Green Energy and Technology

Climate change, environmental impact and the limited natural resources urge scientific research and novel technical solutions. The monograph series Green Energy and Technology serves as a publishing platform for scientific and technological approaches to "green"—i.e. environmentally friendly and sustainable—technologies. While a focus lies on energy and power supply, it also covers "green" solutions in industrial engineering and engineering design. Green Energy and Technology addresses researchers, advanced students, technical consultants as well as decision makers in industries and politics. Hence, the level of presentation spans from instructional to highly technical.

Indexed in Scopus.

Indexed in Ei Compendex.

Călin Iclodean · Bogdan Ovidiu Varga ·
Nicolae Cordoș

Autonomous Vehicles for Public Transportation

Springer

Călin Iclodean
Department of Automotive Engineering
and Transports
Technical University of Cluj-Napoca
Cluj-Napoca, Romania

Bogdan Ovidiu Varga
Department of Automotive Engineering
and Transports
Technical University of Cluj-Napoca
Cluj-Napoca, Romania

Nicolae Cordoş
Department of Automotive Engineering
and Transports
Technical University of Cluj-Napoca
Cluj-Napoca, Romania

ISSN 1865-3529 ISSN 1865-3537 (electronic)
Green Energy and Technology
ISBN 978-3-031-14677-0 ISBN 978-3-031-14678-7 (eBook)
https://doi.org/10.1007/978-3-031-14678-7

This Springer imprint is published by the registered company Springer Nature Switzerland AG
The registered company address is: Gewerbestrasse 11, 6330 Cham, Switzerland

Foreword

The unexamined life is not worth living.
(Socrates)

No reasonable publishing company will permit a book to appear in print without a preface. And even though most readers probably agree that there exists a kind of unavoidable necessity of such introductory texts, with almost equal unanimity they find them superfluous and uninteresting. They tend to spread an impression of bustling prolixity. Good prefaces are rarely written. A notable exception are the famous "Prefaces" of the literary genius (and co-founder of the London School of Economics) George Bernard Shaw, which are entertaining, delightful and full of prolific creativity: His legacy endures seventy years after his death. This will for sure not be the case for the humble preface to this book. However, there is at least one thing that G. B. S. and I have in common: Shaw also prefaced of what his friends wrote!

As vehicle technology progresses into the third decade of the 21st century, many players, including US tech giants like Google, Waymo and Tesla are working on turning the dream of autonomous driving into reality. There is today no car manufacturer who is not in the race to develop driverless technologies. Autonomous driving will become real. And autonomous vehicles are likely to catch on firstly and mainly in urban ("smart city") areas for public transportation, the topic of this book.

Developments in "Autonomous Driving Systems" have brought to the forefront many new challenges in science and engineering. Continued progress involve a very diverse set of knowledge and technology, including localization, sensing, fusion, perception, decision making, actuation, real-time dynamic system (= vehicle) controls, client systems, deep learning, graphic/video processing and much more.

It is the objective of this book to provide an overview to these technological advances and it is, without doubt, a challenge for both, lecturers and students to master all these topics. Since the subject touches upon many new areas, it brings even seasoned automotive engineers out of their comfort zone (including myself).

Autonomous driving is not about mastering one specific problem; it is more than "only" AI or deep learning: it's about the complex integration of many technologies. Integration is the most important technology in autonomous driving. This is very well reflected in the present book, which allows to first grasp the basic concepts of each technology involved as well as their interaction/integration on a vehicle system level.

The selection of topics has been made with great care. Great consideration has been given to the didactic point of view. The authors have tried to present the whole subject-matter in the simplest possible form. This book also satisfies the need of an "easy-to-read" reference book.

Last, but not least: the author of these lines is convinced that the biggest barrier to a rapid development of this field is the very limited talent supply. This is undoubtedly due to several problems upon which we cannot elaborate here. However, teaching young students, and, most importantly, helping them to find the existential pleasures of engineering, is the only way forward to achieve new feats. May this book contribute to this process!

Dipl.-Ing. Dr. Felix Pfister
Business Development Manager
IPG Automotive GmbH
Karlsruhe, Germany

Acknowledgements

First of all, we are deeply grateful to IPG Automotive GmbH, D-76185 Karlsruhe, Bannwaldallee 60, (www.ipg-automotive.com) for their essential logistic support, availability and also for provision of their high-level professional expertise and technical support in the writing process of the book. Their advices definitively increase the scientific and applicative quality of book.

Nevertheless, we would like to thank for their essential logistic support, availability and also for provision of their high-level professional expertise and technical support in the writing process of the book.

At the end but not last, we would like to express our gratitude toward our families for their support, understanding and encouragement, which helped us permanently throughout the preparation of the book.

Călin Iclodean
Bogdan Ovidiu Varga
Nicolae Cordoș

Introduction

The rapid evolution of autonomous driving in the field of information technologies and the automotive industry has made it possible to implement autonomous vehicles for public transport. Although the transport capacities with the small autonomous shuttle buses that are currently used are limited (maximum 15 people), the use of these transport units in large urban areas is beneficial for society.

This book aims to present the current state of implementation of the autonomous technology on autonomous shuttle buses, with direct implications in their evolution in terms of technological progress and with direct links to the legal and social aspects of the public transportation system around the world.

A critical aspect presented in the book is the legal framework of autonomous driving, which is extremely different throughout the world. As autonomous vehicle legislation has some shortcomings in the type-approval, registration, and placement into service of these vehicles on public roads, many countries have found ways to integrate them into testing programs, laying the groundwork for future legislative packages in the field of autonomous driving.

The general term "autonomous vehicle" was used throughout the book to explain the constructive solutions and functional algorithms for autonomous shuttle buses.

At the same time, the book presents the current technological solutions adopted by the main manufacturers of autonomous shuttle buses, resulting in an exhaustive overview of the major components. The operation modalities of these vehicles are presented, which in some countries have been implemented on less-traveled routes where they can travel at speeds of up to 25 km/h (and over in test mode), without hindering public traffic, university schools, industrial areas, airports, sports facilities etc. In other countries, they are used for pilot programs related to public transport, in others for postal transport, and others for scientific research. All these situations show that the first step in autonomous driving has been taken, and these means of transport exist among us.

Public transport is one of the key issues that must be considered when developing a large metropolitan area, namely one of the criteria that define the quality of life in a metropolis, and which is an essential aspect strictly related to the social life of each inhabitant. A model for classifying the public transport based on vehicle dimensions

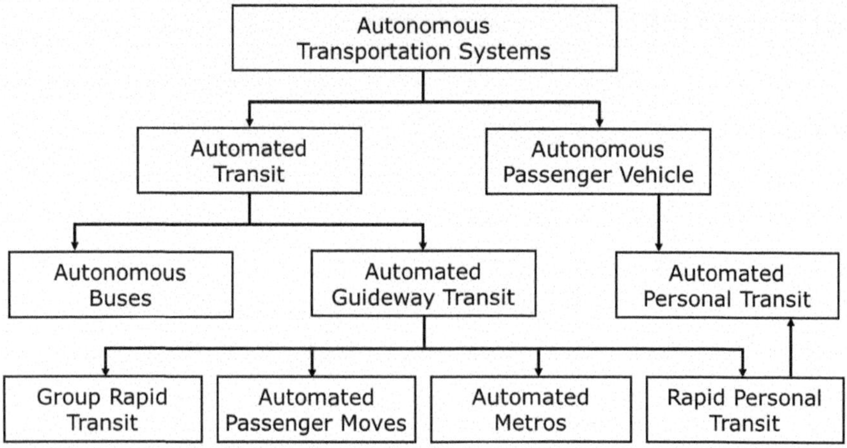

Fig. 1 Model for classifying the public transport

and transport capacity, respectively operational characteristics are shown in Fig. 1 [1].

Public transportation will benefit from the introduction of intelligent vehicles for improve safety in urban areas, reduce the cost for the transportation, decrease congestion and improve the services for the user. Autonomous vehicle technology is not completely mature yet, but it has been attracting economic and industrial interests because those commercial vehicles include increasing levels of advanced driver-assisting systems year after year. The social implications of these huge revolution will change the perception of the transportation systems, increasing the lives quality.

The autonomous vehicles will be equipped with many sensors, and most importantly, safety and reliability will be mandatory, but open requirements. The impact of autonomous vehicles on cities could result in the positive consequences and negative side effects. The autonomous driving will be safer and more accessible for those who cannot drive themselves or do not own a vehicle, following the mobility as a service paradigm. The autonomous vehicle will support public transit and will decrease housing development costs, for elimination of expensive underground garages and space on-ground parking [2].

There are a lot of direct, indirect, and random factors that influence public transport, but the most sensitive factor is human interaction in urban traffic. Drivers' decisions are influenced by many factors and even if they are behind the wheel, driving is not always the only activity they focus on sometimes they do not maintain a constant distance, at the safety limit between vehicles, they do not always focus on the signs of traffic or the color of the traffic lights, do not follow the rules of traffic, and momentary decisions are sometimes influenced by their emotions.

If we eliminate the human factor, traffic jams will be reduced, and road safety will increase. Typically, a driver spends up to an hour in traffic to get from work to home (or vice versa), whether he is using a privately owned vehicle or public transportation

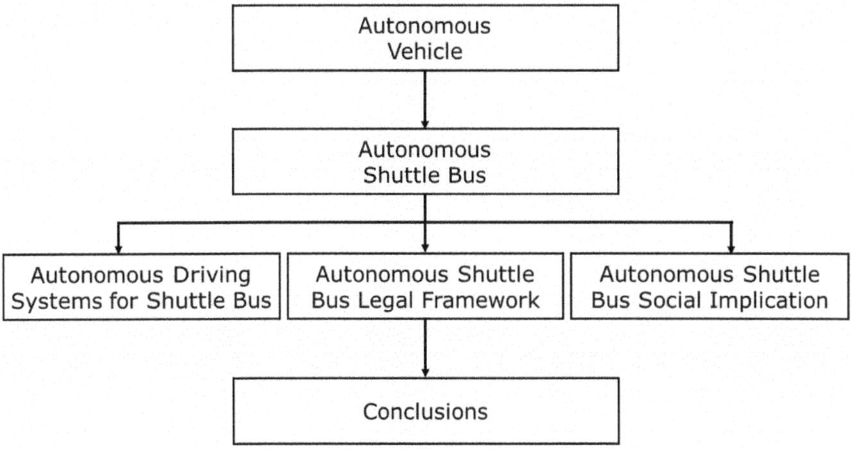

Fig. 2 Main objectives and secondary objectives

because he is traveling in the same traffic. In the case of a dedicated traffic line for public transport, time spent in traffic will be reduced by 40–60% according to studies [3, 4].

The constructive configuration of the autonomous shuttle buses is very similar to the general platform of the autonomous vehicles, however, there are many differences, and these aspects have been highlighted in the following sections of the book.

The structure of this book, respectively main objectives are presented in Fig. 2. The main advantage of autonomous driving is not only the replacement of the human driver with all the emotional and physical limitations, but the possibility for the vehicle itself to make predictions and communicate with the infrastructure and/or other vehicles in traffic. Due to this, autonomous vehicles will become a key component of future smart cities. Cities that will be blocked by traffic jams, especially for public transport, will generate a rapid increase in quality of life, a reduction in energy consumption (in all components related to traffic) and will become more environmentally friendly, generating an increases urban air quality [3].

From the point of view of the autonomous shuttle buses implementation in big cities, the road infrastructure must be developed both as a physical structure (road, lanes, pedestrian shelters etc.) and in terms of digital facilities (sensors, intelligent communication equipment, 4G/5G internet).

The autonomous driving system includes the technological developments of recent years, developments related to processing power, flexibility, and mobility. The autonomous driving level of development has reached the point where an autonomous vehicle is no longer evaluated in terms of performance related to speed, dynamic performance, or braking distance. Autonomous vehicles are assessed by their ability to make decisions about speed, to establish alternative routes before reaching a restricted traffic area, or to reduce speed before reaching a bumpy road section and so on. These decision-makers differentiate autonomous vehicles in general [3].

The main question marks raised by the degree of the autonomous driving system development are the following: is the system able to cross the double line when another vehicle is parked in the direction of travel, or will it require the human factor to take control? Is the system able to recognize the signs made by a police crew in the event of an accident or roadblock? The system can increase its speed in case of medical emergency, or all these decisions cannot be implemented yet the code of the autonomous driving algorithm and the autonomous vehicle cannot perform such tasks that only a human driver would meet?

An autonomous driving system must not only perform its classic driving tasks using sensors as input data generators and actuators to control execution systems, but these systems must also be redundant, flexible, reliable, versatile and in addition to ensure cyber security. All these tasks must be performed in a high-speed communication network, and data processing must be managed by powerful computers [5]. Hence the need for a large amount of computing power, respectively a large amount of electricity [6]. From these examples highlight some evaluation criteria less discussed about autonomous vehicles:

(1) Does processing power influence the choice of an autonomous vehicle?
(2) Will the electronic equipment of the autonomous vehicle accept future software updates?
(3) Will the self-driving system be able to run the software versions that will be developed in the next 3 to 5 years?
(4) Will the cost of these software updates be covered by the benefits it will bring to the operation of the vehicle?
(5) Will there be issues after the first update that will be fixed with the next updates?

References

1. Hasan, U., Whyte, A., Al Jassmi, H.: A review of the transformation of road transport systems: are we ready for the next step in artificially intelligent sustainable transport? Appl. Syst. Innov. **3**, 1 (2020)
2. Ainsalu, J., Arffman, V., Bellone, M., Ellner, M., Haapamäki, T., Haavisto, N., Josefson, E., Ismailogullari, A., Lee, B., Madland, O., Madžulis, R., Müür, J., Mäkinen, S., Nousiainen, V., Pilli-Sihvola, E., Rutanen, E., Sahala, S., Schønfeldt, B., Smolnicki, P.M., Soe, R.-M., Sääski, J., Szymańska, M., Vaskinn, I., Åman, M.: State of the art of automated buses. Sustainability **10**, 3118 (2018)
3. Iclodean, C., Cordos, N., Varga, B.O.: Autonomous shuttle bus for public transportation: a review. Energies **13**, 2917 (2020)
4. Wolf, I.: the interaction between humans and autonomous agents. In: Maurer, M., Gerdes, J.C., Lenz, B. (eds.) Autonomous Driving Technical, Legal and Social Aspects, pp. 103–124. Springer, Berlin/Heidelberg, Germany (2016)
5. https://www.tu-auto.com/nvidia-more-computing-power-makes-for-safer-avs/ (2020)
6. https://www.wired.com/story/self-driving-cars-power-consumption-nvidia-chip/ (2020)

Contents

About the Authors

Călin Iclodean is currently Associate Professor at Technical University of Cluj-Napoca, Romania, Department of Automotive Engineering and Transports. He has a Ph.D. in Mechanical Engineering, since 2013, after studying for doctoral studies in Graz—Austria at AVL List GmbH. He graduated from the Faculty of Electronics and Telecommunications, Department of Applied Electronics of the Technical University of Cluj-Napoca in 1994, and has more than 25 years of working and business experience in IT&C. The principal areas of research of Călin Iclodean are covering the following fields: electric vehicles, autonomous vehicles, fuel cell vehicles, powertrain concept, electronic control unit, in-vehicle communication network, energy efficiency, computer modelling and simulation in the automotive field. His research received financing from European Investment Bank, national and regional authorities, economical private entities. He is also active in the field of peer-reviewed publications where he is a member of a reviewer board for several journals indexed in the Web of Science. At the same time, he is a member of the topic editorial board for several journals indexed in the Web of Science and academic editor for several Special Issues published in journals indexed in the Web of Science. e-mail: calin.iclodean@auto.utcluj.ro

Bogdan Ovidiu Varga is currently Professor at the Technical University of Cluj-Napoca, Romania, Department of Automotive Engineering and Transports and is the coordinator of the Electric and Hybrid Vehicles course and the director of the automotive master programme—Advance Techniques in Automotive Engineering. He received his Ph.D. in mechanical engineering at Technical University of Cluj-Napoca after studying in Germany (University of Hohenheim), Italy (Università di Bologna) and Greece (National Technical University of Athens). Since 2011, he has been Technical Juridical Expert for the Romanian Ministry of Justice in the field of Automotive, Road Traffic and Equipment for Vehicles Control and from 2013 he activates as an associate member of Federation International des Experts en Automobile. The main research activity covers the area of smart and green mobility: electric and hybrid vehicles, fuel cell vehicles, energy efficiency of the vehicles, autonomous vehicles, vehicle evaluation and vehicle diagnosis. He has published several papers in the peer-review journals including Energy Journal from Elsevier and several patterns in the field of electric mobility. e-mail: bogdan.varga@auto.utcluj.ro

Nicolae Cordoş is currently Associate Professor at the Technical University of Cluj-Napoca, Romania, Department of Automotive Engineering and Transports. In 2010, he received his Ph.D. in mechanical engineering at Technical University of Cluj-Napoca after studying in Germany (University of Hohenheim) and Portugal (University of Minho). The relevant activities and the researches interest are in the following fields: vehicle dynamics, intelligent transport systems, autonomous vehicle, and the dynamics of traffic accidents. He has also published scientific papers in the fields of interest mentioned above in the peer-review journals. e-mail: nicolae.cordos@auto.utcluj.ro

Abbreviations

2WD	2-Wheel Drive
3GPP	3rd Generation Partnership
4WD	4-Wheel Drive
5GAA	5G Automotive Association
ABS	Anti-lock Braking System
AC	Alternating Current
ACC	Adaptive Cruise Control
ACK	ACKnowledgement
ADAS	Advanced Driver Assistance Systems
ADF	Angular Dispersion of Features
ADS	Autonomous Driving System
ADSE	Automated Driving System Entity
AEB	Autonomous Emergency Braking
AEBS	Automatic Emergency Braking System
AF	Arbitration Field
AI	Artificial Intelligence
AMCL	Adaptive Monte Carlo Localisation
AMCW	Amplitude Modulated Continuous Wave
APF	Artificial Potential Field
API	Application Programming Interface
AR	Augmented Reality
ASIL	Automotive Safety Integrity Levels
AUTOSAR	AUTomotive Open System ARchitecture
AVA	Autonomous Vehicle Advantage
BCU	Body Control Unit
BCU*	Battery Control Unit
BDS	BeiDou navigation Satellite system
BMS	Battery Management System
BRS	Bit Rate Switch
BS	Base Station
BSD	Blind Spot Detection

BSM	Basic Safety Message
BSW	Blind Spot Warning
BTP/GN	Basic Transport Protocol/Geo Networking
CAN	Controller Area Network
CCD	Charge-Coupled Device
CERT/CC	Computer Emergency Response Team/Coordination Center
CF	Control Field
CFA	Color Filter Array
CGW	Central GateWay
CIT	CarMaker Interface Toolbox
C-ITS	Cooperative Intelligent Transport Systems
CLW	Control Loss Warning
CMOS	Complementary Metal Oxide Semiconductor
CNN	Convolutional Neural Networks
CP	Calculate Parameters
CPU	Central Processing Unit
CRC	Cyclic Redundancy Check
CRCF	Cyclic Redundancy Check Field
CSMA/CA	Carrier-Sense Multiple Access with Collision Avoidance
C-V2X	Cellular V2X
DA	Destination Address
DC	Direct Current
DE	Design Explorer
DEL	Delimiter
DEM	Digital Elevation Model
DFL	Data Field Lenght
D-GNSS	Dynamic Global Navigation Satellite System
DL	Deep Learning
DLC	Data Length Code
DNPW	Do-Not-Pass Warning
DOD	Depth Of Discharge
DoE	Design of Experiment
DOF	Degrees Of Freedom
DOP	Dilution Of Precision
DoS	Denial of Service
DPTC	Véhicule à Délégation Partielle ou Totale de Conduite
DQ	Direct Quadrature
DR	Deep Reinforcement
DR*	Dead Reckoning
DRL	Deep Reinforcement Learning
D-S	Dempster–Shafer
DSM	Digital Surface Model
DSRC	Dedicated Short-Range Communication
DTM	Digital Terrain Model
DVA	Direct Variable Access

D-WEVCS	Dynamic Wireless Electric Vehicle Charging System
E/E	Electrical/Electronic
ECU	Electronic Control Unit
EDL	Extended Data Length
EEBL	Emergency Electronic Brake Light
EIF	Extended Identifier Field
EiL	Electric/Motor-in-the-Loop
EM	Electric Motor
EMC	ElectroMagnetic Compatibility
EMI	ElectroMagnetic Interference
eNB	evolved Node B
EOC	Electrical Optical Converter
EOF	End Of Frame
EP	External Parameters
ERRN	ERRor Not
ESI	Error State Indicator
ESP	Electronic Stability System
EVSE	External Vehicle Supply Equipment
FCTA	Forward Cross Traffic Alert
FCW	Forward Collision Warning
FD	Flexible Data
FD*	Fail-Degraded
FMCW	Frequency Modulated Continuous Wave
FS	Fail-Safe
FTDMA	Flexible Time Division Multiple Access
GIS	Geographic Information System
GLONASS	Global'naya Navigatsionnaya Sputnikovaya Sistema
GNSS	Global Navigation Satellite System
GPS	Global Positioning System
GPU	Graphics Processing Unit
GT	Ground Truth
GUI	Graphical User Interface
HARA	Hazard Analysis and Risk Assessment
HD	High Definition
HDFS	Hadoop Distributed File System
HDM	Human Driver Model
HID	High-Intensity Discharge
Hi-Fi	High Fidelity
HiL	Hardware-in-the-Loop
HMI	Human Machine Interface
HV	High Voltage
HVAC	Heating Ventilation and Air Conditioning
I/O	Input/Output
ID	IDentifier
IDE	IDentifier Extension

IFS	Inter Frame Space
IIDM	Improved Intelligent Driver Model
IMA	Intersection Movement Assist
iMAE	inverse Mean Absolute Error
IMU	Inertial Measurement Unit
INS	Inertial Navigation System
IoT	Internet of Things
IP	Internet Protocol
IP*	Initial Parameters
IPO	Input Process Output
IR	InfraRed
iRMSE	inverse Root Means Square Error
ISO	International Standards Organization
IT	Informational Technology
ITS	Intelligent Transportation System
KML	Keyhole Markup Language
LCA	Lane Change Assistant
LCW	Lane Change Warning
LDM	Local Dynamic Map
LDW	Lane Departure Warning
LED	Light Emitting Diode
LIDAR	Light Detection And Ranging
LIN	Local Interconnect Network
LKAS	Lane Keeping Assist System
LOS	Limited of On-board Sensors
LRR	Long-Range Radar
LSB	Least Significant Bits
LTA	Left Turn Assist
LTE-Uu	Long-Term Evolution User-to-user
LTE-V	Long-Term Evolution for Vehicles
LTE-V2X	Long-Term Evolution for V2X
LTO	$Li_4Ti_5O_{12}$ Battery
LUT	Look-Up Table
LV	Low Voltage
MAC	Media Access Control
MAE	Mean Absolute Error
MCU	Motor Control Unit
MEMS	Micro-Electro-Mechanical System
MiL	Model-in-the-Loop
MIMO	Multi-Input Multi-Output
MITM	Man-In-The-Middle
MLS	Merging Lanes Safely
MOMA	Multi-Objective Memetic Algorithms
MOST	Media Oriented Systems Transport
MPC	Model Predictive Control

MRM	Minimum Risk Maneuver
MRR	Medium-Range Radar
NavIC	Navigation with Indian Constellation
NFC	Near Field Communication
NGSIM	Next Generation SIMulation
NHTSA	National Highway Transportation Safety Association
NIC	Network Interface Controller
NIT	Network Idle Time
NLL	Negative Log Loss
OBD	On-Board Diagnostics
OCV	Open Circuit Voltage
OD	Operational Domain
ODD	Operational Design Domain
OEC	Optical Electrical Converter
OEM	Original Equipment Manufacturer
OGM	Occupancy Grid Map
PAN	Personal Area Network
PI	Proportional Integral
PiL	Processor-in-the-Loop
PLD	Product Liability Directive
pMC	pressure in Master Cylinder
PMSM	Permanent Magnet Synchronous Motor
POI	Point Of Interest
PRM	Probabilistic Roadmap Method
PS	Power Supply
QZSS	Quasi-Zenith Satellite System
RAT	Radio Access Technologies
RCD	Residual Current Device
RCS	Radar Cross Section
RCTA	Rear Cross Traffic Alert
ReLU	Rectified Linear Unit
RF	Radio Frequency
RFI	Radio Frequency Interference
RGB	Red-Green-Blue
RKE	Remote Keyless Entry
RMSE	Root Means Square Error
RNN	Recurrent Neural Networks
ROI	Region Of Interest
ROS	Robotic Operating System
RP	Resulting Parameters
RSI	Raw Signal Interface
RSU	Road-Side Units
RT	Real Time
RTK	Real-Time Kinematic
RTOS	Real Time Operating System

RTR	Remote Transmission Request
Rx	Reception
SA	Source Address
SAE	Society of Automotive Engineers
SaFAD	Safety First for Automated Driving
SAV	Shared Autonomous Vehicle
SCI	Serial Communication Interface
SiL	Software-in-the-Loop
SLAM	Simultaneous Localisation And Mapping
SMO	Shuttle Modellregion Oberfranken
SNPD	Slave Node Position Detection
SNR	Signal-to-Noise Ratio
SOC	State Of Charge
SOC*	Start Of Cycle
SOF	State Of Function
SOF*	Start Of Frame
SOH	State Of Health
SPA	Sound Pressure Amplitude
SRR	Short-Range Radar
SRR*	Substitute Remote Request
SSD	Single Shot multibox Detector
SSL	Secure Sockets Layer
STBN	STandBy Not
S-WEVCS	Static Wireless Electric Vehicle Charging System
TCMV	Technical Committee on Motor Vehicles
TCP	Transport Control Protocol
TCU	Transmission Control Unit
TCU*	Telematic Control Unit
TOF	Time-Of-Flight
TOPS	Theoretical Operations Per Second
TPMS	Tire Pressure Monitoring System
TSE	Traffic State Estimation
TSL	Transport Layer Security
Tx Data	Transmit Data
Tx	Transmission
TxEN	Transmit Data Enable Not
UART	Universal Asynchronous Receiver Transmitter
UE	User Equipment
UN	United Nations
USB	Universal Serial Bus
UTM	Universal Transverse Mercator
V2I	Vehicle-to-Infrastructure
V2N	Vehicle-to-Network
V2P	Vehicle-to-Pedestrian
V2V	Vehicle-to-Vehicle

V2X	Vehicle-to-Everything
ViL	Vehicle-in-the-Loop
VIN	Vehicle Identification Number
VIS	VIsible Spectrum
VPN	Virtual Private Network
VRU	Vulnerable Road Users
VVE	Virtual Vehicle Environment
WAVE	Wireless Access in Vehicular Environments
WDR	Wide Dynamic Range
WHO	World Health Organization
WEVCS	Wireless Electric Vehicle Charging System
WPT	Wireless Power Transfer
YOLO	You Only Look Once

Chapter 1
Autonomous Driving Basics

1.1 Autonomous Driving History

History has shown that long before the beginnings of autonomous vehicle development, the aeronautical, naval and railway industries successfully (partially or completely) replaced human intervention in the control of airplanes, submarines, trains, and subways. The beginning of self-driving was a complicated process, it was a technology that developed in the background, without concrete applications for the transport economy or for the industry in general.

In the mid-1920s, Francis P. Houldina, an electrical engineer with a military background, equipped a Chandler vehicle with an antenna that received radio signals from a second vehicle moving in front of the first. The control of the Chandler was performed by an electric motor that performed tasks depending on the movement of the first vehicle [1, 2]. Another visionary of autonomous driving was Norman Bel Geddes, an industrial designer who, with the support of General Motors, presented in 1939 during the Futurama exhibition at the World's Fair, a vehicle powered by magnetic fields generated by circuits incorporated in the rolling track [1, 3]. In 1957, RCA Labs in the United States introduced to the public two vehicles supplied by General Motors, which were equipped with receivers that interpreted the signals from the road. Thus, the vehicles were able to manage automatic steering, acceleration, and automatic braking [1, 4]. Dr. Robert L. Cosgriff, who worked at Ohio State University in 1960 on the Communications and Control Systems Laboratory, presented the idea that driver automation based on road information would become a reality in the next 15 years. In 1986 the Carnegie Mellon University Institute of Robotics developed one of the first self-driving vehicles to ever be controlled by a computer (Navlab) [1, 5].

In Europe, the pioneers of autonomous driving were members of a team in a road research laboratory that, in 1960, successfully tested a Citroen DS traveling at speeds of over 130 km/h without being assisted by a driver. This vehicle operated following the magnetic field of a series of electric wires mounted on the rolling track [1, 6]. In 1980, Mercedes-Benz developed a driverless van, which operated on public roads.

C. Iclodean et al., *Autonomous Vehicles for Public Transportation*, Green Energy and Technology, https://doi.org/10.1007/978-3-031-14678-7_1

This project was carried out by Prof. Ernst Dickman, from the Bundeswehr University in Munich. In 1995, Mercedes-Benz completed the Eureka Prometheus project, one of the largest research and development programs for autonomous vehicles in the history of that time. The project redefined the state of the art for self-driving vehicles. Many universities and car manufacturers participated in this pan-European project between 1987 and 1995. Schiphol Airport, Amsterdam, Netherlands in December 1997, was the first public entity to use a Shuttle Park vehicle to travel between terminals. This technology was the basis for future developments in the field, being for the first time when self-driving skills were publicly tested [1, 7].

Since 1997, the United States Department of Transportation has issued the first regulations regarding the "demonstration of an automated system of motor vehicles and highways", so that the legislation began to be favorable to the development of vehicles equipped with autonomous driving systems. Since 2000, there has been a revolution in self-government. Independent researchers, car companies, software companies etc. they tried to claim ownership of the autonomous vehicle, which generated fierce competition that led to the rapid development of this technology. In 2009 Google began to develop the project of autonomous vehicles [1, 8].

1.2 Levels of Driving Automation

From the point of view of the autonomous driving standardization, SAE International (Society of Automotive Engineers) has established a standard SAE J3016™: Taxonomy and Definitions for Terms Related to Driving Automation Systems for On-Road Motor Vehicles", which defines a framework regulates and facilitates the approach of good practices by manufacturers in the design, development, testing, and safe use of vehicles with various levels of automation.

The autonomous vehicle is equipped with hardware and software technology that ensures the ability to drive without the need for human intervention on the control mechanisms, with or without remote monitoring of the vehicle by specially qualified personnel for monitoring. The short- and medium-term benefits of self-driving will have a clear impact on public transport once this solution is widely used. Autonomous vehicles currently circulating on public roads around the world have reached levels 3 and 4 of automation, according to the specifications of SAE J3016™ (Fig. 1.1) [1, 9].

Level 0 (Driver Only) is characterized by a simple technology that does not present any active automation, the only thing that can be done is to provide visual or audible warnings. The role of the driver is to monitor the driving environment (eyes on) and to perform the dynamic driving tasks (acceleration, braking) and lateral driving (steering/hands on) [9].

Level 1 (Assisted) is characterized by a technology that provides driver assistance, which can be activated, and which performs dynamic driving tasks in the longitudinal or transverse plane. The role of the driver is to monitor the driving

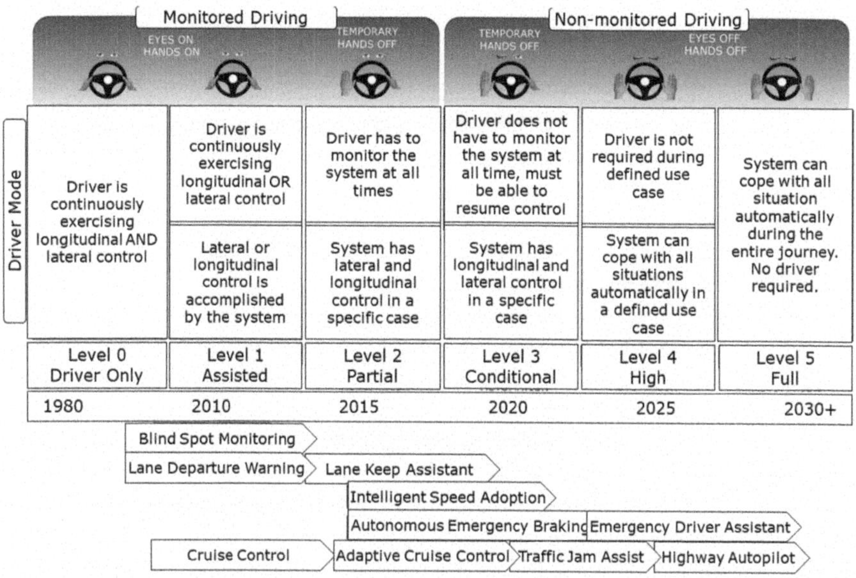

Fig. 1.1 The levels of self-driving (according to SAE J3016™)

environment (eyes on) and to perform dynamic driving tasks longitudinal (acceleration, braking) or transverse (steering/hands on). The driver constantly monitors the dynamic driving task performed by the autonomous driving system and has the responsibility to determine when it is necessary to assist or deactivate it, except for systems that automatically intervene in cases of emergency [9].

Level 2 (Partial Automation) is characterized by a partially automated technology, which can be activated or deactivated, and which performs dynamic driving tasks in longitudinal and transverse plane (at least two automation functions). The role of the driver is to monitor the driving environment (eyes on) and to constantly monitor the dynamic longitudinal (acceleration, braking) and transverse (steering/temporary hands off) driving tasks that can be performed by the autonomous driving system. and has the responsibility to determine when it is necessary to assist or deactivate it, except for systems that automatically intervene in cases of emergency [9].

Level 3 (Conditional Automation) is characterized by an automated conditional technology, which can be activated or deactivated depending on the driving conditions. The autonomous driving system may require the driver to take control in certain critical situations, or the driver may intervene if he deems it necessary to take control. The driver must not monitor the driving environment (temporary eyes off) and must not constantly monitor the dynamic driving tasks (temporary hands off) but must be permanently able to intervene and take control if the situation requires it. The autonomous driving system performs the tasks of dynamic longitudinal driving (acceleration, braking) and transverse (steering), can assess the severity of extreme situations, when over the driver's decisions, moves to minimum risk conditions and can act automatically in emergency situations [9].

Level 4 (High Automation) is characterized by a technology with a high level of automation. The autonomous driving system is activated or deactivated automatically, according to the implemented autonomous driving algorithm; the driver can request to take control only if this is necessary. The driver must not monitor the driving environment (eyes off) and must not supervise dynamic driving tasks (hands off), he may be absent from the autonomous vehicle, which through the minimum risk conditions can handle any emergency [9].

Level 5 (Full Automation) is characterized by a technology with a total level of automation, which for the moment is developed only as a theoretical principle, still implemented on no vehicle. The autonomous driving system controls all dynamic driving tasks, it can handle any extreme situation to avoid any danger. The driver may be permanently absent, all persons inside the level 5 autonomous vehicle being passengers.

According to SAE J3016™ specifications, the ODD (Operational Design Domain) is a description of the operating conditions under which an autonomous management system is specifically designed to operate, including, but not limited to environment, relief restrictions during the day and night, characteristics of the road or traffic. Self-driving systems from level 1 to level 4 are directly dependent on the limitations of the ODD (Table 1.1), and level 5 systems are not affected by these limitations. ODD attributes can have a real or virtual infrastructure, and the mode of operation can be static or dynamic [9, 10].

The standardization of the driving automation levels of a motor vehicle according to SAE J3016™ has the following purposes [9]:

(1) Establishing the role of the driver throughout the operation of the self-driving system.
(2) Establish the scope of laws, regulations, standards, and norms regarding self-driving.

Table 1.1 ODD attributes type

ODD attributes	Infrastructure type	Operating
Road (rolling track)	Physical	Static
Road (HD map)	Virtual	Static
GPS positioning	Virtual	Static
Road markings	Physical	Static
Traffic signs	Physical	Static
Road traffic	Physical	Dynamic
Weather conditions	Physical	Dynamic
Communications	Virtual	Dynamic
Traffic management	Virtual	Dynamic
Fleet surveillance	Virtual	Dynamic

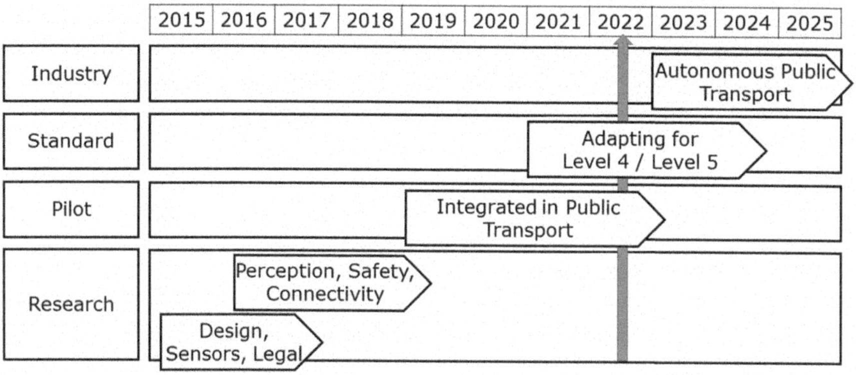

Fig. 1.2 Evolution of the public transport by autonomous buses

(3) Establish an appropriate framework for the specifications and technical requirements of self-driving.

The six levels of driving automation according to SAE J3016™ [9] are descriptive and informative and indicate the minimum capacities for each level.

Autonomous shuttle bus-type buses for public passenger transport are present and operational in many of the world's major metropolises, where they have been implemented on urban routes, in traffic, operating in public, in various testing regimes, being monitored. permanently by a human operator (level 3, according to SAE J3016™ in Fig. 1.2) [1, 11, 12].

Level 4 of automation (according to SAE J3016™) is being tested in various locations around the world (e.g., Waymo (Google) in Phoenix, AZ, USA, Apollo (Baidu) in China, Evo (Navya) in France [1, 13]) and involves the use of autonomous vehicles in urban traffic without being accompanied by a human operator. These autonomous vehicles travel only based on the perceptions and decisions of the autonomous driving system, using Machine Learning or Artificial Intelligence technologies [1, 14].

UN (United Nations) Regulation no. 79 revision 3 of November 30, 2017: Uniform provisions concerning the type-approval of vehicles about steering equipment, lays down uniform provisions for the arrangement and performance of steering systems fitted to the vehicles. In previous revisions of the regulation, it is specified that the main steering system contains a mechanical link between the steering control and the wheels to determine the trajectory of the vehicle [15]. Due to the expansion of autonomous technology and requirements on the functional safety of motor vehicles, the elimination of the mechanical steering column has led to a revision of the traditional approach of UN Regulation no. 79 which has been modified to adapt to the requirements of new technologies.

For the purposes of this revision of the regulation, the autonomous steering system is defined as a system that incorporates a complex electronic control function that causes the vehicle to follow a defined path or change its route in response to signals transmitted by on board the vehicle. The advanced driver assistance steering system according to the specifications of UN Regulation no. 79 is an additional system to the main steering system, which aids the driver in controlling the direction of the vehicle and which includes the automatically commanded steering function.

This function intervenes in the command and control of the direction in situations classified in the following categories [15, 16]:

Category A assists the driver in low-speed maneuvers or parking when the vehicle is traveling at a maximum speed of 10 km/h.

Category B1 assists the driver in maintaining the vehicle on the selected lane, by controlling the lateral movement.

Category B2 is activated by the driver and keeps the vehicle on the lane by controlling the lateral movement without requiring the subsequent command of the driver.

Category C is activated by the driver and can perform a single lateral maneuver (change of lane) when is commanded by the driver.

Category D is activated by the driver and indicates the possibility of performing a single lateral maneuver (change of lane) but performs the function only after confirmation of the action by the driver.

Category E is activated by the driver and can continuously determine the possibility of a maneuver (change of lane) without other driver commands.

1.3 Autonomous Driving Small/Medium Vehicles

Road vehicles are classified according to Regulation (EC) No 661/2009 of the European Parliament and of the Council of July 13, 2009 concerning type-approval requirements for the general safety of motor vehicles, their trailers and systems, components and separate technical units intended therefor in the following categories [17]:

Category M motor vehicles with at least four wheels, with a maximum weight exceeding 1000 kg, used for the transport of persons:

Category M1 motor vehicles used for the transport of persons, with a maximum of eight seats, outside the driver's seat.

Category M2 motor vehicles used for the transport of persons, with more than eight seats, outside the driver's seat, with a maximum weight of less than 5000 kg.

Category M3 motor vehicles used for the transport of persons, with more than eight seats, outside the driver's seat, with a maximum weight exceeding 5000 kg.

Category N motor vehicles with at least four wheels, with a maximum weight exceeding 1000 kg, used for the goods transport:

Category N1 motor vehicles used for the goods transport, with a maximum weight of less than 3500 kg.

Category N2 motor vehicles used for the goods transport, with a maximum weight between 3500 kg and 12,000 kg.

Category N3 motor vehicles used for the goods transport, with a maximum weight exceeding 12,000 kg.

Another classification of road vehicles according to the purposes they serve according to the Bosch Automotive Handbook [18] is the following:

Passenger car (vehicles for up to nine people):

(a) Saloon (sedan) closed body, maximum four side doors.
(b) Convertible (sedan) saloon convertible ceiling, fixed side panels.
(c) Coupé closed body coupe, maximum two side doors.
(d) Station wagon body with spacious interior and a loading area.
(e) Truck station wagon delivery truck.
(f) Special passenger car ambulance, recreational vehicle.
(g) Multipurpose passenger car.

Commercial vehicles (passenger or freight vehicles):

(a) Bus car for transporting more than nine people.
(b) Minibus vehicles for the transport of a maximum of 17 people (this category includes the autonomous shuttle bus-type analyzed in the book).
(c) Special bus (people with disabilities etc.).
(d) Truck (freight vehicle).
(e) Multi-purpose truck (multifunctional cargo vehicle).
(f) Special truck (special purpose goods vehicle).

Among the models of small/medium class autonomous vehicles, existing in circulation on public roads, the following can be exemplified [19]:

Tesla S model equipped with the following sensors (Fig. 1.3):

(1) Forward facing radar (1).
(2) Forward facing cameras (3).
(3) Forward looking side camera (1).
(4) Rearward looking side camera (1).
(5) Rear view camera (1).
(6) Ultrasonic sensors (12).

Fig. 1.3 Tesla S Model (Photo author May 31, 2022)

Volvo-Uber XC90 model equipped with the following sensors:

(1) Lidar 360° (1).
(2) Front radar (1).
(3) Rear radar (1).
(4) Forward facing camera roof (3).
(5) Forward facing camera front spoiler (3).
(6) Forward looking side camera (1).
(7) Rearward looking side camera (1).
(8) Rear view camera (1).
(9) Front and back ultrasonic sensors (2).

Waymo Chrysler Pacifica model equipped with sensors:

(1) Long range lidar 360° (1).
(2) Front mid-range Lidar (1).
(3) Rear mid-range Lidar (1).
(4) Forward looking side short-range Lidar (1).

1.4 Autonomous Driving for Public Transportation

The integration of autonomous vehicles in public passenger transport is recent, but experimental research in the field has shown that these vehicles can be safely implemented in most common road scenarios in all major cities [1]. The introduction of autonomous vehicles in public transport in large cities has as main objectives: the safety of people transported, the reduction of congestion in traffic (during peak hours), respectively a sustainable driving that leads to reduced energy consumption, with effects direct and on pollutant emissions.

Regarding the role of autonomous vehicles in the traffic streamlining, especially in large urban areas, it has the width of traffic lanes, respectively the road in the situation of autonomous transport implementation. At present, most lanes have a standard width between 2.5 and 4 m, for vehicles widths between 2 and 2.5 m. This tolerance is intended for the driver's behavior as a parameter of traffic safety in rectilinear movement in general and in corners. Autonomous vehicles that have the travel route stored in the autonomous driving system allow lower tolerances for deviations from the road. This may lead in the future to the narrowing of traffic lanes for the vehicle's movement and to the organization of several lanes on the same direction of travel (two lanes of 3.75 m can become three lanes of 2.5 m) leading to fluidization of the road traffic [10].

Two of the biggest challenges to be solved before the widespread implementation of autonomous vehicles are harmonization and finalization of zonal legislation on their traffic regime on public roads, respectively the development of autonomous technology up to levels 4 and 5 of automation. Until then, autonomous vehicles will only be able to be used in testing on public roads, on private property, and the existing technology will require strict criteria for monitoring and assistance during their use.

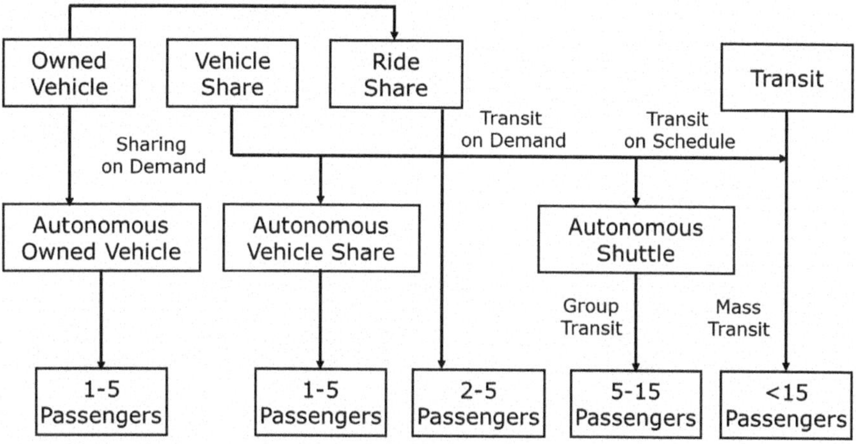

Fig. 1.4 Automated urban mobility solutions

Autonomous vehicles intended for public passenger transport belong to the emerging group of automated urban mobility solutions [1, 20, 21] (Fig. 1.4). This solution involves four concepts regarding to the access to this mode of transport, namely: the self-driving vehicle owned used by a family (owned car), the self-driving vehicle shared with others (ride share), the self-driving vehicle that runs in taxi mode (car share), respectively the autonomous shuttle bus-type shared by a larger number of people (mass transit).

Following previous comparison (see in Fig. 1.4), it is possible to define four technical and organizational automated mobility solutions presented in Fig. 1.5. These are: autonomous automobiles, self-driving car-sharing and ridesharing or carpooling (self-driving, car-on-demand) and driverless shuttles [20].

The diagram presents four solutions for automated modes of transport defined according to the infrastructure needed, number of riders, route flexibility and objective of travel. The self-owned mode, needs parking at home, work due to the more often use of self-driving car-sharing, few parking bays and few charging points for electric vehicles are necessary, and self-driving ridesharing is used, pick-up and drop-off bays need to be prepared. The first two solutions, providing individual transport, or up to four riders similarly, self-driving ridesharing provides 1–4 places for riders, but the vehicle will exchange passengers often, providing para-transit mode and the driverless shuttle provides space for about 10 exchanging riders. Most of the presented mobility solutions rides on flexible routes, but it is possible that driverless shuttles will ride on a fixed route [20].

Currently, there are several companies worldwide that have already produced autonomous shuttle bus-type for public transport. These companies initially developed pilot programs, validated their products, adapted to the specific needs of various parts of the world, and now produce these vehicles in series production. Among these most notable companies are Apollo Baidu (China) [1, 21], EasyMile (Toulouse,

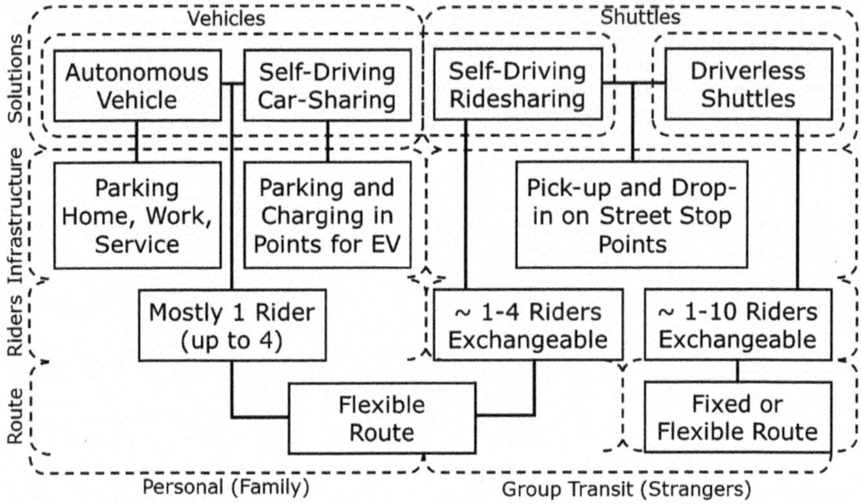

Fig. 1.5 Automated mobility solutions

France) [1, 22], Navya (Lyon, France) [1, 23], respectively Olli (National Harbor, MD, USA) [1, 24].

Table 1.2 presents an extract of the current fleet of autonomous buses in active service worldwide, namely: car manufacturer, construction model, number of vehicles on the routes, cost of transport, respectively the type of routes on which the listed vehicles are in operation.

The book is strongly focused about transporting people with autonomous vehicles, and the diversity of the subject in terms of research brings together scientists from all branches of engineering.

Understanding the environment and its dynamic changes in relation to autonomous vehicles, their integration into the current road infrastructure, as well as the risks in terms of cybersecurity, propulsion systems, sensors, processing power of integrated computers, Machine Learning elements and Artificial Intelligence, as well as the social aspects that appear with the evolution of this technology, are all topics that must be evaluated and presented as key elements in the future development of autonomous vehicles in particular, respectively of autonomous buses for public transport in particular.

Table 1.2 Autonomous shuttle bus-type for public transportation (Routes were operational at the time of documentation)

Shuttle	Location	Operator	Units	Launch	Track (km)	Fee	Environment	Link
Navya	Confluence Lyon, France	Keolis	2	09/2016	1.350	Free	Pedestrian[1]	[25]
Navya	Parc Olympique Lyonnais, France	TCL Lyon	2	11/2019	1.400	Free	Public[2]	[26]
Navya	ZAC des Gaulnes Lyon, France	Berthelet	1	03/2019	1.200	Free	Public	[27]
Navya	l'Abbaye Fontevraud, France	Keolis	1	05/2018	0.800	Free	Pedestrian	[25]
Navya	Universite de Lille, France	Keolis	2	12/2018	1.400	Free	Public	[25]
Navya	Universite de Rennes, France	Keolis	2	01/2018	1.300	Free	Public	[25]
Navya	Meyrin, Geneve, Switzerland	TPG Geneve	2	09/2018	2.100	Free	Public	[28]
Navya	Place de la Planta Sion, Switzerland	CarPostal	2	06/2016	3.540	Free	Public	[29]
Navya	Neuhausen Rheinfall, Switzerland	VB/SH	1	03/2017	1.500	Free	Public	[25]
Navya	l'Ancienne Marly, Switzerland	TPF	2	08/2017	1.300	Taxă	Public	[25]
Navya	Sylt Schleswig–Holstein, Germany	SVG	1	05/2019	2.700	Free	Public	[25]
Navya	Ilse-Arlt-Straße Wien, Austria	Wiener	2	06/2019	4.000	Free	Public	[30]
Navya	Groningen, Netherlands	Arriva (DB)	1	08/2018	1.000	Free	Public	[25]
Navya	Contern, Luxembourg	Sales-Lentz	1	09/2018	1.000	Free	Public	[31]
Navya	Oslo Waterfront Oslo, Norway	Holo	4	05/2019	2.200	Free	Public	[25]
Navya	Science Park Gothenburg, Sweden	Autonomous	2	04/2019	1.400	Free	Public	[25]
Navya	Helsinki, Finland	Metropolia	2	06/2019	2.000	Free	Public	[25]
Navya	St Perth Esplanade Perth, Australia	RAC	2	07/2016	3.500	Tax	Public	[25]
Navya	Adelaide, Australia	Flinders	1	06/2018	1.200	Free	Public	[25]
Navya	Ann Arbor, Michigan, US	MCITY	2	12/2016	1.600	Free	Public	[25]

(continued)

[1] Pedestrian area: type of traffic mixed with pedestrians and bikes.

[2] Public road: type of traffic mixed with pedestrians, bikes, and motorized vehicles.

Table 1.2 (continued)

Shuttle	Location	Operator	Units	Launch	Track (km)	Fee	Environment	Link
Navya	Lake Nona Orlando, Florida, US	Beep	2	09/2019	3.540	Free	Public	[25]
Navya	Las Vegas, Nevada, US	Keolis	1	11/2017	1.000	Free	Public	[25]
Navya	Montcalm, Montreal, Canada	Keolis	1	08/2018	2.000	Free	Public	[25]
Navya	Masdar City, Abu Dhabi, UAE	Navya	3	09/2018	0.900	Free	Pedestrian	[25]
Navya	West Kowloon, Hong Kong	Kowloon	1	07/2017	0.300	Free	Pedestrian	[25]
EasyMile	Velizy-Villacoublay, Paris, France	RATP/SCA	1	06/2018	–	–	Government[3]	[32]
EasyMile	Sorigny, France	TLD	1	11/2018	1.500	Free	Public	[33]
EasyMile	Satory Versailles, France	Transdev	1	12/2018	1.000	Free	Public	[33]
EasyMile	InnoZ Campus, Berlin, Germany	BVG	1	12/2017	0.600	Free	Public	[33]
EasyMile	Bad Birnbach, Germany	DB	2	10/2017	1.400	Free	Public	[33]
EasyMile	Project See-Meile, Berlin, Germany	BVG	1	08/2019	1.200	Free	Public	[33]
EasyMile	GreenTec, Enge-Sande, Germany	BMVI	1	06/2018	2.700	Free	Public	[33]
EasyMile	Bernmobil Demo Bern, Switzerland	AVOC	1	06/2019	2.000	Free	Public	[33]
EasyMile	Salzburg Research, Koppl, Austria	Digibus	1	04/2018	1.400	Free	Public	[33]
EasyMile	Austin Airport, Austin, Texas, US	AUS	1	08/2019	0.700	Free	Pedestrian	[33]
EasyMile	Virginia Tech, Virginia, US	NRV	1	05/2019	0.800	Free	Public	[33]
EasyMile	Calgary Zoo Calgary, Canada	PWT	1	09/2018	0.557	Free	Public	[33]
EasyMile	Renmark Aged Care, Australia	TAG	1	08/2019	4.500	Free	Public	[33]
EasyMile	Toormina, Australia	Busways	1	06/2019	1.000	Free	Pedestrian	[33]
EasyMile	Southeast, Nanjing, China	NJNDTIG	1	10/2018	1.400	Free	Public	[33]
EasyMile	National University of Singapore	Comfort	1	07/2019	1.600	Free	Public	[33]

(continued)

[3] Government property: legal designation for the ownership of property by governmental legal entities.

Table 1.2 (continued)

Shuttle	Location	Operator	Units	Launch	Track (km)	Fee	Environment	Link
Apollo	Software Park Xiamen, China	Baidu	1	04/2018	–	Free	Public	[34]
Apollo	Kink Long Xiamen, China	Baidu	1	03/2018	–	Free	Pedestrian	[34]
Apollo	Haidian Park Xiamen, China	Baidu	1	05/2019	–	Taxă	Public	[34]
Apollo	Xiongan New Area, China	Baidu	5	12/2017	4.000	Free	Public	[34]
Apollo	Shenzhen, China	Shenzhen	4	12/2017	1.200	Free	Public	[34]
Apollo	Yangquan, Shanxi, China	Baidu	22	01/2019	26.00	Free	Public	[34]
Apollo	Wuhan, China	Baidu	10	10/2018	5.000	Free	Pedestrian	[34]
Olli	ITCILO Campus, Turin, Italy	ITC-ILO	1	01/2020	–	Tax	Public	[35]
Olli	Colmar-Berg, Luxembourg	Sales-Lentz	3	03/2019	–	Free	Pedestrian	[35]
Olli	Concord, California, US	CCTA	1	10/2019	–	Taxă	Pedestrian	[35]
Olli	National Harbor, Maryland, US	MDOT	1	10/2019	–	Taxă	Public	[35]
Olli	Goodyear Tire, Akron, Ohio, US	Goodyear	1	02/2019	–	Taxă	Pedestrian	[35]
Olli	University of Buffalo, New York, US	UB	1	09/2019	–	Free	Pedestrian	[35]
Olli	King Abdullah, Saudi Arabia	SAPTCO	3	12/2019	–	Free	Public	[35]

References

1. Iclodean, C., Cordos, N., Varga, B.O.: Autonomous shuttle Bus for public transportation: a review. Energies **13**, 2917 (2020)
2. http://content.time.com/time/magazine/0,9263,7601250810,00.html (July 24, 2020)
3. https://archive.org/details/magicmotorways00geddrich/page/n9/mode/2up (July 24, 2020)
4. https://www.wired.com/2009/11/autonomous-cars/ (July 24, 2020)
5. https://trid.trb.org/view/1289421 (July 24, 2020)
6. http://uk.news.yahoo.com/how-the-first--driverless-car--was-invented-in-britain-in-1960-093127757.html (July 24, 2020)
7. http://faculty.washington.edu/jbs/itrans/parkshut.htm (July 24, 2020)
8. Bishop, R.: Intelligent vehicle technology and trends; Artech House: Norwood, MA, USA, pp 7–24 (2005)
9. https://www.sae.org/standards/content/j3016_202104/ (May 17, 2022)
10. https://www.its-platform.eu/wp-content/uploads/ITS-Platform/AchievementsDocuments/AutomatedDriving/EU%20EIP_SA42_%20Deliverable%20_Task_3_2020_Road_Map_and_Action_Plan_v6.0.pdf (April 22, 2022)

11. https://navya.tech/wp-content/uploads/documents/Brochure_Shuttle_EN.pdf (July 24, 2020)
12. https://www.ertrac.org/uploads/documentsearch/id57/ERTRAC-CAD-Roadmap-2019.pdf (July 24, 2020)
13. https://www.thedrive.com/tech/15848/waymo-is-already-running-cars-with-no-one-behind-the-wheel (July 24, 2020)
14. Amini, A., et al.: Learning robust control policies for end-to-end autonomous driving from data-driven simulation. IEEE Robot Autom Lett **5**, 1143 (2020)
15. https://www.unece.org/fileadmin/DAM/trans/main/wp29/wp29regs/2017/R079r3e.pdf (October 02, 2020)
16. https://www.vie-publique.fr/sites/default/files/rapport/pdf/184000287.pdf (October 02, 2020)
17. https://eur-lex.europa.eu/eli/reg/2009/661/2019-04-24 (March 30, 2020)
18. Bosch, R.: Bosch Automotive Handbook, 10th Edition (2019)
19. https://www.wevolver.com/article/2020.autonomous.vehicle.technology.report (May 02, 2021)
20. Ainsalu, J., et al.: State of the art of automated buses. Sustainability **10**, 3118 (2018)
21. http://apollo.auto/minibus/index.html (July 24, 2020)
22. https://easymile.com/vehicle-solutions/ez10-passenger-shuttle (July 24, 2020)
23. https://navya.tech/shuttle/ (April 04, 2020)
24. https://techcrunch.com/2022/01/13/local-motors-the-startup-that-created-the-olli-autonomous-shuttle-has-shutdown/ (April 22, 2022)
25. https://navya.tech/en/autonom-shuttle/#storeLocator__bottomHalf (April 12, 2020)
26. https://www.leprogres.fr/rhone-69-edition-est-lyonnais/2019/11/15/la-navette-autonome-n1-inauguree-a-decines (April 22, 2022)
27. https://navya.tech/fr/inauguration-de-lexperimentation-mia/ (April 22, 2022)
28. https://h2020-avenue.eu/?portfolio=geneva (July 24, 2020)
29. https://www.postauto.ch/en/project-smartshuttle (July 24, 2020)
30. https://www.wienerlinien.at/eportal3/ep/channelView.do/pageTypeId/66533/channelId/-4400687#4400868 (July 24, 2020)
31. https://h2020-avenue.eu/portfolio-item/luxembourg/ (July 24, 2020)
32. https://www.auto-infos.fr/article/les-navettes-autonomes-francaises-a-la-conquete-du-monde.218134 (April 22, 2022)
33. https://easymile.com/application-map-easymile/ (July 24, 2020)
34. https://github.com/ApolloAuto/apollo-platform (May 17, 2022)
35. https://corporate.goodyear.com/us/en/media/news/goodyear-local-motors-join-forces-on-autonomous-shuttles.html (April 22, 2022)

Chapter 2
Autonomous Driving Technical Characteristics

2.1 Powertrain Architecture

2.1.1 Overview

In recent years, autonomous driving has become not only a hypothetical driving solution, but a reality. The vehicle models used in the implementation of this concept are similar to the pre-existing, classic solutions, but in addition to the current systems they also include a complex network of sensors. Depending on the degree of automation, these classic control elements begin to disappear and will disappear completely once the degree of automation reaches level 5. In parallel, new decision management solutions have begun to appear (tablets and/or joysticks) for the control of autonomous vehicles in cases of emergency or unforeseen situations.

The complexity of the direct mechanical control by the driver will decrease as the degree of autonomous control over the dynamic behavior of the vehicle increases, until the driver will no longer have any control over the behavior of the vehicle, except for the indication of the destination address. This process of physical elimination works in parallel with the modernization of the level of automation. However, the complete replacement of a driver requires a series of upgrades in the sensor system and an increased processing power of the control units of the autonomous driving system.

2.1.2 Driveline

The traction power of the autonomous vehicle's wheels is provided by the electric propulsion engine. The torque for the left and right wheels is provided by a differential that has the same transmission ratio as the electric motor shaft at maximum angular speed. The torque and angular speed of the electric motor are controlled by the inverter

C. Iclodean et al., *Autonomous Vehicles for Public Transportation*, Green Energy and Technology, https://doi.org/10.1007/978-3-031-14678-7_2

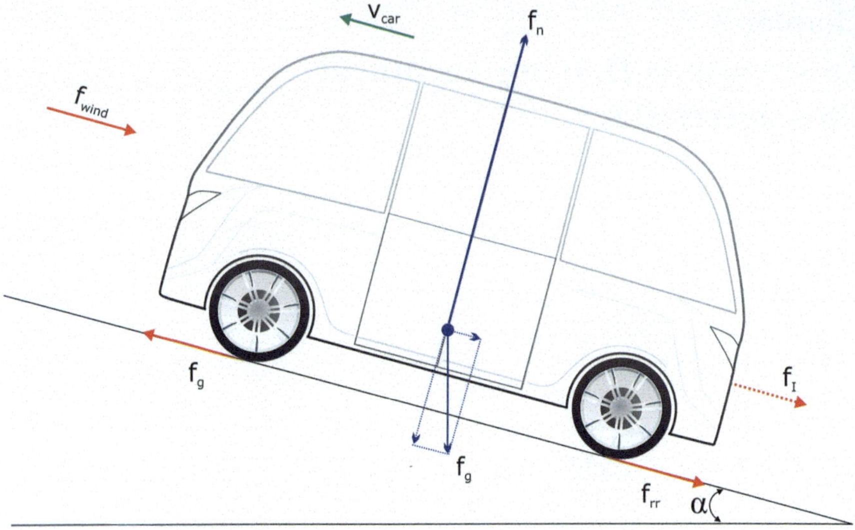

Fig. 2.1 Forces acting on the moving vehicle

which converts the DC (Direct Current) voltage of the battery into alternating voltage to supply the electric motor.

The forces that the electric motor must overcome for the vehicle to move forward are those due to rolling resistance, resistance due to the ramp/slope, air resistance, respectively those due to inertia (Fig. 2.1). The tractive force of a vehicle is described by the equations [1]:

$$\begin{aligned} f_t = \underbrace{M_{vehicle}\dot{v}_{vehicle}}_{f_I} + \underbrace{M_{vehicle}g}_{f_g}\sin(\alpha) + \operatorname{sign}(v_{vehicle})\underbrace{\overbrace{M_{vehicle}g\cos\alpha}^{f_n}c_{rr}}_{f_{rr}} \\ + \operatorname{sign}(v_{vehicle}+v_{wind})\cdot\underbrace{\frac{1}{2}\rho_{air}C_{drag}A_{front}(v_{vehicle}+v_{wind})^2}_{f_{wind}} \end{aligned} \quad (2.1)$$

$$C_{rr} = 0.01\left(1 + \frac{3.6}{100}v_{vehicle}\right) \quad (2.2)$$

where: f_t(N) vehicle tractive force, $M_{vehicle}$(kg) vehicle mass, $\dot{v}_{vehicle}$ (m/s^2) vehicle acceleration, f_I(N) vehicle inertial force, g(m/s^2) gravitational acceleration (9.8), f_g(N) vehicle gravitational acceleration, α (rad) angle of longitudinal inclination of the road, $v_{vehicle}$(m/s) vehicle velocity, f_n(N) the normal force of the vehicle, c_{rr}(-) the rolling resistance coefficient, f_{rr}(N) the rolling resistance force, v_{wind}(m/s) wind speed, ρ_{air}(kg/m^3) air density at 20 °C (1.2041), C_{drag}(-) the aerodynamic coefficient,

$A_{front}(m^2)$ the surface of the front section of the vehicle, $f_{wind}(N)$ force due to air resistance.

The propulsion system of autonomous vehicles is equipped with asynchronous/synchronous traction motors powered by a DC/AC (Alternating Current) converter (according to the options made available by the manufacturer). The equations that describe how a permanent magnet synchronous motor PMSM (Permanent Magnet Synchronous Motor) works in a direct quadrature DQ-frame (Direct Quadrature), which rotates synchronously with the mains voltage vector so that three-phase signals that vary over time are transformed into direct current signals, are given by the following relations [1]:

$$V_d = R_s I_d + L_d \frac{dI_d}{d_t} - \omega_e L_q I_q \tag{2.3}$$

$$V_q = R_s I_q + L_q \frac{dI_q}{d_t} - \omega_e L_d I_d + \omega_e \lambda_{pm} \tag{2.4}$$

$$P_{EM} = \frac{3}{2}\left(V_d I_d + V_q I_q\right) \tag{2.5}$$

where: $V_d(V)$ D axis voltage, $R_s(\Omega)$ stator phase resistance, $I_d(A)$ D axis current, $L_d(H)$ D axis inductance, $\omega_e(rad/s)$ the angular frequency of the stator, $L_q(H)$ Q axis inductance, $I_q(A)$ Q axis current, $V_q(V)$ Q axis voltage, $\lambda_{pm}(Wb)$ permanent magnetic flux, $P_{EM}(W)$ electric input power.

The mechanical torque of the PMSM motor is expressed by the following relation [1]:

$$\tau_e = J_s \frac{d\omega_s}{dt} + B_V \omega_s + \tau_c + \tau_s = \frac{3}{2}\frac{P}{2}\left(\lambda_{pm} I_q + \left(L_d - L_q\right) I_d I_q\right) \tag{2.6}$$

where: $\tau_e(Nm)$ electromechanical torque, $J_s(kgm^2)$ the moment of inertia of the shaft, $\omega_s(rad/s)$ the angular frequency of the shaft, $B_v(Nms/rad)$ viscous friction coefficient, $\tau_c(Nm)$ is the Coulomb friction torque, $\tau_s(Nm)$ shaft torque, $P(-)$ the number of poles of the motor.

The inverter converts the DC voltage of the electric battery into alternating voltage to supply the electric motor and is characterized by the circuit diagram shown in Fig. 2.2 [1]. The inverter transmits the electrical power between the electric motor (with phase voltages V_A, V_B, and V_C) and the battery by starting and stopping the commutators Q_A+, Q_A-, Q_B+, Q_B-, Q_C+ and Q_C-. The commutators have a starting resistance $R_{Q,Inv}$ and $R_{D,Inv}$. The parallel diodes of each commutator open a path for the motor currents during the time the circuits of both sides are open. The average power losses of a commutator $P_{Q,Inv}$ and of a diode $P_{D,Inv}$ in a period are given by the relations [1]:

$$P_{Q,Inv} = \left(\frac{1}{8} + \frac{m_i}{3\pi}\right) R_{Q,Inv} \hat{I}_p^2 + \left(\frac{1}{2\pi} + \frac{m_i}{8}\cos(\Phi_{EM})\right) V_{Q,th,Inv} \hat{I}_p \tag{2.7}$$

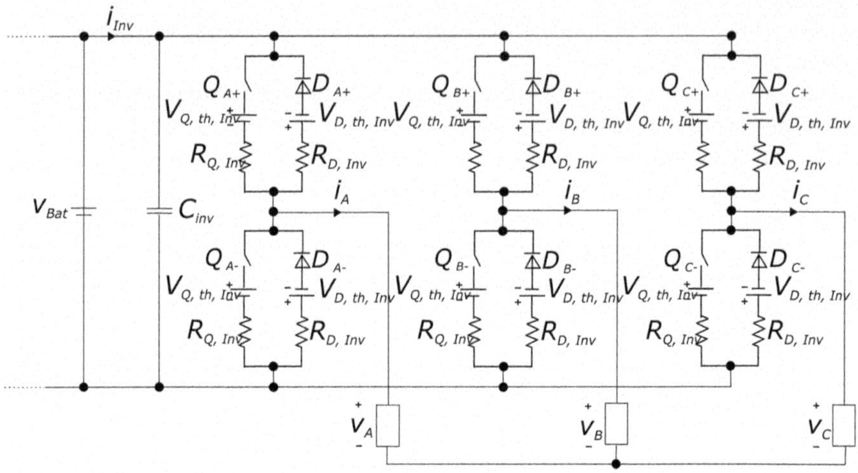

Fig. 2.2 Inverter electrical diagram

$$P_{D,Inv} = \left(\frac{1}{8} + \frac{m_i}{3\pi}\right) R_{D,Inv} \hat{I}_p^2 + \left(\frac{1}{2\pi} + \frac{m_i}{8}\cos(\Phi_{EM})\right) V_{D,th,Inv} \hat{I}_p \tag{2.8}$$

where: $P_{Q,Inv}(W)$ commutator power loss, m_i modulation index, $R_{Q,Inv}(\Omega)$ inverter commutator resistance, $\hat{I}_p(A)$ the peak of the current phase, $\Phi_{EM}(rad)$ power factor angle, $V_{Q,th,Inv}(V)$ inverter commutator threshold voltage, $P_{D,Inv}(W)$ diode power loss, $R_{D,Inv}(\Omega)$ inverter diode resistance, $V_{D,th,Inv}(V)$ inverter diode threshold voltage.

If we assume that the voltage drops threshold of the commutators and diodes is equal, i.e., $V_{th,Inv} = V_{Q,th,Inv} = V_{D,th,Inv}$ and that the resistances of the commutators and diodes are also equal, i.e., $R_{Inv} = R_{Q,Inv} = R_{D,Inv}$, the inverter total power loss $P_{Inv,loss}$ and efficiency η_{Inv} are given by the following relations [1]:

$$P_{Inv,loss} = 6\left(P_{Q,Inv} + P_{D,Inv}\right) = \frac{3}{2} R_{Inv} \hat{I}_{Inv}^2 + \frac{6}{\pi} V_{th,Inv} \hat{I}_{Inv} \tag{2.9}$$

$$\eta_{Inv} = \begin{cases} \frac{P_{EM}}{P_{Inv}}, & P_{EM} \geq 0 \\ \frac{P_{Inv}}{P_{EM}}, & P_{EM} < 0 \end{cases} \tag{2.10}$$

where: $I_{Inv}(A)$ is the input current to the inverter, $P_{Inv}(W)$ is the input power to the inverter, η_{Inv} is the efficiency of the inverter.

The electric traction motors generate the motor moment at the wheels of the autonomous vehicle according to the control algorithm of the electric vehicle and according to the limitations imposed by the BMS (Battery Management System) [2].

Electric motors also operate as electric generators performing regenerative braking, in which case they will recover a maximum amount of braking energy

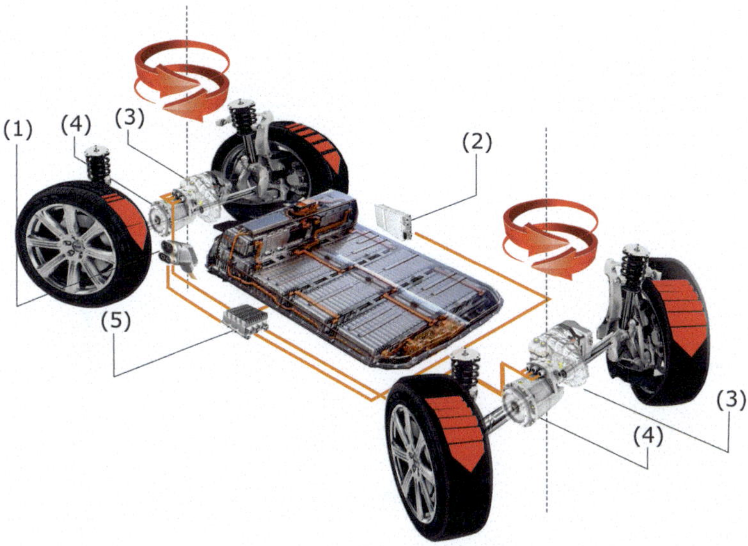

Fig. 2.3 Propulsion system architecture

(energy that will be stored in a battery through an AC/DC converter and/or a supercapacitor). Another solution is to direct the recovered kinetic energy through the power supply (AC/DC and DC/DC) to the low voltage electrical consumers of the autonomous vehicle (Fig. 2.3), where (1) standard electrical charging connector, (2) main inverter, (3) driveline, (4) electric motor/generator, (5) power electronics [3].

The command and control of the operation of the propulsion system is performed by the electronic control unit ECU (Electronic Control Unit), which is integrated in the autonomous driving system of the vehicle.

2.1.3 High Voltage Battery

The high voltage electric battery that powers the autonomous vehicle consists of a series of low voltage cells (2.2–3.7 V) grouped in modules and connected in series and/or in parallel to increase the power and voltage of the electric battery. The complete battery pack has a voltage of 80 V (depending on the manufacturer) with a power between 20 and 35 kWh. The battery pack is continuously monitored by the battery management system to meet the criteria [4]:

(1) Charges/discharges at equal voltage for each SOC (State Of Charge).
(2) Optimal operating temperature for the base cells.
(3) Optimal C rate (speed at which the battery is charged or discharged) for the entire battery pack.
(4) Complete SOH (State Of Health) of the battery.

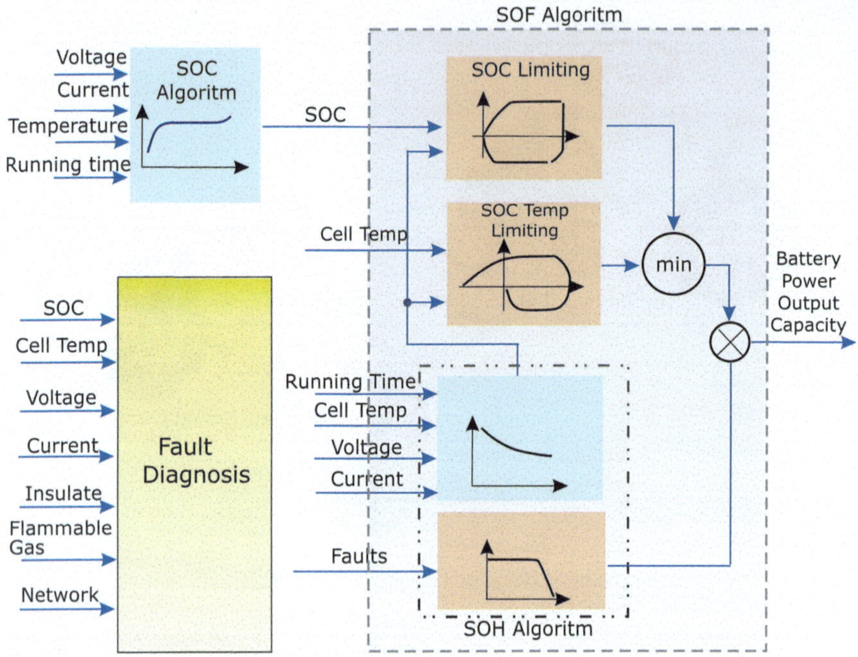

Fig. 2.4 Battery Management System algorithm

Battery power output capacity is calculated based on the operating algorithm of the BMS system and takes into account the following performance criteria: SOC describing the difference in battery charge considered compared to a 100% charged battery, SOH describing the difference in battery health considered compared to a battery with a 100% state of health, respectively SOF (State Of Function) which describes the difference between the energy available from the battery and the maximum energy that could be stored in the battery (Fig. 2.4)[5].

Electric batteries have a capacity that will ensure the operation of autonomous vehicles on the selected routes in a predetermined time (generally the duration of a working day, respectively 8 to 10 h) at maximum load with passengers, with the temperature conditioning systems in the passenger compartment in operation, in all possible situations and in all environmental conditions.

The batteries use Lithium technology, with a high density of stored energy, respectively with a minimum volume and mass, necessary to obtain the required autonomy, with a maximum safety in operation in the climatic conditions in which autonomous vehicles operate. Electric batteries allow a charging regime from 0 to 99% in a well-defined time interval depending on their capacity, regardless of the ambient thermal regime.

There are models of autonomous electric vehicles that are equipped with solutions to extend the autonomy using an additional battery with a maximum capacity of

Table 2.1 Electric batteries that equip autonomous shuttle buses

Elements	Apollo Baidu	EasyMile EZ10	Navya Evo	Olli
Capacity	no data	20.0 kWh	33.0 kWh	18.5 kWh
Model	no data	LiFePO4	LiFePO4	Lithium

10% of the capacity of the main battery (or a supercapacitor), which is used in situations where the capacity of main batteries falls below 20% to ensure additional autonomy. This battery/supercapacitor is separated from the main power supply and is permanently maintained at a maximum charge level.

The main characteristics of the high voltage electric battery that is equipped on the main models of autonomous shuttle buses are presented in Table 2.1 [3, 6–9].

Several types of batteries are used in the construction of electric vehicles. Studies and research conducted in recent years have shown which of these types of batteries best meets the performance criteria (energy storage efficiency, constructive characteristics, price, safety, and durability).

Lithium batteries are currently the most widespread technology that equips electric vehicles, due to the high density of stored energy (100–250 Wh/kg) and due to the high power per unit mass (800–2000 W/kg), which allows the development of lighter and smaller battery packs at competitive prices [10]. Another advantage of Lithium batteries is the lack of memory effect (gradual loss of maximum energy storage capacity in case of repeated recharging without being completely discharged), resulting in an increased life cycle. The disadvantages of Lithium batteries are the operating temperature, the short lifespan (charging-discharging cycles) and the operational safety, factors that require the existence of a BMS battery management system, necessary to monitor and control the internal temperature of the cells [10].

Ni-MH batteries are among the first types of batteries used to equip the electric vehicles in the 2000s. The advantages of these batteries are the high density and energy power, which allow an extended autonomy of electric vehicles (for the same capacity of the battery pack compared to Lithium ones, but at a lower mass than them), the long lifespan, the very good thermal properties (operation at temperatures from -30 °C to $+70$ °C), and the safety of charging and using these batteries. The main disadvantage of Ni-MH batteries is the outdated technology and the high cost due to reduced production [10].

$Na\text{-}NiCl_2$ are used to equip some prototypes of vehicles or electric buses used in public transport. The main advantages are the high density of stored energy (100–120 Wh/kg), the low cost compared to other existing technologies and the long lifespan. The main disadvantage of $Na\text{-}NiCl_2$ batteries is the internal operating temperatures that have high values (250–350 °C). Thus, when the vehicle is not in use, the batteries are kept at operating temperature by an external heating system, which consumes electricity [10].

Table 2.2 The characteristics of the studied electric batteries

Name	Li-Ion	Ni-MH	Na-NiCl$_2$	M.U
Capacity	24	24	24	kWh
Nominal voltage	323	288	289	V
Maximum charge	75	85	84	Ah
Minimum voltage	308	274	275	V
Maximum voltage	339	302	304	V
Cells/unit	17	20	30	–
Operating temperature	33	36	270	°C
Thermal capacity	0.40	0.40	6.00	W/K
Caloric capacity	795	677	950	J/kgK
Battery mass	318	534	457	kg
Battery price	300	400	500	Euro/kWh

The electrical characteristics of the three types of batteries considered and presented above are found centralized in Table 2.2. The storage capacity considered for each type of battery was 24 kWh, this being an acceptable average value for powering an autonomous shuttle bus for public transportation [10].

2.1.4 Charging System

The charging modes of electric and autonomous vehicles are defined according to IEC 61851-1 depending on the type of power, the voltage level, the bidirectional communication between the vehicle and the charging station, as well as the existence of earthing and protection devices. The loading modes are classified as follows (Fig. 2.5) [11]:

(1) AC mode-1 consists of the SAE J1772 standard socket and an extension cable that attaches to the charging port of the autonomous vehicle and to the standard power supply socket. For this charging system there is no protection against electric shocks, the only protection is provided by the earthing system of the electrical installation. AC mode-1 charging can be supplied to the autonomous vehicle with a built-in charger of up to 1.9 kW from a single-phase station of 250 VAC or from a three-phase station of 480 VAC and at a maximum current of 10–16 A. AC mode-1 is a slow charging mode and for this reason, it is preferred mainly overnight, the average charging time being 10–15 h.

(2) AC mode-2 consists of a slow charging process from a general-purpose socket with RCD (Residual Current Device) on cable. In this mode the charging is done from a standard socket, but with EVSE (External Vehicle Supply Equipment) which is supplied by the manufacturer together with the autonomous vehicle. This cable provides protection against electric shocks residual current device

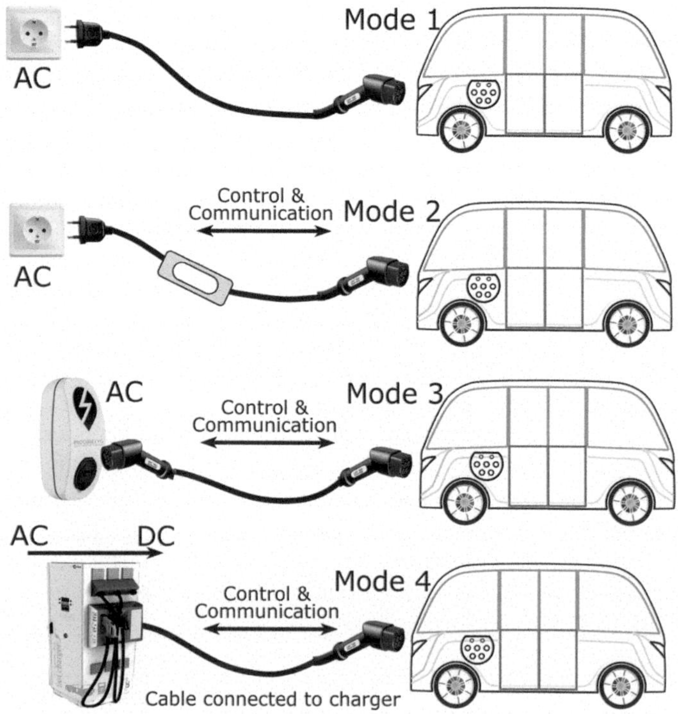

Fig. 2.5 Charging modes (according to IEC 61851-1)

RCD, maximum overcurrent protection, overtemperature protection, as well as connection to the earthing system of the electrical installation. As well as for AC mode-1, the power is provided by a 250 VAC single-phase station or a 480 VAC three-phase station and at a maximum current of 32 A.

(3) AC mode-3 consists of a semi-fast charging process from a special socket. This charging mode allows faster charging than the modes 1 and 2, depending on the power supplied by the grid and the power of the vehicle charger, and the communication is established between the vehicle and the special socket via a pilot line.

(4) DC mode-4 consists of a fast-charging process through an external DC charger that uses an AC to DC converter. The autonomous vehicle is recharged in a maximum of one hour from a direct current source with a maximum value of up to 400 A. The external charger is powered by a three-phase circuit of up to 575 VAC. The vehicle communicates with the external charger for complete control of the autonomous vehicle's battery charge and for maximum protection against electric shocks [11].

The autonomous vehicle charging strategy is based on the fact that there is a direct link between the SOH of the electric battery and used the charging solution.

Table 2.3 Charging levels (SAE J1772)

Level	Voltage (V)	Phases (~)	Current (A)	Power (kW)	Time
Level 1	120	1*AC	12–16	1.4–1.9	≤17 h
	200–450	DC	≤80	≤36	≤1.2 h
Level 2	240	1*AC	≤80	≤20	≤7 h
	200–450	DC	≤200	≤90	≤20 min
Level 3	208–600	3*AC	No data	≤20	No data
	200–600	DC	≤400	≤240	≤10 min

The continuous use of fast charging stations will generate a direct influence on its reliability and usability over time.

Charging stations are equipment that charges electric batteries that equip autonomous vehicles in a well-defined time frame, depending on their capacity, through special or dedicated connectors/couplers [12, 13]. The charging stations used to power electric vehicles are characterized by the following specifications (Table 2.3) [14]. Charging levels are used to classify the power, voltage, and nominal current of charging stations according to the specifications defined by SAE J1772 [3, 11, 15, 16]:

(1) AC level-1 uses a 250 VAC socket and a maximum current of 16 A, which ensures a maximum charging power of up to 1.9 kW. The type of charging is slow charge, and it is done through a SAE J1772 charging cable, the charger being built into the vehicle.
(2) AC level-2 consists of a charge from a 250 VAC single-phase station, with a maximum current of 30 A and a maximum charging power of 7.2 kW. The charging cable is of the SAE J1772 type, and the maximum output power is of 19.2 kW, the charger being incorporated in the vehicle. If a three-phase 480 VAC station is used at a maximum current of 80 A, charging will be done through a cable with a type 2 connector (Mennekes) (IEC-62196) and charging times are shorter. Levels AC-1 and AC-2 use a charger integrated in the autonomous vehicle, the power of the charger being proportional to the level of charge and to the power supply from the grid (single-phase or three-phase).
(3) AC level-3 consists of a semi-fast charge from a special socket that provides a power of up to 130 kW, making the SOC very fast and using three-phase alternative current at 400 VAC.
(4) DC level-4 consists of a fast charge through an external DC charger that uses an AC to DC converter. This way the autonomous vehicle is recharged in a maximum of one hour from a direct current source with a maximum value of up to 400 A. The external charger is powered by a three-phase circuit of up to 575 VAC. The vehicle communicates with the external charger for complete control of the autonomous vehicle's battery charge and for maximum protection against electric shocks.

Charger types that equip the sockets used in the power supply of the autonomous shuttle buses charging system are according to the specifications defined by the international standard IEC 62196-2 [3]. For example, Apollo Baidu uses a 50 kW DC type 2 CHAdeMO charging connector type [3, 17], EasyMile uses a type 2 AC charging connector (Mennekes) [3, 18], Navya uses a type 2 (Mennekes) (IEC-62196) [3, 10] and Olli uses a type 2 AC charging connector (Mennekes) [3, 19].

Electric batteries are charged by charging stations by connecting to the on-board connector, or by wireless charging. S-WEVCS (Static Wireless Electric Vehicle Charging System) works on the principle similar to that of transformers, where electricity is transferred from a primary coil to a secondary coil (Fig. 2.6) [11].

The operating principle of an S-WEVCS is the principle of the transformers, where the energy is transferred from a primary to a secondary coil. AC voltage is converted to DC using an AC/DC rectifier. Next, a high frequency DC/AC converter provides high frequency current to the primary coil. The current in the primary coil creates an alternating magnetic field, which induces an alternating voltage on the secondary coil.

The secondary coil is installed below the vehicle. To increase the system's power transfer capability, a resonant capacitor is used in the secondary coil. Finally, an AC/DC rectifier converts the alternating voltage of the secondary coil to continuous voltage so that the battery of the vehicle is charged. The wireless power transfer WPT (Wireless Power Transfer) system is activated when the vehicle is in the charging area. The charging time depends on the power source level, the charging coil sizes

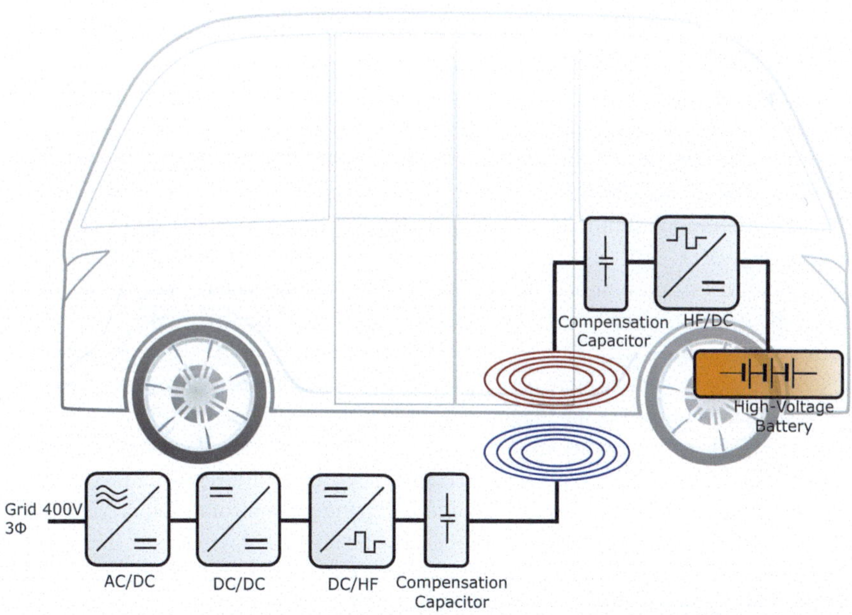

Fig. 2.6 Static wireless electric vehicle charging system

and the air gap distance between the two coils. The average distance between two coils is about 150–300 mm.

Using WEVCS (Wireless Electric Vehicle Charging System) the power transfer is limited compared to traditional plug-in chargers. The WEVCS classes can reach 22 kW according to the SAE J2954 electromagnetic compatibility topic [11].

Dynamic electric vehicle wireless charging D-WEVCS (Dynamic Wireless Electric Vehicle Charging System) is a technology that reduces the problems associated with the range and the cost of electric vehicles. This method charges the batteries, while the vehicle is in motion, and the vehicle's autonomy is increased.

In the case of D-WEVCS technology, when the electric vehicle passes over the primary coils, it receives a magnetic field through a secondary coil and converts the coil current to DC to charge the batteries [11, 20–22, 23].

2.1.5 Steering Systems

The steering of autonomous vehicles is a key factor, and due to the lack of steering wheel, direct control over the direction of travel is missing. Steering control decisions are managed by a decision algorithm supervised by the autonomous driving system. The steering of an autonomous vehicle is a reactive action generated by a series of external factors, rather than the decision of a virtual driver that is reflected in a proactive way based on events on the road, in traffic or in the surroundings. The steering in this case is the means, not the method by which the vehicle follows its path established by the autonomous driving system. This is why an autonomous vehicle operates on the basis of a virtual/optimal trajectory. This way there is no freedom for the autonomous vehicle to cut the demarcation lines between lanes nor the possibility of encountering dangerous situations in traffic, but there is always a safe method of driving and a controlled balance between the position of the vehicle's center of gravity and the three axles of the forces acting on each of the steering wheels. Autonomous shuttle buses can be equipped with two or four-wheel steering systems (Fig. 2.7) [3].

In the case of autonomous shuttle buses, the steering wheels on the rear axle turn in the opposite direction to the steering wheels on the front axle. Four-wheel steering systems have major advantages. One of these advantages is the reduction of the steering angle, which increases safety by reducing the response time required. This system also improves the maneuverability of a vehicle during tight cornering or parking [3, 21]. Autonomous shuttle buses with four steering wheels are equipped with independent steering controllers on each wheel. This allows greater flexibility in controlling the vehicle's movement via the independent control of the driving force and the rotation angle (δA, δB, δC, δD) of each wheel [3, 24, 25]. The steering system of the autonomous shuttle buses is built around a rack-type pinion mechanism and a steering linkage mechanism controlled by a 24 VDC motor. In Fig. 2.8 is presented a kinematic diagram of the steering mechanism: (0) chassis, (1) steering mechanism

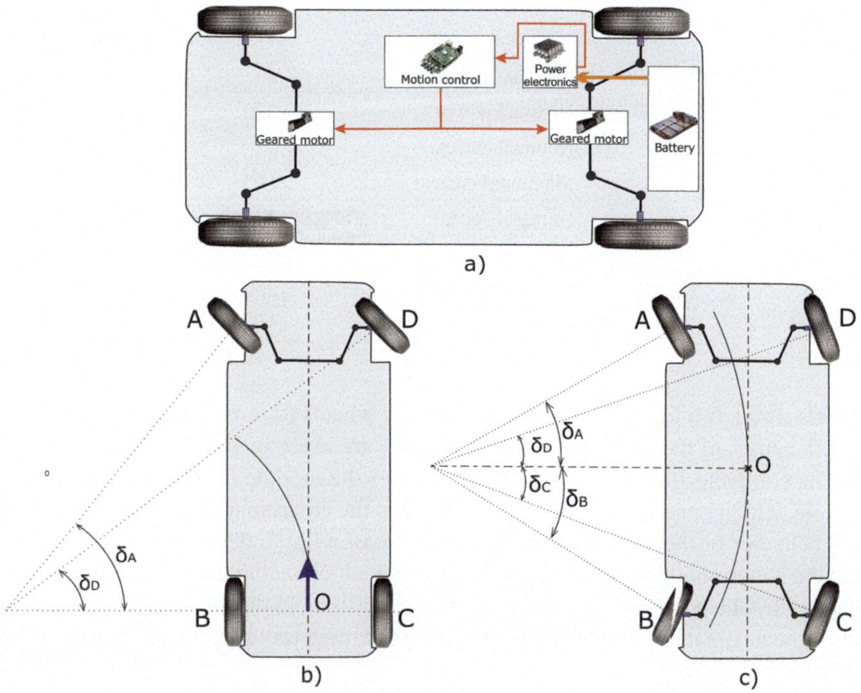

Fig. 2.7 Steering system: **a** overview of the steering system; **b** the steering system with two steering wheels; **c** the steering system with four steering wheels

driven by gear motor, (2) rack, (3) left link, (4) right link, (5) left wheel, (6) right wheel [3, 7].

The steering system of the autonomous shuttle bus is powered by an electric gear motor that drives the steering pinion (1), which in turn drives the rack (2). The electric gear motor is controlled by the motion control block in the autonomous control system. The pinion-and-rack gear converts the rotational motion of the electric gear motor into translational movement that is transmitted to the hubs of the

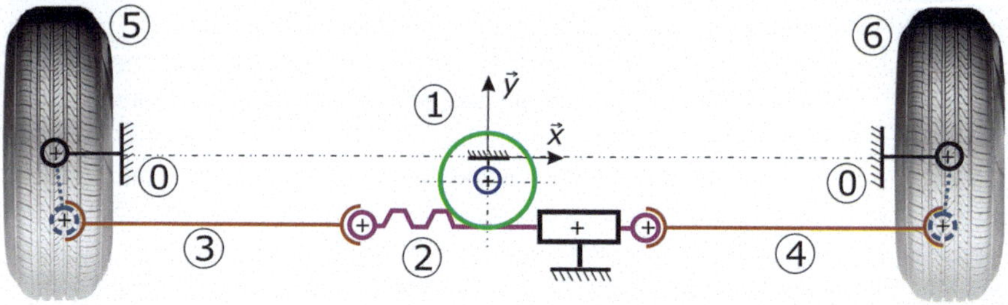

Fig. 2.8 Kinematic diagram of the front wheel steering system

Table 2.4 Specifications of the electric gear motor

Technical specifications	Values	M.U
Nominal voltage	24	V
Nominal power	320	W
Nominal current	13.5	A
Maximum current	105	A
Nominal speed	30	1/min
Nominal torque	45	Nm
Maximum torque	350	Nm
Reduction ratio	50	–

wheels through a link (3, 4), which causes the wheels to turn (5, 6). The technical specifications of the type of electric gear motor are given in Table 2.4 [3, 26].

The complete steering control chain includes the module for the behavior of the vehicle, which generates a planned trajectory in the environment in which it travels. The behavior of the vehicle is based on the use of a file to define the road network that the autonomous vehicle must travel on, and which includes the streets and a file that defines the points that the autonomous vehicle will go through while traveling to a specific destination. The planned and generated trajectory is entered in the steering controller [3, 27].

The task of the steering controller (Fig. 2.9) is to compare the current position of the autonomous vehicle with the alignment of the planned trajectory (where: (X_{def}, Y_{def}) is the route's geographical coordinates—the X and Y, (V_{def}) is the vehicle's speed on the route, (ω_{ro}) is the required angular speed of the wheel in order to follow the route, ($\sum F_x$) is the sum of the forces in the longitudinal direction, ($\sum F_y$) is the sum of the forces in the transverse direction, ($\sum F_z$) is the sum of the forces in the vertical direction, (F_{xi}) is the longitudinal forces on the wheel in the longitudinal direction, (F_{yi}) is the transverse forces to the wheel in the longitudinal direction, (M_i) is the torque at the driving wheels, (δ_i) is the steering angle of the wheels, (v_x) is the vehicle's longitudinal speed, (v_y) is the vehicle's transverse speed, (ω_r) is the angular speed of the wheel). Depending on the current speed of the autonomous vehicle, the controller generates the actuation command on the steering system. The centralized control structure is based on the characteristics of the vehicle dynamics, and the subsystems are controlled directly by a centralized controller.

The design of the controller is generally based on linear or non-linear models, and the design method of a MIMO (Multi-Input Multi-Output) system is adopted to solve the problems related to the vehicle's longitudinal and lateral dynamics [3, 28, 29].

The stability and maintenance of the desired direction of travel of the autonomous vehicle is ensured in critical situations—due to loss of grip—by the ESP (Electronic Stability Program) which intervenes in remedying these critical situations by commanding and controlling the braking system and by adapting the torque generated by the powertrain. The operating principle of the ESP is based on information

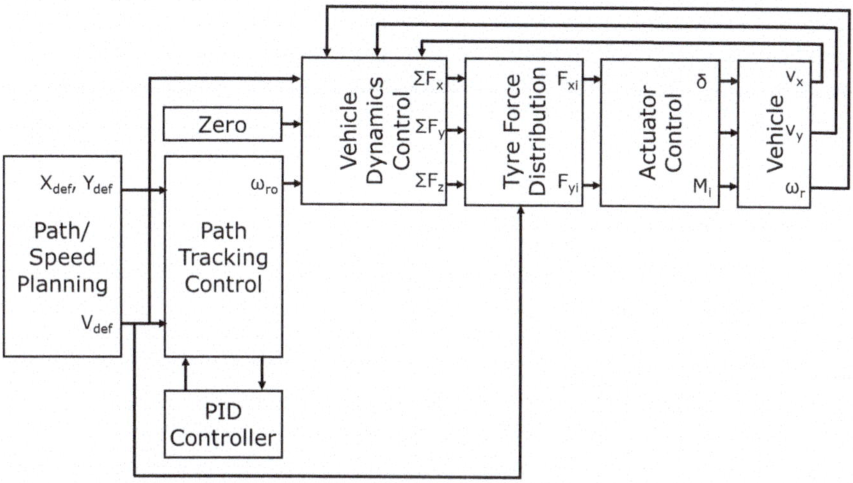

Fig. 2.9 Block diagram of the steering controller

that comes from sensors. Thus, the ESP compares the planned driving direction with the actual direction of the autonomous vehicle at the time of the comparison.

The planned direction of travel of the vehicle and the intention to maintain this direction by the autonomous steering system is determined on the basis of information on the steering angle of the steering system, wheel speed, acceleration value and pressure in the hydraulic braking circuit (Fig. 2.10, where (1) electrohydraulic element with integrated electronic control module, (2) steering angle sensor, (3) transverse gyration and acceleration sensor, (4) wheel speed sensor).

The actual driving direction of the autonomous vehicle is calculated based on the information provided by the gyration sensor (which monitors the rotation around

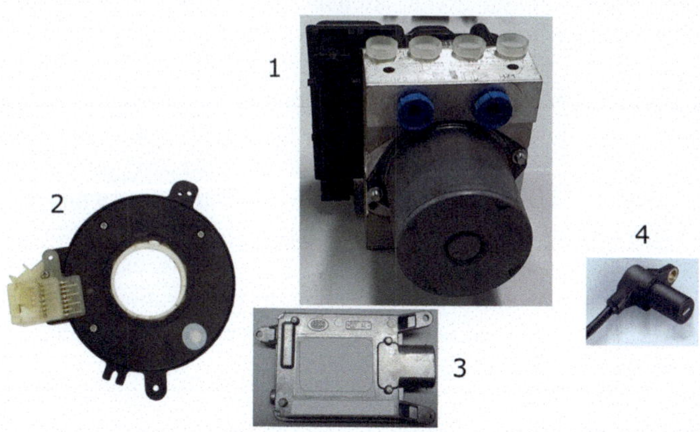

Fig. 2.10 Electronic Stability Programe

the vertical axis), respectively by the value of the lateral acceleration. Thus, through individual wheel braking, combined with the reduction of torque, the ESP has the role of bringing and keeping the vehicle in the direction of travel planned by the autonomous driving system [3].

The operating principle of the ESP is as follows: the steering angle sensor (2) generates electrical signals based on the magneto resistive effect (Hall) which transmit to the control unit information on the position of the steering system at any time. The gyration sensor and side acceleration sensor are integrated in the same unit (3). The gyrometer is mechanical and it detects the rotation of the autonomous vehicle around the vertical axis in the event of loss of stability. The lateral acceleration sensor (Hall with effect) transmits to the control unit the information on the intensity of the lateral acceleration. The wheel speed sensors (4) transmit information regarding their rotational speed. Based on the information received from the sensors, the electronic control unit of the autonomous control system acts on the valves of the electro-hydraulic pump (1) in order to activate/deactivate the brake system individually on each wheel, respectively on the propulsion system to change the torque.

2.1.6 Braking System

In the case of classic vehicles, the driver uses the senses, experience, and skills he/she has in order to make predictions in road traffic. Usually when breaking the vast majority of drivers use a combination of engine brake and brake pedal. Truck and bus drivers use a retarder before using the brake pedal. In the end, braking is done partly instinctively and partly through awareness. Taking into account all these variables, vehicle manufacturers do not provide a certain braking distance depending on the different speeds of the vehicle, but they equip the vehicles with braking systems capable of ensuring a minimum deceleration measured in m/s^2. The braking distance for vehicles at different travel speed values is shown in Fig. 2.11. This distance depends on certain factors, for example road surface, tire wear, weather conditions, the driver's reaction time (thinking distance) etc.

In the case of autonomous vehicles, the same rules apply as laid down in the legislation, so that these vehicles can always reach the desired level for deceleration. A real problem experienced on autonomous shuttle buses traveling with passengers in traffic is how this deceleration is distributed relative to the braking distance and how the passengers in the bus, especially those traveling standing, experience this deceleration. Consequently, the algorithm controlling the deceleration must be adaptable and able to distinguish between normal braking, deceleration and smooth braking, respectively emergency braking [3, 21].

Depending on the value of the required deceleration, respectively depending on the distance the vehicle has to travel to stop safely, the control algorithm of the braking system may decide to use the hydraulic braking system or the regenerative braking system, independent one from another or at the same time. ABS (Anti-lock Braking

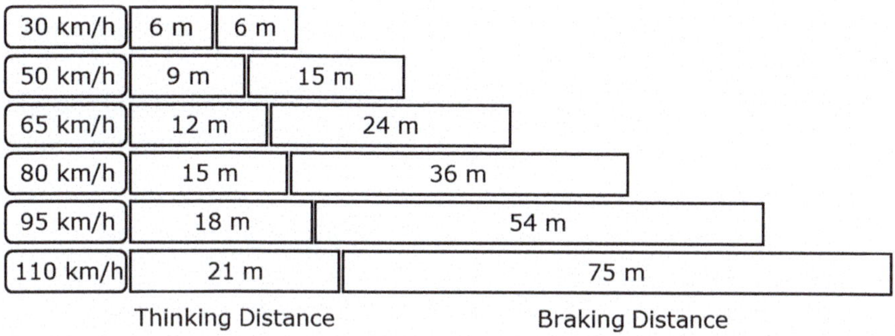

Fig. 2.11 Typical braking distance

System) has the role of preventing the locking of the wheels while the autonomous vehicle is braking and allows maintaining control of the steering during braking and shortening the braking distance.

Similar to the ESP, the ABS (Fig. 2.12, where (1) brake disc, (2) wheel speed sensor, (3) hydraulic modulator, (4) ABS control unit, (5)-(6) brake circuit pipes) consists of a control unit, four speed transducers (one for each wheel) and a number of hydraulic valves on the braking circuit. The control unit constantly monitors the rotational speed of each wheel. When it detects that one of the wheels is rotating more slowly than the others (a condition that will bring it to a standstill), it operates the valves to lower the pressure in the braking circuit, reducing the braking force on that wheel. The ABS constantly monitors the rotational speed of all the wheels and compares the speed of each wheel with the average of the reference speeds for the other wheels, determining which of the vehicle's wheels rotate at a speed higher or lower than the reference speed [3].

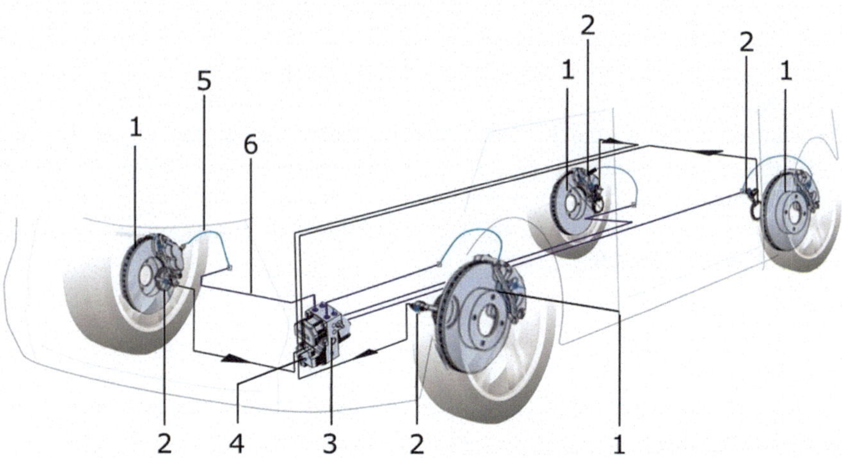

Fig. 2.12 Anti-lock Braking System

The ABS allows the control of the braking of the autonomous vehicle according to the following criteria [3]:

(1) Braking below the road grip limit when the ABS does not intervene, and the braking torque is maintained at the maximum value set by the autonomous driving system.
(2) Braking at the road grip limit when the ABS detects the tendency of the brake wheel to lock and controls the maintenance or decrease of the pressure in the braking system so that maximum road grip is used. When the wheel lock tendency occurs, the ABS controls the isolation of the corresponding brake cylinder from the rest of the braking system. Depending on the accentuation or disappearance of the tendency to lock the wheels, the pressure in the brake cylinder will be ordered to be reduced or increased, thus executing a succession of braking-relaxation cycles that will keep the wheel in the optimal grip area.

Combined braking that occurs when passing areas with different coefficients of adhesion, when the ABS ensures the prevention of wheel lock-up - on low-grip sections, and maximum braking—on high-grip sections. The ABS also works efficiently if the vehicle crosses a road section with different adhesion coefficients.

The operating principle of the ABS is as follows: on each wheel is installed a detection ring and a speed sensor that sends an electrical signal to the control unit. By analyzing the frequency of this signal, the control unit can monitor the rotational speed of each wheel and can calculate the acceleration or deceleration. During braking, the control unit may at any time determine whether the rotational speed of any wheel differs from the reference speed. If one of the wheels is on a slippery surface and cannot sustain the required braking level, then the wheel speed will decrease compared to the reference speed, meaning the sliding and deceleration of that wheel will start to increase. The control unit will detect these situations throughout the braking process, thus avoiding the locking of any of the wheels. When the control unit detects excessive slip and requires a rapid change of deceleration on one of the wheels, it sends an electrical signal to the pressure modulator valve to release the excess pressure in the service brake actuator on that wheel [3].

The AEB (Autonomous Emergency Braking) system has the task to decelerate safely the autonomous vehicle to the velocity of the target frontal object. For this action, the system compares the time-to-collision with a time-threshold-brake to decide if a braking intervention is required.

For a stationary or a very slow-moving target object, the time-to-collision $t_{tc}(s)$ is calculated as follows [30]:

$$t_{tc} = \frac{d}{v_{rel}} \tag{2.11}$$

where d(m) is relative distance and $v_{rel}(m/s)$ is relative velocity between the vehicle and the target object. If the target object is decelerating, the time-to-collision is calculated as follows:

$$t_{tc} = \frac{\sqrt{v_{rel}^2 - 2dD_{rel}} - v_{rel}}{D_{rel}} \tag{2.12}$$

where $D_{rel}(m/s^2)$ is relative deceleration. The time-threshold-brake for a stationary target object is calculated as follows:

$$t_{tb} = \tau_B + \frac{v_{rel}}{2D_{\max}} \tag{2.13}$$

where τ_B (s) is brake loss time. For a non-stationary target object, the time-threshold-brake is calculated with:

$$t_{tb} = \tau_B + \frac{v_{rel} + D_{rel}\tau_B}{2(D_{\max} - D_{obs})} \tag{2.14}$$

where $D_{max}(m/s^2)$ is maximum allowed deceleration of vehicle and $D_{obs}(m/s^2)$ is the actual target object deceleration [30].

2.2 Sensor Systems

2.2.1 Overview

One of the main topics of current researches on autonomous shuttle buses is whether the sensors that equip these vehicles are able to replace a human driver, especially for a public transportation vehicle in urban traffic. Replacing the decisive factor of the human brain with a series of control algorithms—algorithms based on information received from sensors—presents a number of unknowns. In the case of classical driving, the decision-making element is the brain of the driver who receives information based on the senses (sight and hearing), but these human qualities are not always sufficient. There are people who have very good eyesight and hearing, but do not have the ability to drive because they do not have the knowledge and reflexes developed enough to control a vehicle. However, in addition to the perception based on sight and hearing together with acquired knowledge, there is another important element that is also necessary for a person to have the ability to drive a vehicle, namely the practical experience of driving it. This means that the driver's reactions are accumulated by the human brain leading to improved driving skills in terms of reflexes and reactions (driving style in different conditions related to traffic or weather).

This means that once the driver's brain is trained to drive, the vast majority of a vehicle's driving functions will be performed almost automatically (similar to typing without looking at the computer keys), which means that the mental effort will become negligible. For example, for a driver who travels the same route for a long time, the human brain becomes so familiar with the route that it is no longer able to remember the events that took place on that route except in situations where these events were exceptional [3].

In the process of developing autonomous driving, specialists used artificial brain training methods through a similar approach to training a person who drives automatically, based on information gained from previous experiences. The problems in decision-making by artificial brains can occur in the event of unpredictable events, moments when the processing power of the human brain increases exponentially, and it can make decisions on the spot.

Due to the lack of processing power of the human brain, the autonomous vehicle must rely on the information received from the sensor system. Some of the sensor's abilities far exceed human sensory abilities, for example: people do not have good night vision, do not have superior skills to detect movement through hearing, cannot see through fog or heavy rain etc. However, humans can compensate for these deficiencies through the processing power of the brain and through interaction and reaction to environmental disruptors. The autonomous vehicle algorithm does not have the ability to reach the level of processing power of the human brain but compensates for this lack through sensory skills and communication facilities between the vehicle and the V2X (Vehicle-to-Everything) environment, respectively between the vehicle and another vehicle V2V (Vehicle-to-Vehicle). Thus, autonomous vehicles can compensate for the lack of processing power to perform a task independent to the decision-makers related to the driving process.

Sensors are technical devices that react to specific environmental characteristics. An autonomous vehicle, which lacks the observation and perception of a human operator, must be able to perceive its surroundings in order to operate safely without the intervention of a human operator. Therefore, autonomous vehicles are equipped with a large number of sensors (Fig. 2.13) [3, 31–33] that scan the environment, identifying what is around the vehicle, road markings, traffic signs, static and dynamic objects:

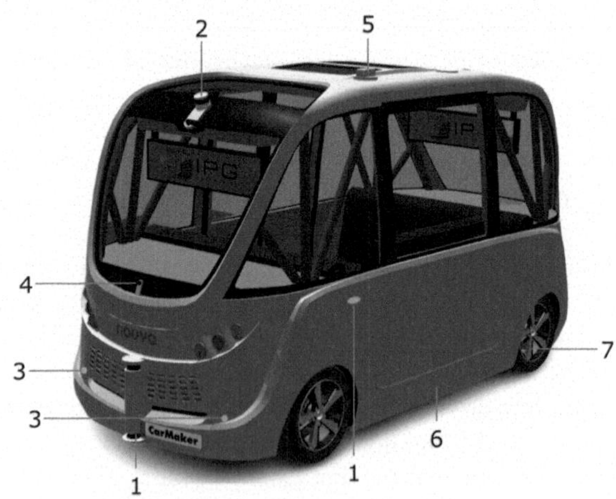

Fig. 2.13 Sensor system

(1) Lidar 2D 6-layer which provides a 360° visibility by accurately measuring the distance to static or dynamic objects.
(2) Lidar 3D 64-layer which scans the area within a radius of 120 m to create a 3D map and a real perception of the environment.
(3) Front radar/rear radar medium range radar with a coverage of up to 60 m (45°) and long-range radar with a coverage of up to 200 m (10°).
(4) Video camera for detecting and classifying objects based on information extracted from object textures.
(5) GPS/GNSS (Global Positioning System/Global Navigation Satellite System) sensor that provides the position relative to the geographical coordinates of the autonomous vehicle.
(6) IMU (Inertial Measurement Unit) inertial sensor, consisting of an accelerometer and a three-axis gyroscope which provide the position and orientation of the vehicle.
(7) Wheel-mounted odometer or speed sensor which measures the distance traveled by the vehicle and indicates its position on the HD (High Definition) map.

The great disadvantage of optical sensors (camera and Lidar) is the decrease in the detection performance of objects in the environment in case of unfavorable weather conditions (rain, snow, fog etc.). These factors reduce the performance of the optical sensors by affecting the detection and classification accuracy of the static and dynamic objects. Vargas Rivero et al. reported in [86] a reduction of up to 40% in the accuracy of Lidar sensors in case of adverse weather conditions, respectively a reduction of up to 5 m in the maximum detection distance. The sensors that equip the Apollo Baidu autonomous shuttle buses are presented in Table 2.5, according to public data on the Github.com/ApolloAuto platform [34–39].

The characteristics of the sensors equipping the autonomous vehicles (the maximum values for: field of view, range, accuracy, number of frames, resolution, color perception, and the minimum values for: meteorological phenomena, visibility, maintenance requirements and production cost) are presented in Table 2.6 and Fig. 2.14 (where the sensor capabilities are expressed using the following values: 0 is a very low score, 1 is a low score, 2 is an almost correct score, 3 is a correct score, 4 is a good score and 5 is a very good score) [3, 40–44].

Table 2.5 Sensor models that equip the platform Baidu Apollo

Type	Manufacturer	Model	M.U
Lidar	Velodyne	Alpha Prime VLS-128	1
Lidar	Velodyne	HDL-64E S3	1
Lidar	Hesai	Pandora (3D 40 channels)	1
Radar	Continental	ARS-408-21 (LRR 77 GHz)	1
Camera	Leopard	LI-USB30-AR023ZWDR	2
IMU	NovAtel	IMU-IGM-A1 (200 Hz)	1
RTK	NovAtel	ProPak6 (GNSS Receiver)	1
GNSS	Navtech	PwrPak 7D (GNSS Receiver)	1

Table 2.6 Comparisons between sensor capabilities

Performances	Human	Lidar	Radar	Camera	V2X
Object detection	4/5	4/5	4/5	1/2	4/5
Object classification	4/5	2/3	0/1	4/5	4/5
Image resolution	4/5	3/4	2/3	4/5	4/5
Distance estimation	1/2	4/5	4/5	3/4	4/5
Proximity estimation	3/4	2/3	3/4	4/5	4/5
Line tracking	4/5	1/2	1/2	4/5	4/5
Speed detection	2/3	4/5	4/5	1/2	4/5
Area of visibility	4/5	2/3	4/5	3/4	4/5
Color/contrast detection	4/5	0/1	0/1	4/5	4/5
Weather conditions	2/3	2/3	4/5	1/2	4/5
Low light	2/3	4/5	4/5	1/2	4/5
High light	2/3	4/5	4/5	3/4	4/5
Sensor dimensions	–	1/2	4/5	4/5	4/5
Sensor price	–	1/2	4/5	4/5	1/2

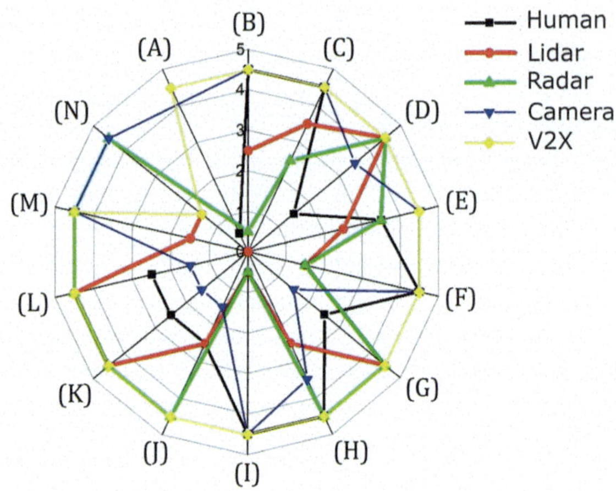

Fig. 2.14 Comparisons between sensor capabilities

Frost and Sullivan presented in [45] an evolution of the sensors that equip a (partially) autonomous vehicle from level 1 to level 5. It can be observed that as the degree of autonomy increases, the number of sensors that scan and interpret everything what happens in the environment surrounding the autonomous vehicle also increases (Table 2.7). The signals from the sensors fuse both in the sensor processor and in the autonomous driving system according to an architecture with a horizontal structure that manages several specific applications in parallel (Fig. 2.15) [46].

Table 2.7 Evolution of sensors depending on the level of autonomy

Level 1		Level 2		Level 3		Level 4		Level 5	
Sensor	No	Sensor	No	Sensor	No	Sensor	No	Sensor	No
Ultrasound	4	Ultrasound	8	Ultrasound	8	Ultrasound	8	Ultrasound	10
Radar LRR	1	Radar LRR	1	Radar LRR	2	Radar LRR	2	Radar LRR	2
Radar SRR	2	Radar SRR	4	Radar SRR	4	Radar SRR	4	Radar SRR	4
Camera SR	1	Camera SR	4	Camera SR	2	Camera SR	3	Camera SR	3
Total	8	Total	17	Bi-Camera	1	Bi-Camera	1	Bi-Camera	2
				Infra-Red	1	Infra-Red	1	Infra-Red	2
				Lidar	1	Lidar	4	Lidar	4
				GNSS	1	GNSS	1	GNSS	1
				Total	20	Total	24	Total	28
2012		2016		2018		2020		2025	

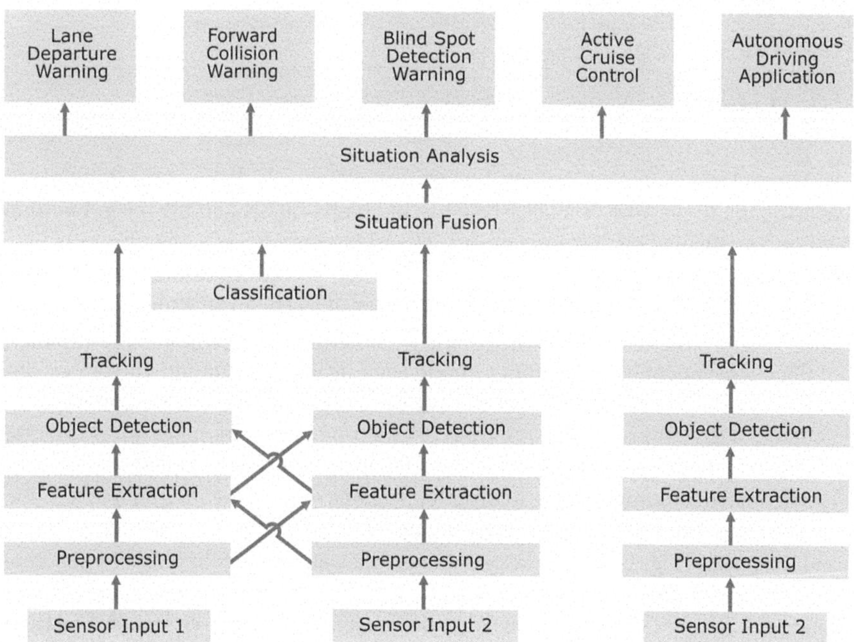

Fig. 2.15 Fusion of the signals received from sensors

The characteristics of the electromagnetic spectrum (structured on wavelengths and frequencies) that the sensors with which autonomous vehicles are equipped with use actively or passively, reflect the advantages and disadvantages of these sensors in degraded environments or in case of adverse weather conditions (Fig. 2.16) [40].

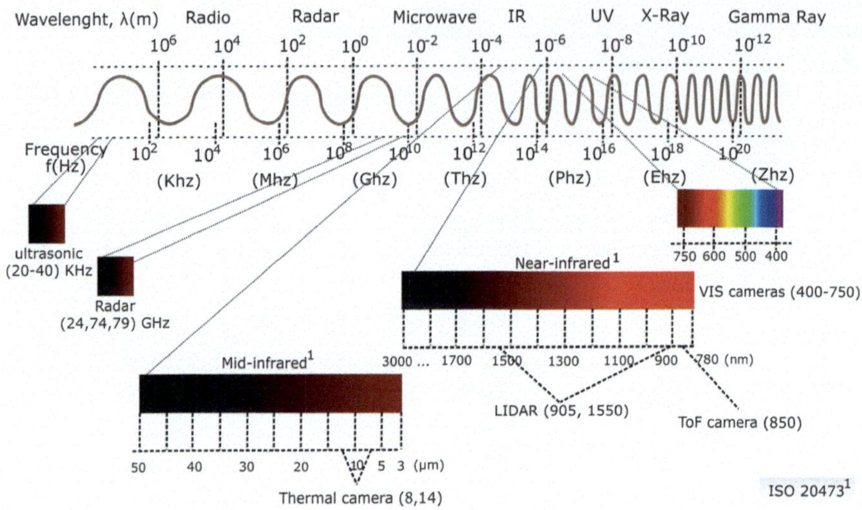

Fig. 2.16 Electromagnetic spectrum characteristics

2.2.2 *Lidar*

Lidar (Light Detection And Ranging) systems operate on the basis of the principle of measuring the flight time of a laser beam to the target object and the reception time of this beam reflected by the target object. The Lidar emission is in the infrared range (905 nm or 1550 nm). Lidar systems are classified according to the type of information they obtain from the environment in Lidar 2D systems or Lidar 3D systems, or according to the construction: rotating or fixed. Lidar 2D collects information from the environment by projecting a single laser beam on a rotating mirror perpendicular to the axis of rotation. Lidar 3D collects information from the environment by projecting a beam of laser rays from a group of diodes mounted on a rotating mechanism and generates high-precision 3D maps for the environment.

Lidar sensors are affected by weather conditions (rain, snow, fog, dust etc.) due to the diffraction of light in these environments [40].

Lidar sensors (Fig. 2.17) use laser distance measurement technology to generate a virtual image of the environment, respectively to determine its precise position and to detect the static or dynamic objects on the road for a 2D map (Lidar 2D) or for a 3D map (Lidar 3D) [3, 47].

The distance measuring principle used by Lidar sensors is the TOF (Time-Of-Flight), where the time traveled by the pulsed laser, projected on a target object and the reflection of the beam back to the source is measured in order to calculate the distance. Lidar sensors emit a cloud of individual pulses over a wide range of intervals, with peaks of high instantaneous power, which perform a scan on the surface of target objects (Fig. 2.18). Lidar sensors allow programmable control of the position of the laser beam using the MEMS (Micro-Electro-Mechanical System)

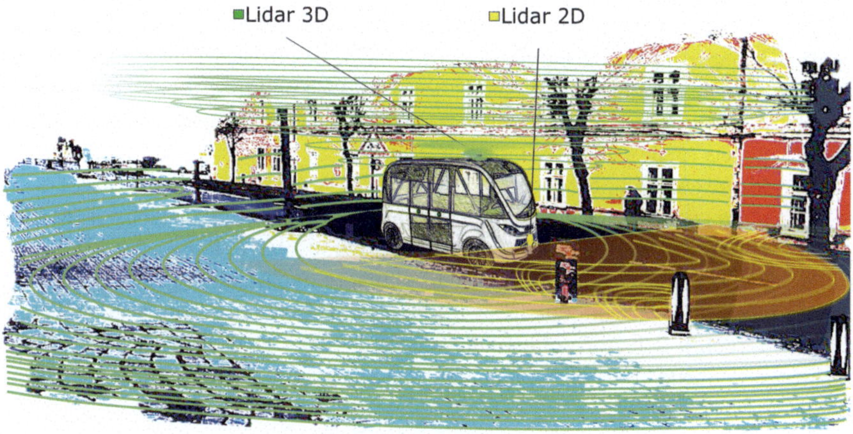

Fig. 2.17 Lidar sensors (Photo author January 01, 2020)

consisting of miniature mirrors, whose angle of inclination varies in the presence of a stimulus, so that the angular direction of the incident beam is modified, and the laser beam is directed to a specific point of the target object. This scan is performed in a 2D plane ensuring the scanning of a complete area of the target [48].

Lidar 2D sensors capture information from the environment by projecting a single laser beam on a flat surface, circular at 180°, perpendicular to the axis of rotation. Lidar 2D sensors are generally mounted in pairs of two units so as to minimize areas without coverage and ensure continuous visibility over the entire circular surface of the autonomous vehicle [3, 44].

Lidar 3D sensors use a set of laser diodes (between 4 and 128 laser channels) mounted on a rotating device that scans the environment in a 360° horizontal field

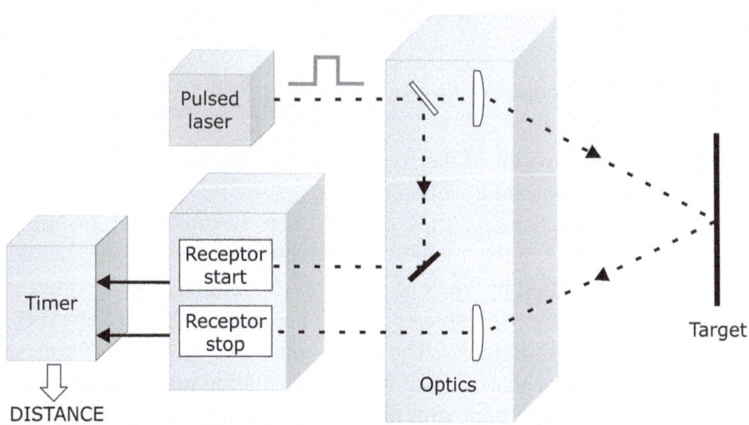

Fig. 2.18 Lidar principle of operation

of view and a 20°–45° vertical field of view. The accuracy of the images that make up the 3D map is given by the number of laser beams used. Lidar 3D sensors are mounted so as to minimize the areas without coverage and to scan the area of travel on a circular radius in order to create a 3D map of the virtual environment including existing objects (pedestrians, vehicles etc.) [3, 40].

The AMCW (Amplitude Modulated Continuous Wave) approach consists of using the intensity modulation of a continuous light wave instead of the laser pulses. This principle is known as amplitude-modulated continuous-wave. It uses the phase-shift induced in an intensity-modulated periodic signal to obtain the range value.

The optical power is modulated with a constant frequency and after reflection from the target, a detector collects the received light signal. Measurement of the distance is deduced from the phase shift occurring between the reflected and the emitted signal $\Delta\Phi$ [48]:

$$\Delta\Phi = k_M d = \frac{2\pi f_M}{c} 2R \geq R = \frac{c}{2}\frac{\Delta\Phi}{2\pi f_M} \tag{2.15}$$

where R(m) is the range to the target and c(m/s) is the speed of light in the free space, k_M(1/s) is the wavenumber associated to the modulation frequency, d(m) is the total travelled distance and f_M(Hz) is the modulation frequency of the amplitude of the signal.

For the FMCW (Frequency Modulated Continuous Wave) approach, the emitted instantaneous optical frequency is periodically shifted, usually by varying the power applied to the source. The reflected signal is mixed with the emitted source, creating a beat frequency that is a measure of the probed distance. The source is a laser diode to enable coherent detection. The signal is sent to the target, and the reflected signal that arrives at the receiver, after a TOF is mixed with a reference signal built from the emitter output. For a static target, the delay between the collected light and the reference causes a constant frequency difference. Frequency f_r(Hz) is directly proportional to t_{oF}(s) and hence proportional to the target range [48]:

$$f_r = \text{slope } \Delta\tau = \frac{B}{T} t_{oF} = \frac{B}{T}\frac{2R}{c} \geq R = f_r \frac{cT}{2B} \tag{2.16}$$

where B(Hz) is the bandwidth of the frequency sweep, τ (s) denotes the period of the ramp, and t_{oF}(s) is the total travelled time.

The main technical characteristics of the Lidar sensors used in automotive applications are presented in the Table 2.8 [3, 35, 49–51].

Recent studies, presented by Yu et al. [52] show a combination of video camera (optical passive), radar (electromagnetic active) and Lidar (optical active) that can build 3D images of the environment. Thus, the signals provided by the three sensors merge into a virtual sensor that transforms the information provided into a detailed representation of the environment with the passive and active objects that are in the vicinity of the autonomous vehicle. Lidar sensors have a precise operating range, a higher power and frequency, a shorter wavelength than radio waves, which ensures a

Table 2.8 Technical data of Lidar sensors

Lidar sensors	Lidar 2D	Lidar 3D
(Horizontal) detection angle	270°	360°
(Vertical) detection angle	–	30°
Horizontal resolution	0.25°–0.5°	0.1°–0.4°
Vertical resolution	–	2°
Range	0.50–50 m	1–100 m
Number of lines	1	16
Scan area (10% reduction)	18–30 m	–
Scan area (90% reduction)	20–50 m	–
Scan frequency	25 Hz/50 Hz	20 Hz

better reflection rate in non-metallic environments, and they work the same regardless of ambient light or climatic conditions (rain, snow, fog etc.).

2.2.3 *Radar*

Radar systems are used for military and civilian applications: air or ground detection systems, airports, weather systems, etc. The principle of operation is as follows: the radar system measures the distance between the transmitter and the target object by calculating the flight time of the transmitted signal and the time of receiving the echo reflected by the target object. Another feature of radar systems is that they can accurately provide the direction and speed of the targeted objects.

Radar systems are used in the following automotive applications: BSD (Blind Spot Detection), LCA (Lane Change Assistant), RCTA (Rear Cross Traffic Alert), FCTA (Forward Cross Traffic Alert) or video fusion radar. Radar waves have a higher penetrability because they offer good characteristics in all weather conditions and can accurately detect targets with a short range in front, side, and rear of the autonomous vehicle [40].

Radar sensors that equip autonomous vehicles are set at the front and at the rear of the vehicles to measure the distance to the detected objects and to calculate their speed and estimate the direction of travel (Fig. 2.19) [53–55].

Automotive radars are divided into three categories: LRR (Long-Range Radar), MRR (Medium-Range Radar), and SRR (Short-Range Radar). LRR is used to measure the distance and speed of other vehicles, MRR is used in the wider field of view, and SRR is used to detect objects near the vehicle. Two main frequency bands are used in the automotive radar systems, 24 GHz, and 77 GHz.

SRR for vehicles uses the 24 GHz frequency because the band can detect objects at short and medium distances. A 24 GHz radar is also used to detect an object that can be obstructed or is located very close. Radar systems with the same repeatability can also be used to detect dead spots, which directly involves avoiding collisions.

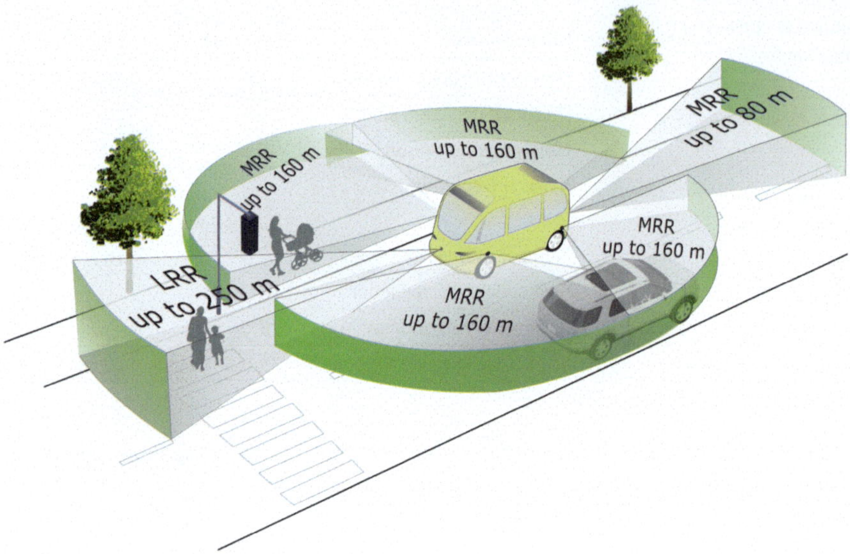

Fig. 2.19 Radar sensors

SRR sensors are not used to measure the angle of the detected objects and have a very wide side coverage. They are operated in pulse mode or in continuous wave mode. Small range radars require a controllable antenna with a large scan angle, creating a wide field of view.

LRR uses the higher permitted transmission power 77 GHz to obtain better performance. By producing a better resolution of objects, a greater accuracy is provided.

The bandwidths available in the 77 GHz band is 4 GHz and, the bandwidth available in the 24 GHz band is 200 MHz. The difference between the frequency of the signal emitted by the transmitter and the frequency of the received reflected signal is linearly dependent on the distance from the transmitter to the object. The resolution is the minimum distance between the objects so that they can be distinguished as different. The transition from 24 to 77 GHz results in a 20 times better performance in terms of resolution and accuracy. The resolution of the 77 GHz radar range can reach 4 cm. The 24 GHz radar achieves a resolution of 75 cm. Therefore, the advantage of the 77 GHz system is that it can detect objects that are at a short distance one from another.

The wavelength of 77 GHz signals is a third the frequency of the 24 GHz system, which enables significantly smaller modules in the spatially limited areas of the vehicle [3, 32, 33, 56].

The cruising speed control system for an ACC (Adaptive Cruise Control) autonomous vehicle uses the radar sensor as an active element and is shown in Fig. 2.20, where: (1) is radar sensor (77 GHz), (2) ECU, (3) ESP, (4) vehicle dashboard, (5) acceleration control, (6) sensor, (7) TCU (Transmission Control Unit).

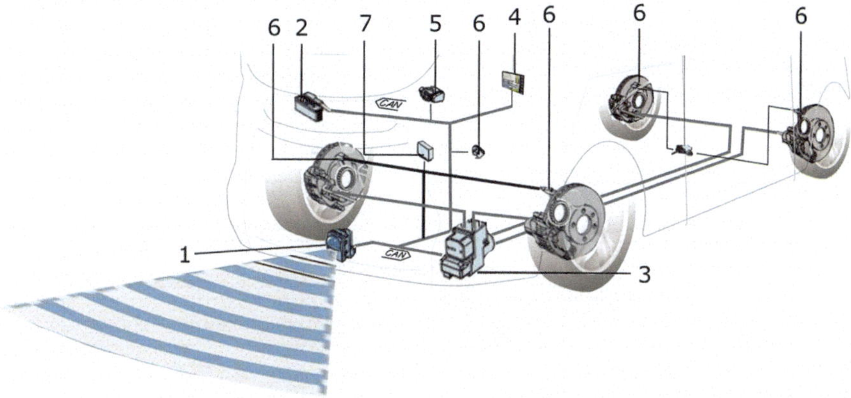

Fig. 2.20 Adaptive Cruise Control system

The role of ACC has automatically adapted the acceleration of a vehicle so that it maintains a constant velocity, one that is preselected by the autonomous driving system. This control system has a linear feature when the vehicle is moving in a straight sector, without other vehicles in the vicinity. When the vehicle engages in a turn or approaches another vehicle traveling at a slow speed, the intervention of a sensor is required to monitor and adapt the cruising speed to the existing traffic conditions [57].

The ACC measures the distance to the vehicle in front and controls the acceleration and deceleration in order to maintain a minimum safety distance according to the velocity preset by the autonomous driving system.

Based on the signals from the radar sensor, the ACC calculates the distance to the vehicle in front and the velocity in relation to it. If there are several vehicles in the radar sensor's field of view at the same time, the minimum distance is the parameter used by the autonomous driving system to decide which of these vehicles should be tracked [58].

If the autonomous vehicle approaches a slow or stationary vehicle, the ACC shall request to the autonomous driving system to reduce its velocity. If the required deceleration in this situation exceeds 30% of the maximum stopping power of the vehicle, visual and audible warning signals shall be generated which shall require the human operator to apply the braking system until the vehicle has come to a complete stop. The structure of the ACC on functional levels is shown in the Fig. 2.21 [3]:

Level 1 has the role of measuring the signals generated by the radar sensors and wheel speed sensors, yaw rate sensor, etc.

Level 2 has the role of defining the operation of radar sensors based on the signals received from the objects in the active field of the autonomous vehicle, characterized by distance, velocity, position, trajectory, etc.

Level 3 has the role of selecting the object (target vehicle) detected by the radar sensor which is the most significant for the ACC.

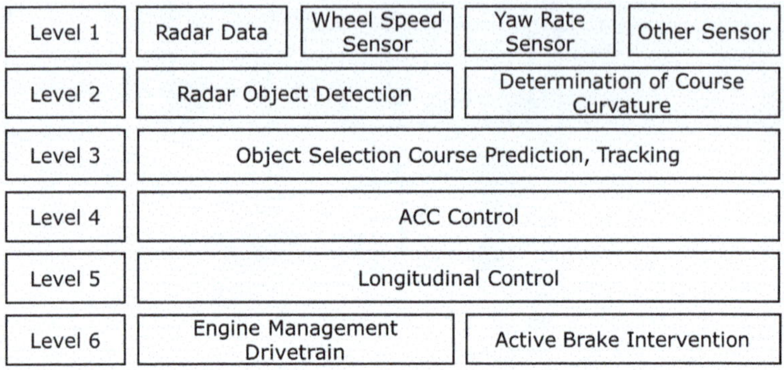

Fig. 2.21 Block diagram of the ACC

Level 4 has the role of establishing the travel characteristics (velocity, trajectory etc.) based on the travel characteristics of the target vehicle.

Level 5 has the role of determining the acceleration/deceleration of the autonomous vehicle for driving control according to the travel characteristics of the target vehicle.

Level 6 has the role of establishing the operating regime of the electric motor for accelerating the autonomous vehicle, respectively to establish the intervention on the braking system in case of deceleration.

The main technical characteristics of the radar sensors (SRR, MRR and LRR) equipping autonomous vehicles are presented in Fig. 2.22 and in Table 2.9 [3, 59, 60].

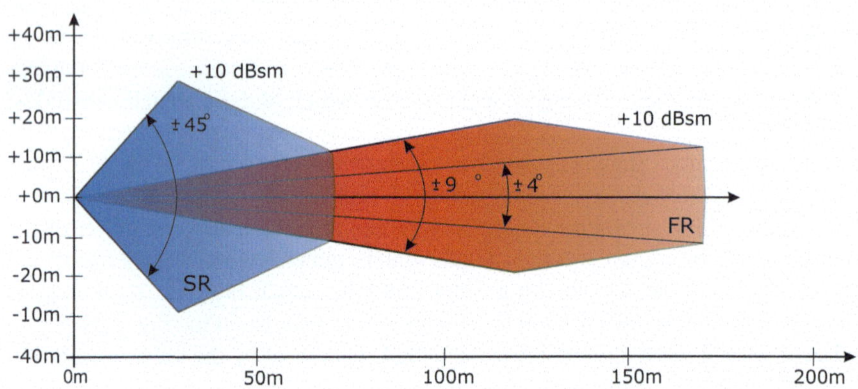

Fig. 2.22 Technical data of radar sensors

Table 2.9 Technical data of radar sensors

Radar sensor	LRR	MRR	MRR rear
Bandwidth	77 GHz	77 GHz	77 GHz
Detection range	0.36–250 m	0.36–160 m	0.36–80 m
Measuring range	6° (200 m)	6° (160 m)	5° (70 m)
	10° (100 m)	9° (100 m)	75°
	15° (30 m)	10° (60 m)	–
	20° (5 m)	25° (36 m)	–
	–	42° (12 m)	–
Distance accuracy	±0.12 m	±0.12 m	±0.12 m
Velocity accuracy	0.11 m/s	0.11 m/s	0.14 m/s
Angle accuracy	±0.3°	±0.3°	±0.8°
Separation objects: distance	0.72 m	0.72 m	0.72 m
Separation objects: velocity	0.4 m/s	0.66 m/s	1.4 m/s
Separation of objects: angle	4°	7°	7°
Maximum detected objects	24	32	32
Time cycle	60 ms	60 ms	60 ms

2.2.4 *Video Camera*

The video camera system that equips autonomous vehicles can be classified according to the wavelength in video cameras that operate in the VIS (VIsible Spectrum) or video cameras that operate in the IR (InfraRed) spectrum. The optical element used by the video camera to capture a scene is called an imaging sensor and it was developed based on two constructive technologies: CCD (Charge-Coupled Device) and CMOS (Complementary Metal Oxide Semiconductor).

The cameras operating in the VIS capture scenes at wavelengths between 400 and 780 nm (similar to the human eye) and are used in the perception systems of autonomous vehicle to obtain information about the vehicle's surroundings. These video cameras are used in the following applications: blind spot detection, side view control, accident recording, object identification, lane change assistant and sign detection. VIS cameras are strongly affected by variations in lighting conditions, respectively weather conditions (rain, snow, fog, dust etc.) and for this reason they are used together with radar and Lidar technologies.

The combinations of two VIS cameras ensure their operation with stereoscopic view that provides a 3D representation of the scenes around the autonomous vehicle.

Cameras operating in the IR spectrum are passive sensors operating at wavelengths between 780 nm and 1 mm. IR cameras are used: in situations where there are areas with very low lighting (tunnels, traveling with the sun in front of the vehicle etc.), in detecting bodies that emit heat (pedestrians, animals, other vehicles in operation) [40].

Autonomous vehicles are equipped with a system of high-resolution stereo video cameras, located at the front and rear, respectively, to provide a circular image of the environment. These video cameras have the role of permanently monitoring/recording the movement of the autonomous vehicle, the route followed, the traffic signs etc.

Monocular video cameras capture 2D images without generating precise details about distances to the observed objects and without estimating information to help establish scenarios to avoid these objects. Stereo video cameras have the ability to estimate the distance to the observed objects by measuring the difference between two images from different angles [3, 20]. Video cameras (Fig. 2.23) perform the following tasks for autonomous vehicles: detecting the lane, detecting the distance of other objects or vehicles in traffic, and detecting traffic signs.

A video camera works based on the light passing through the lens, that is filtered by a color filter matrix Bayer CFA (Color Filter Array), and which provides information about the light intensity in different regions (area) depending on the wavelength. The raw data of the image captured by the video camera is then transformed into color images by an algorithm that converts photons into photoelectrons [61–63]. After the image processing step, the visual perception (property) is used to recognize and understand the objects detected in the environment using the classification algorithm (class) (Fig. 2.24) [3, 64].

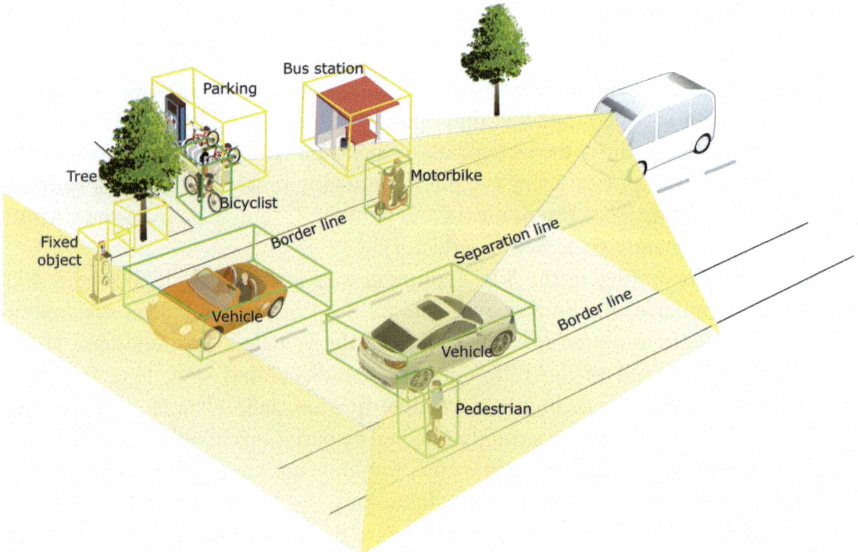

Fig. 2.23 Video camera

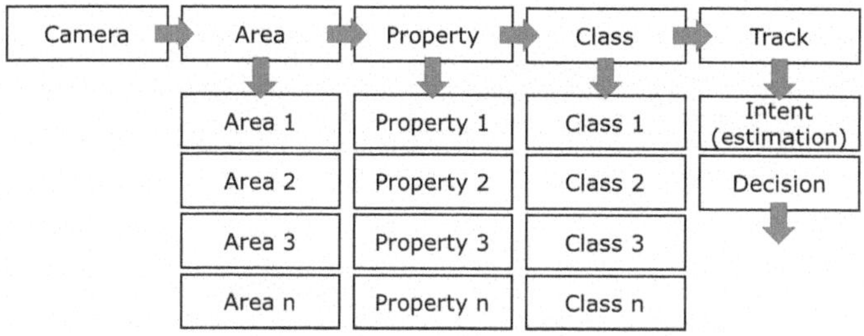

Fig. 2.24 Object classification algorithm

2.2.5 GPS/GNSS

The Global Navigation Satellite System (GNSS) comprises all global and regional satellite navigation systems and operates on the principle of detecting the position of a receiver (autonomous vehicle) against a fixed reference consisting of a group of geostationary satellites that orbit more than 20,000 km away from the earth's surface. Satellites emit signals containing information about their position, orbital parameters etc. Based on these signals, the reception systems extract information about their geographical position, velocity etc. [20, 65, 66]. The principle of operation of the GNSS is based on the transmission time of the extrapolated signal emitted by at least four visible satellites and received by the receiver, from which it extracts information about the geographical coordinates and time (x, y, z, t) [3, 40]. The signals transmitted by satellites are influenced by numerous errors [40]:

(1) Synchronization of atomic clocks of the orbiting satellites.
(2) The influence on the signal in the ionosphere and troposphere.
(3) The noise of the receptor equipment.
(4) Multipath effect due to signal reflection.
(5) Particular geographical features.

GNSS technology is scalable and allows access to unlimited independent users to the satellite infrastructure. Satellites work as radio wave generators and permanently transmit their position using specific signals. Satellites transmit signals in the 2–40 MHz frequency band, with a spectral range between 1.2 and 1.6 GHz. Transmissions of signals from satellites are synchronized, and the receivers receive these signals and calculate the position of the object based on the time of reception of the signals [67].

The D-GNSS (Differential Global Navigation Satellite System) has been developed to mitigate errors affecting the measurement of the signal from satellite constellations. D-GNSS consists of two GNSS receivers, a base station, and a mobile rover. The base station has a position characterized by coordinates and continuously communicates signal corrections to the rover. The signal corrections ensure an accuracy between 0.7 m and 3 m.

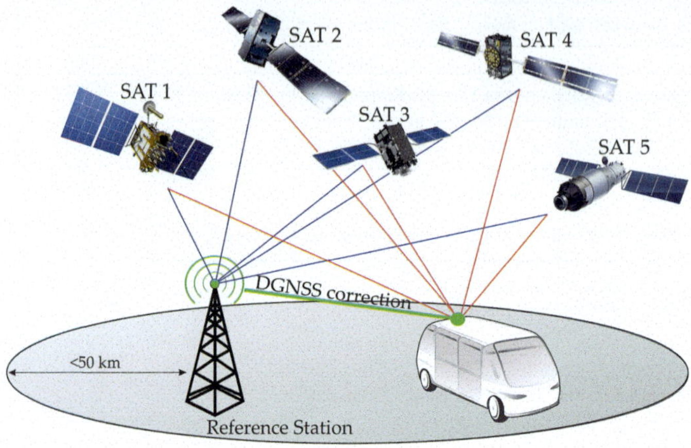

Fig. 2.25 GNSS/GPS sensors

The most commonly used GNSS is the GPS, which uses a RTK (Real-Time Kinematic) correction tool to position the autonomous vehicle with maximum accuracy (Fig. 2.25).

The RTK system uses the signal received from the satellites to improve the positioning accuracy of the rovers. The base station retransmits the signal received from the satellites to one or more rovers which compare these signals with the signals received directly from the satellites and accurately calculate the position. These systems have an accuracy of up to 2 cm and usually use a radio modem for the communication between the base stations and the rovers within a maximum radius of 50 km [40].

The GNSS consists of segments interconnected through radio waves and transmissions. These segments are the space segment consisting of the satellite network, the user segment consisting of GPS receivers and the control segment consisting of ground control and monitoring stations. The control segment communicates with satellites though bidirectional transmissions, and it has the following roles: to tracking the satellites, orbital control, to monitor the operation and integrity and to maintain the stability and accuracy of orbit positioning [67].

There are situations conditioned by external factors in which GNSS offers an accuracy that can deviate by a few meters, although the autonomous vehicle requires an accuracy of a few centimeters. The correction, respectively the achievement of this precision of a few centimeters is obtained with the help of a base station located on the ground. The GNSS measures the distance from the satellites and from the base station, thus obtaining a positioning with maximum accuracy for the autonomous vehicle.

The transfer of GNSS/RTK differential corrections from the reference base station to the autonomous vehicles is performed via mobile communications or the Internet (4G/5G) [3]. There are currently several operational GNSS whose technical characteristics are presented in Table 2.10 [67].

Table 2.10 GNSS (Global Navigation Satellite System)

System	GPS	GLONASS	BDS	Galileo	NavIC	QZSS
Owner	USA	Rusia	China	EU	India	Japan
Coverage	Global	Global	Global	Global	Regional	Regional
Launch	1993	1993	2012	2016	2016	2018
Altitude (km)	20,180	19,130	21,150	23,222	36,000	32,600
Pitch (°)	56	64.8	55	56	29	43
Period (hours)	11.97	11.26	12.63	14.08	23.93	23.93
Satellites	24	24	35	30	7	4
Accuracy (m)	5.00	4.50	1.00	1.00	1.00	1.00

Where: GPS Global Positioning System, GLONASS GLObal NAvigation Satellite System, BDS BeiDou navigation Satellite system, NavIC Navigation with Indian Constellation, QZSS Quasi-Zenith Satellite System.

The INS (Inertial Navigation System) can be used to continuously positioning the autonomous vehicles during GNSS signal losses around urban canyons, bridges, tunnels, and trees. GNSS with INS motion model follows the position of the autonomous vehicle is usually done relative to a reference. The INS's measurements usually provided in the body (sensors) frame would need to be transformed into the navigation frame for tracking purposes. The transformation matrix R^{nb} from the body frame to navigation frame is as shown in equation [68]:

$$R^{nb} = \begin{bmatrix} \cos\theta\cos\psi & -\cos\theta\sin\psi + \sin\phi\sin\theta\cos\psi & \sin\phi\sin\psi + \cos\phi\sin\theta\cos\psi \\ \cos\theta\sin\psi & \cos\theta\cos\psi + \sin\phi\sin\theta\sin\psi & -\sin\phi\cos\psi + \cos\phi\sin\theta\sin\psi \\ -\sin\theta & \sin\phi\cos\theta & \cos\theta\cos\psi \end{bmatrix} \quad (2.17)$$

where ϕ is the roll, θ is the pitch and ψ is the yaw. However, as our study is limited to the two-dimensional tracking of vehicles, f and q are thus considered to be zero, thus the rotation matrix R^{nb} becomes [68]:

$$R^{nb} = \begin{bmatrix} \cos\psi & -\sin\psi & 0 \\ \sin\psi & \cos\psi & 0 \\ 0 & 0 & 1 \end{bmatrix} \quad (2.18)$$

2.2.5.1 Inertial Measurement Unit

Inertial Measurement Unit (IMU) includes a three-axis accelerometer and a three-axis gyroscope, respectively, and its role is to measure the acceleration of the autonomous vehicle and to specify its location and orientation (Fig. 2.26).

When a vehicle is in motion, it can perform linear and rotational movements around each of the three axes: lateral, longitudinal, and vertical (pitch, roll, yaw).

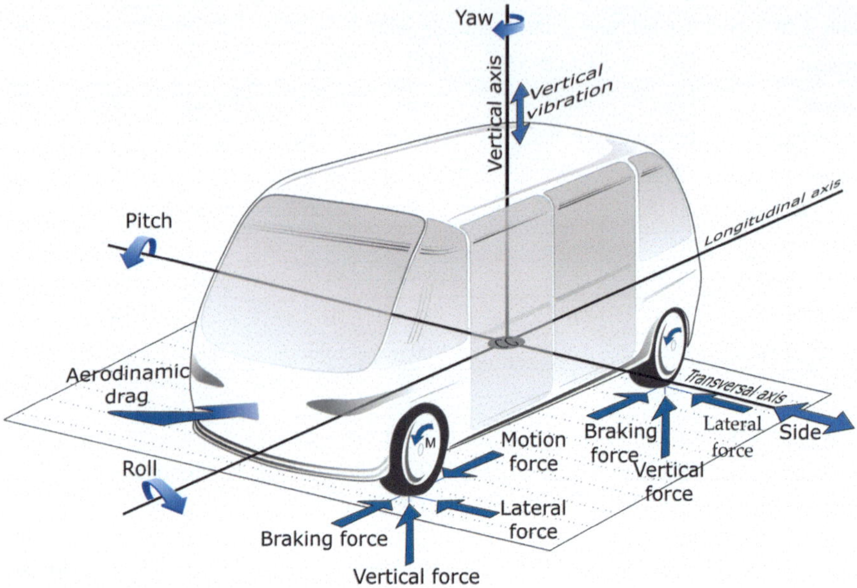

Fig. 2.26 IMU sensor

Inertial measurements, which can be made without reference to an external point, include linear acceleration, angular velocity, and angular acceleration. This information can be used to increase or improve external measurements (GPS coordinates) [3, 69].

IMU is a device that contains an accelerometer that measures a force vector specific to the body, a gyroscope that measures the vector of inertial angular velocity, and a magnetometer that measures the vector of the magnetic field around the device. These three sensors are mounted so that their measuring axes each constitute distinct orthogonal systems, leading to the existence of six degrees of freedom. IMU provides basic information about vehicle dynamics (acceleration, rotation, and orientation).

2.2.6 *Ultrasonic Sensors*

Ultrasonic sensors are proximity sensors that have the role of signaling obstacles that may occur in the immediate vicinity of the autonomous vehicle. Ultrasonic sensors emit acoustic pulses that are reflected by the obstacles and then received by the sensors and transmitted to a control unit that calculates the distance to the obstacle. The system warns the autonomous driving system of the proximity of an obstacle, and the reduction of the distance to it is signaled by increasing the frequency of the emitted signal, including the acoustic frequency until a continuous tone is reached which indicates a predefined minimum distance.

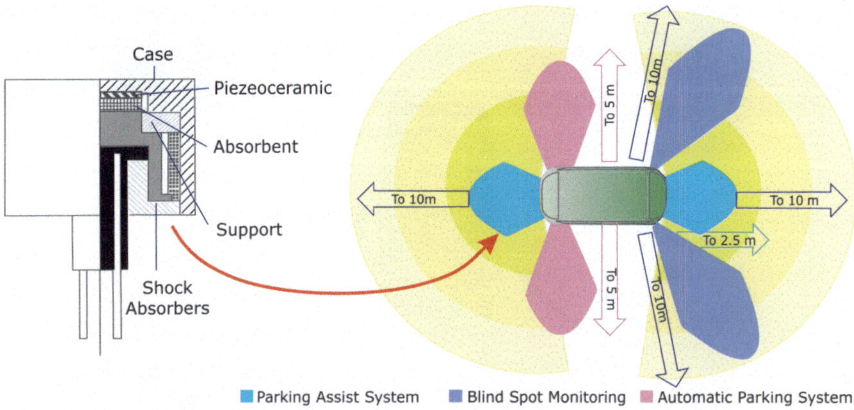

Fig. 2.27 Ultrasonic sensors

Ultrasonic sensors can be mounted both in the front, on the sides and in the rear of the autonomous vehicle (Fig. 2.27).

Ultrasonic sensors use sonic waves, in the range of 20 kHz to 40 kHz, generated by a magneto-resistive membrane, to measure the distance d(m) to an object. Its principle of operation is based on the TOF (s) measurement of the sonic wave from the transmitter to the reception, where c (m/s) velocity of the wave [40, 70]:

$$d = \frac{c}{2} \mathrm{ToF} \tag{2.19}$$

2.3 Vehicle Connections

2.3.1 *Vehicle-to-Everything (V2X)*

2.3.1.1 Overview

The concept of V2X (Vehicle-to-Everything) in general and the concepts of V2V (Vehicle-to-Vehicle), V2I (Vehicle-to-Infrastructure) and V2P (Vehicle-to-Pedestrian) in particular are communication links between vehicles and the intelligent environment (vehicles, infrastructure, pedestrians etc.), designed to increase traffic safety and efficiency.

V2X communications (Fig. 2.28) are complex and depend on the level of development of intelligent infrastructure along the route (V2I), but mostly they depend on the number of vehicles that have the technology and interconnection capabilities (V2V). The V2X concept aims to provide connected vehicles with information on

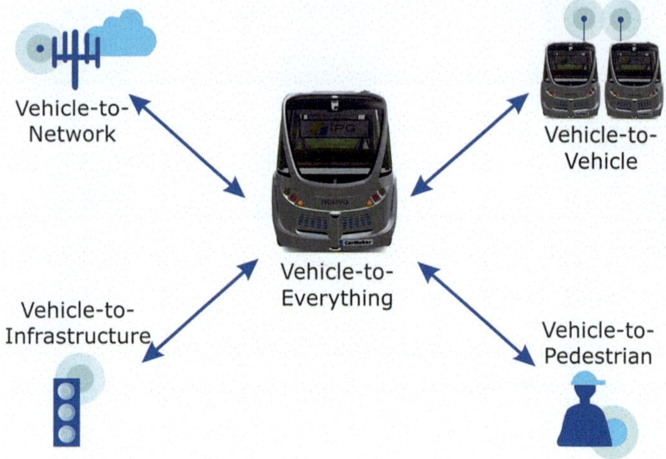

Fig. 2.28 V2X communications

local road traffic, helping to increase traffic safety, to make traffic more efficient, to reduce the incidence of accidents, and to develop a smart transportation system.

The V2X concept uses two communication technologies: DSRC (Dedicated Short-Range Communication) and LTE-V2X (Long-Term Evolution for V2X) on long distance.

DSRC uses 802.11p technology, which is a standard WAVE (Wireless Access in Vehicular Environments), which allows connected vehicles to exchange relevant safety information directly with each other, or with the road infrastructure [71].

LTE-V2X is a mobile communication technology for V2X that ensures the transfer of large volumes of data [72]. Applications developed based on the V2X concept are classified into:

(I) Applications to improve traffic safety (road hazard warnings, collision, and speed warnings).
(II) Applications to increase transportation efficiency (optimizing vehicle driving to ensure fluid traffic flow).
(III) Applications for traffic information services (current traffic information, route recommendations, identification of free parking spaces).

Inter-vehicle communication is based on 802.11p technology and C-V2X (Cellular V2X), respectively. V2X communications can be achieved through direct communication (vehicles connect with the intelligent entities around them in a peer-to-peer connection), network-assisted communication (vehicles connect via RSU (Road-Side Units) or BS (Base Station)) and hybrid communication (vehicles communicate with each other through a combination of direct communication and network-assisted communication using cellular network components) [73]. The main technologies used in V2X communications are the following [74].

2.3.1.2 DSRC Based V2X

DSRC (5.9 GHz) based V2X provides the following cooperative applications: vehicle warning, emergency brake lighting and vehicle platooning. Because many vehicles drive on the road, the V2X applications need an high capacity, high bandwidth, and very low latency in a densely populated environment. In the case of critical and emergency warning notifications, latency plays a very important role—for example in which regards the security of traffic, it prevents collisions. Some of the features and requirements of future V2X are high bandwidth, low delay, high scalability, availability, reliability, and high data speeds. DSRC has limitations for all the specifications of the future V2X requirements in regard to the physical layer, the absence of collision and the interference management. DSRC is an asynchronous system that is based on the CSMA/CA (Carrier-Sense Multiple Access with Collision Avoidance) protocol [74].

2.3.1.3 Cellular V2X

The C-V2X (Cellular V2X) was developed by the 3GPP (3rd Generation Partnership) project based on V2X RAT (Radio Access Technologies). which in release 14 that gives the highest priority for radio access modifications suitable for V2X. LTE-V (Long-Term Evolution for Vehicles) was introduced in vehicle networks as an alternative technology for intelligent transport, in addition to DSRC.

C-V2X can use V2N (Vehicle-to-Network) mode based on existing licensed cellular networks. The cellular network provides access to the cloud through licensed spectrum using a network architecture for the vertical industry. C-V2X describes the transmission modes that allow for direct V2X connections over the PC5 interface through the side link channel. The PC5 refers to the point of contact where the vehicular nodes connect with other nodes directly over the same channel, where it does not require connection with the base station.

Based on the specifications of release 14 and release 15, the 3GPP announced the C-V2X technology. The release 14 included two additional transmission side link modes to enable V2X communications with low latency.

The C-V2X can perform in both within-coverage and outside-of-coverage environments and is can also work with both the conventional LTE air interface and the sidelink air interface.

V2X based on LTE-Uu air interface is the standard air interface for connecting UE (User Equipment) and an eNB (evolved Node B). Each UE that supports the LTE-Uu (Long-Term Evolution UE to UE) protocol relays its signal on uplink to the eNB, and the eNB transmits the signal to the destination UE on downlink. The eNB will use scheduling to lower the scheduling overhead that is involved with V2X uplink transmission. In scheduling, the July B allocates resources to a user over several subsequent transmissions, as most of the traffic is periodic and has similar packet sizes.

V2X based on PC5 air interface: according to the 5GAA (5G Automotive Association), the device-to-device communication mode must be enabled in C-V2X for direct safety message communication.

In C-V2X side-link mode 3, the scheduled mode works only in the presence of a base station or an eNB.

C-V2X side-link mode 4 works independently in the absence of eNB support by directly interacting using the PC5 side-link radio interface using a 5.9 GHz frequency band like the DSRC. It is also known as the autonomous mode. It can interact with other nearby vehicles in a decentralized manner without depending on central cellular network connections. The mode 4 in C-V2X showed better performance compared to the 802.11p technology in several situations during an initial comparison.

Communications based on C-V2X technologies (Fig. 2.29) are based on the host computer and C-V2X router in both the construction architecture of the autonomous vehicle and the construction architecture of the ITS (Intelligent Transportation System) smart infrastructure [22].

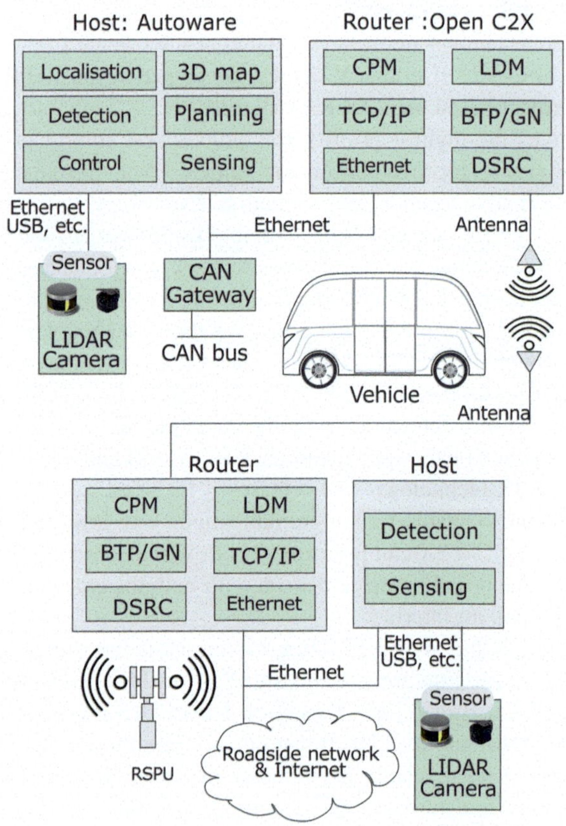

Fig. 2.29 C-V2X communications

Host computer from the construction of the autonomous vehicle consists of the following functional blocks: detection, sensing, planning, 3D maps, localization, control.

C-V2X router both in the construction of the autonomous vehicle and in the constructive architecture of the ITS consists of the following functional blocks: antenna, DSRC, BTP/GN (Basic Transport Protocol/Geo Networking), LDM (Local Dynamic Map), TCP/IP (Transport Control Protocol/Internet Protocol), Ethernet.

Host computer in the constructive architecture of the ITS intelligent infrastructure consists of the following functional blocks: detection, sensing.

There are two technologies that allow V2X communication, 802.11p and C-V2X technologies. The information rates necessary for the sensor data sharing between the vehicular nodes is not supported in 802.11p or LTE networks. The 5G communications provide higher data transfer rates for a specific vehicle using multiple antennas.

In general, V2X communications can be direct communication, fully network-assisted communication, and hybrid communication, as shown in Fig. 2.30. In direct communication, vehicles connect with neighboring vehicles—peer-to-peer. Communication must be carried out via the RSU or BS in fully network-assisted communications. In hybrid communications, vehicles communicate with each other through a combination of direct communication and network-assisted communication by using 802.11p and cellular networks components [74].

2.3.2 *Vehicle-to-Vehicle (V2V)*

Vehicle-to-Vehicle communication is a cooperative driving application that involves transmitting messages from a host vehicle to other remote vehicles (Fig. 2.31) [75].

Conceptually, vehicles can improve traffic awareness by exchanging situational information directly. Most V2V applications are based on determining the status of other vehicles situated away from the vehicle that is about to be informed through messages. If a vehicle in the vicinity of the V2V system fails to connect to the system, it is invisible to the application. As such a vehicle can be a potential danger to traffic safety, the responsibility for driving at the safety limit lies with the autonomous driving system, which in this situation is guided by the information received from the sensor system (Lidar, radar, video camera etc.).

The main element of the V2V system is the BSM (Basic Safety Message), a message defined by the SAE J2735 standard [76], which provides information on the status of the autonomous vehicle (current position, velocity etc.). The current position of the autonomous vehicle and its velocity, respectively, determines its longitudinal trajectory of movement in relation to the other vehicles in traffic. This trajectory can be influenced by the velocity or by the steering of the autonomous vehicle and is controlled by the autonomous driving system based on the information received from the sensors, and from the V2V systems, respectively [77].

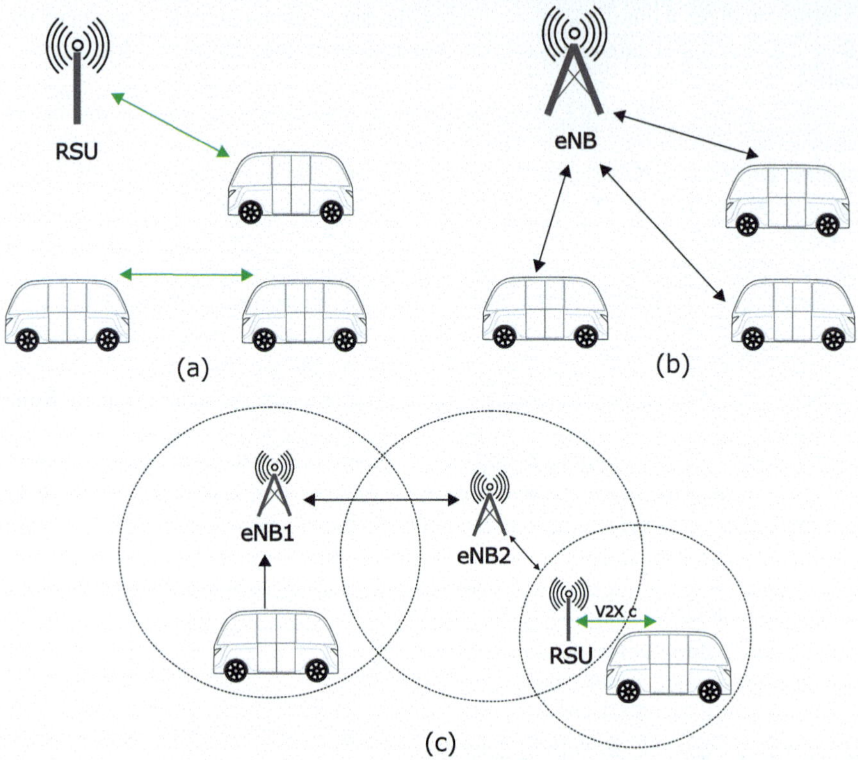

Fig. 2.30 V2X communications: **a** direct (802.11p/PC5); **b** full network assisted; **c** hybrid (5G + PC5)

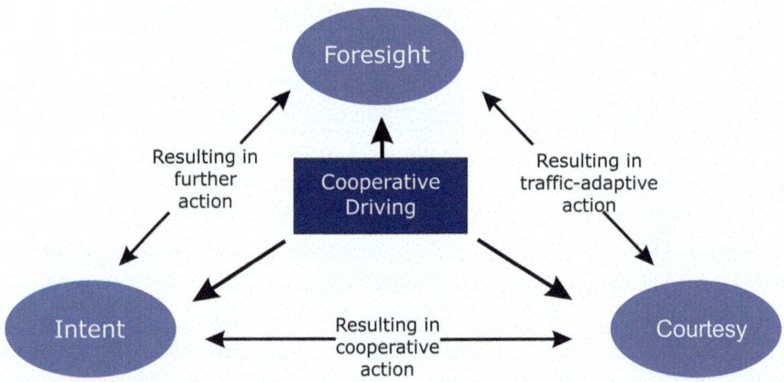

Fig. 2.31 Cooperative driving

The ACC controls the velocity of the autonomous vehicle to maintain its limit at the level set by the autonomous driving system and for the safety of the vehicle, automatically adjusting the distance to the nearest remote vehicle (Fig. 2.32) [77, 78].

The FCW (Forward Collision Warning) application has the role of warning the autonomous driving system in case there is a risk of a frontal collision with a vehicle moving in the same direction, on the same lane (Fig. 2.33) [79, 80].

The EEBL (Emergency Electronic Brake Light) application aims to warn vehicles driving behind the autonomous vehicle about the situation of a possible braking of the autonomous vehicle. The EEBL system transmits this status information via the BSM to nearby vehicles (Fig. 2.34) [78].

The DNPW (Do-Not-Pass Warning) application uses information about the velocity and location of vehicles in the vicinity of the autonomous vehicle in order to determine if it can safely pass slower vehicles on the same lane and traveling in

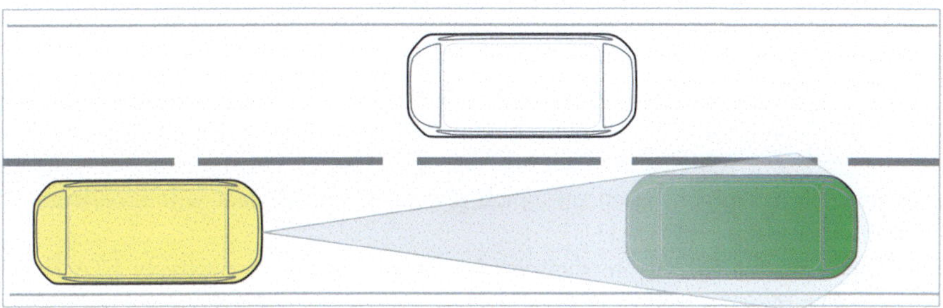

Fig. 2.32 Adaptive Cruise Control

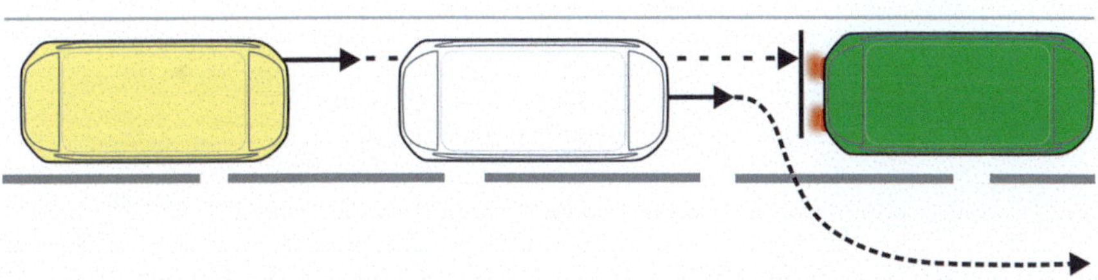

Fig. 2.33 Forward Collision Warning

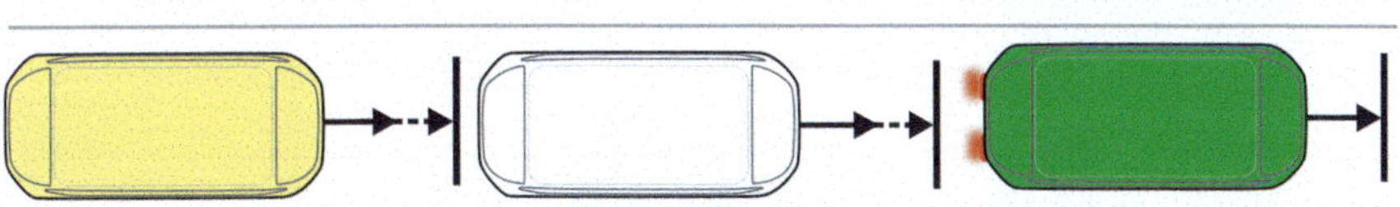

Fig. 2.34 Emergency Electronic Brake Light

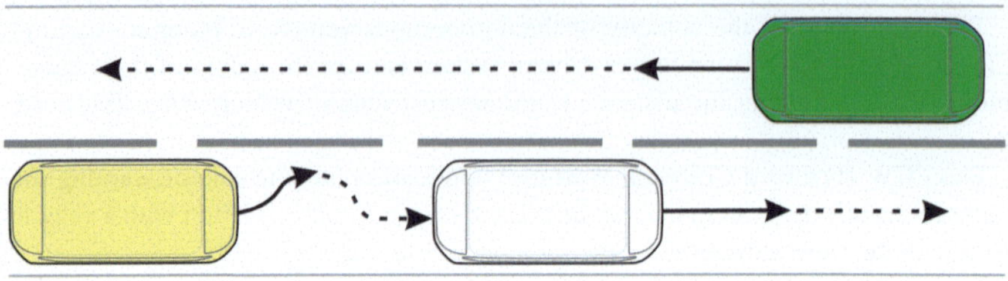

Fig. 2.35 Do-Not-Pass Warning

the same direction, in order to overtake them without endangering vehicles traveling in the opposite direction (Fig. 2.35) [79].

The application BSW (Blind Spot Warning), respectively LCW (Lane Change Warning) performs an assessment and provides information on the side trajectories of vehicles that are in the vicinity of the autonomous vehicle and that travel on adjacent lanes. If these trajectories coincide, the autonomous driving system will change the trajectory of the vehicle (Fig. 2.36) [80].

The MLS (Merging Lanes Safely) application involves a lane change warning and a rear-end collision warning. The autonomous vehicle receives information about the side trajectories of vehicles in its vicinity and traveling on the same lane and in the same direction of travel, based on which it determines when it will accelerate to merge the lanes (Fig. 2.37) [81].

The applications IMA (Intersection Movement Assist), respectively LTA (Left Turn Assist) aim to determine all the trajectories of the vehicles equipped with V2V systems located in the vicinity of the autonomous vehicle (Fig. 2.38) [42, 77, 78].

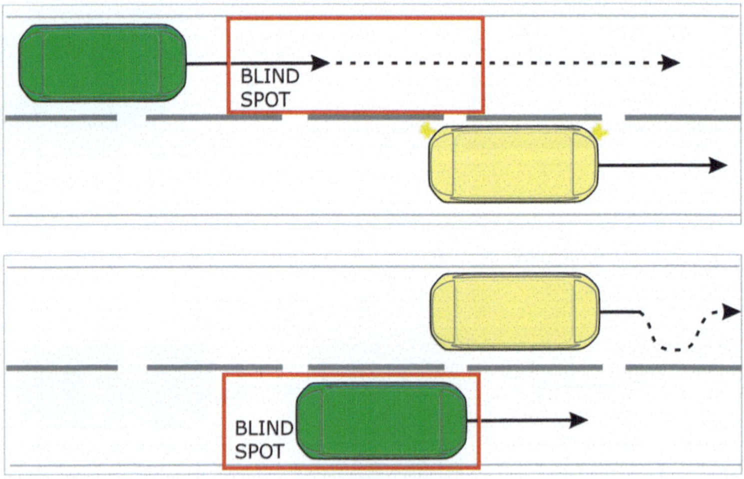

Fig. 2.36 Blind Spot Warning and Lane Change Warning

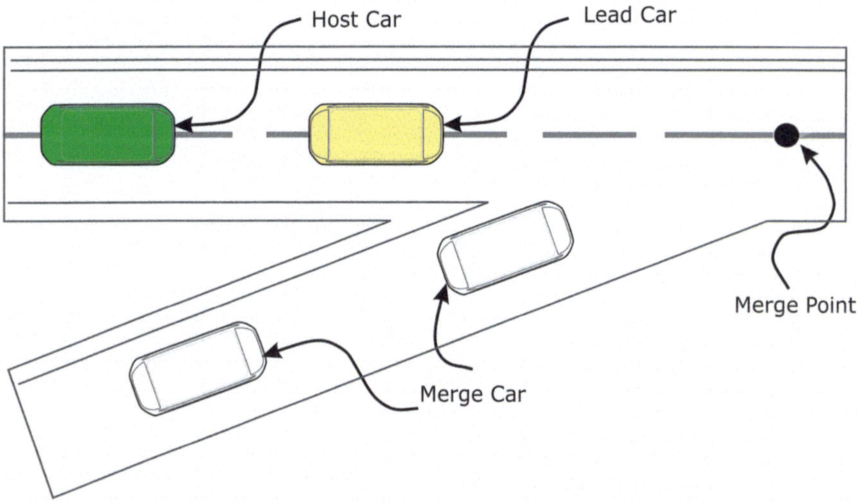

Fig. 2.37 Merging Lanes Safely

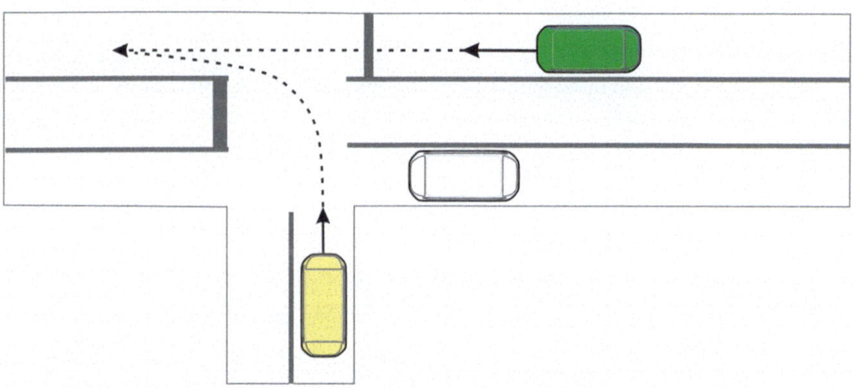

Fig. 2.38 Intersection Movement Assist and Left Turn Assist

The CLW (Control Loss Warning) application intervenes in the situation when a vehicle in the vicinity of the autonomous vehicle loses control, in which case the application will send a warning message to the remote vehicles. Upon receipt of the message, the autonomous vehicle will compare the trajectory of the uncontrolled vehicle with its trajectory, and if the trajectories intersect, the autonomous driving system will determine the change of the trajectory of the autonomous vehicle (Fig. 2.39) [79].

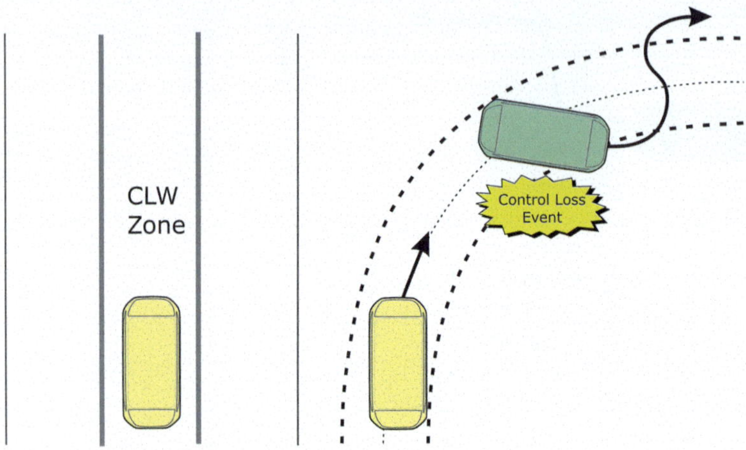

Fig. 2.39 Control Loss Warning

2.3.3 Vehicle-to-Infrastructure (V2I)

Vehicle-to-Infrastructure (V2I) communications support a wide range of applications for both traffic safety and flow. Safety applications generally involve alerts and warnings, while mobility applications may involve the collection of data from vehicles in order to better understand and be aware of the general situation of the road traffic in a given area.

V2I systems generally address a wide range of information at both zonal and regional levels. In general, road information is generated by one or more sources and is communicated to the vehicles, or the information is generated by the vehicles and is transmitted to the environment in which they travel (Fig. 2.40) [71].

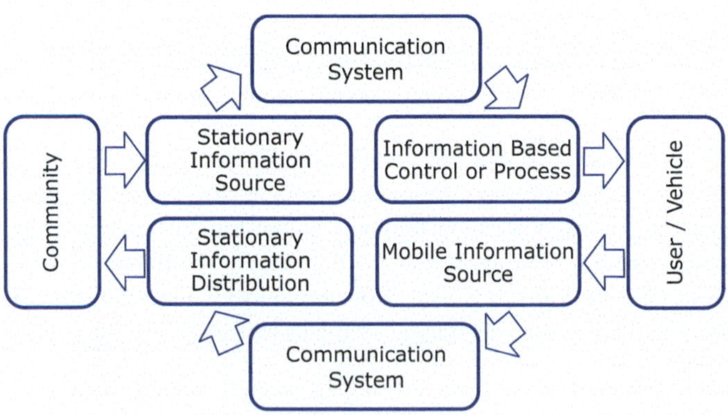

Fig. 2.40 V2I communication and I2V communication

The information transmitted by a V2I system will include geographical identification elements so that the information can be used in a context based on the location of the vehicle. After identifying the geographical location, the system will provide information on the state of the road, so that the autonomous driving system is permanently informed through alerts about all road events that take place on the road sector in question. These road events can be static or dynamic, and the V2I system must monitor, evaluate, and report them accordingly.

(I) Static Events are permanently applicable or applicable for certain periods of time (hours, days, weeks etc.). These events are also marked by permanent or temporary traffic signs. A typical static event is a potential unforeseen road hazard. For these situations, the V2I system generates or retrieves a risk warning message from a road authority. The message related to a potential danger is transmitted to vehicles in the immediate vicinity of the monitored area, to allow the autonomous driving system to reduce planning perception times [72].

(II) Dynamic Events occur temporarily for short periods of time, so potential hazards are intermittently present. If the information about static events is transmitted only once, the information about dynamic events is transmitted regularly, periodically, because it changes over time, and the information must reflect the current state and evolution of the monitored event. Potentially dangerous dynamic events can occur in variable locations or at fixed locations in the road traffic [72].

V2I systems are based on information, alert, warning, and control applications that are transmitted to the autonomous driving system before the autonomous vehicle reaches a maximum decision point where, based on the information received, the system decides to take measures to allow braking until the vehicle has been stopped on a minimum distance (Fig. 2.41) [72]. V2I communication is performed through a local communications system (located in a region on a road infrastructure), namely the DSRC type, and the exchange of information is performed when the vehicle is in the radio coverage area.

2.3.4 Vehicle-to-Pedestrian (V2P)

V2P (Vehicle-to-Pedestrian) systems meet the needs of different groups of VRU (Vulnerable Road Users). These V2P systems use communication technologies and mechanisms to interact with users. V2P technology has the advantage of warning the autonomous driving system, but also the pedestrians, about possible potential traffic hazards.

The VRU are defined in the ITS regulations as "non-motorized users, such as pedestrians and cyclists, and those on two wheels, people with disabilities or reduced mobility" [82]. The V2P system involves a regular exchange of safety messages between vehicles in traffic and the VRU. This communication is done directly through ad-hoc communication (802.11p technology), or indirectly through infrastructure-based communication.

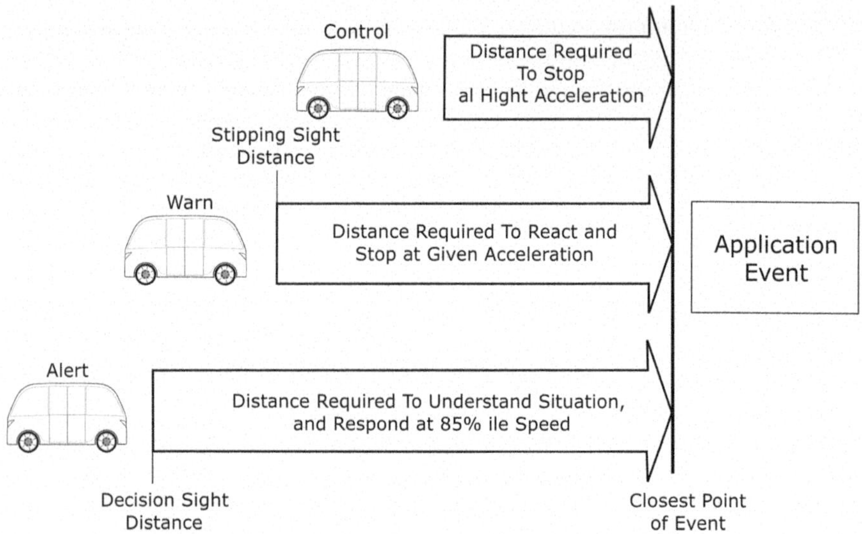

Fig. 2.41 Information, alert, warning, and control

In the case of direct communication, the system consists of two components: the device for the vehicle and the device for the user, which performs the three phases of the protocol: detection, tracking and prediction of the trajectory. In the case of indirect communication (through infrastructure), the central information processing unit is responsible for the implementation of the three phases of the protocol (Fig. 2.42) [83, 84].

Direct communication establishes a communication channel between the vehicle and the user, and these entities communicate with each other directly. This type of communication requires the least latency in sending messages but requires that all devices be equipped with the same type of communication technology.

Indirect communication establishes a communication channel between the vehicle and the user that communicates with each other indirectly through the infrastructure [85]. The devices can be equipped with the same cellular technology, or with 802.11p and Wi-Fi technologies [84]:

(1) 802.11p operates in the 5.9 GHz frequency, and it has been designed for V2X communications. These technologies support the exchange of safety messages reliably and with low latency even under the typical high vehicular mobility conditions. 802.11p based systems support a communication range of up to 1 km.
(2) Cellular-based V2P systems have a larger communication range due to the use of central infrastructure using 3G/LTE for communication and smartphones. C-V2X technology is part of the proposed 5G architecture.
(3) Wi-Fi-based V2P safety systems use a smartphone as a VRU device and have 100–150 m of range for communication. This range may be enough in urban

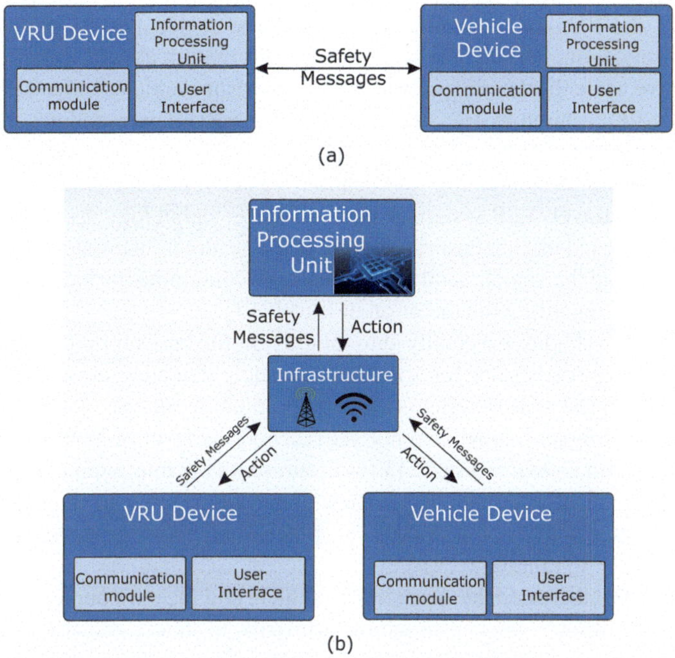

Fig. 2.42 V2P communication system: **a** direct communication, **b** indirect communication

areas with typical vehicle speeds of up to 50 km/h. Wi-Fi-based V2P systems can be deployed without the infrastructure support.

(4) Co-operative pedestrian localization system that operates at 2.44 GHz and 5.768 GHz. It uses a tag as a VRU device which can directly communicate with the vehicle's device without any infrastructure. This V2P system achieves a communication range of up to 100 m. Scalability and latency performance of this system need to be researched further due to its requirement of fixed tag identification numbers.

(5) Bluetooth-based V2P safety system for cyclists. This system uses iBeacon as a VRU device to communicate with the vehicle directly. The communication range is up to 50 m.

(6) V2P communication in 700 MHz ITS band—it evaluates various channel access mechanisms for V2P system.

(7) 802.15.4 technology can achieve the communication range of up to 80 m and can be helpful only for unilateral notification of the collision.

Indirect communication establishes a communication channel between the vehicles and VRU that communicates with each other indirectly through the infrastructure. The devices can be equipped with the same cellular technology, or with different technologies 802.11p and Wi-Fi [85].

The role of VRU devices can be active, when the device actively participates in V2P communication, periodically transmitting information about location, velocity etc. or passive when the device receives information from vehicles [84]. V2P systems based on DSRC technology use a smartphone as a VRU device and use its sensors to optimize the secure transmission of messages. Both the VRU and the vehicle send safety messages containing the location, velocity, and information regarding the direction of travel. V2P systems based on DSRC technology use a Smartphone as a VRU device and use its sensors to optimize the secure transmission of messages. Both the VRU and the vehicle send safety messages containing the location, velocity, and information regarding the direction of travel.

Autonomous vehicles, especially autonomous shuttle buses for public transportation, aim to make roads safer for VRU by accurately detecting these traffic participants. Although the sensor systems that equip autonomous vehicles have become more and more efficient, they have not yet reached the level of human perception. The logic diagram shown in Fig. 2.43 [64] represents the minimum conditions that the sensor systems of autonomous vehicles must meet in order to safely detect and estimate the future actions of the VRU. This process will allow autonomous vehicles to navigate safely in relation to VRU.

The interaction between the autonomous vehicle and the environment is achieved through sensors that collect information primarily to detect and track static and dynamic objects. The quality of the information collected by sensors can be influenced by external sources such as ambient light, weather conditions, sensor coverage etc. In order to increase the accuracy of detection and tracking of the static and dynamic objects, autonomous vehicles use different sensor systems that—by merging the recorded information—increase the accuracy of the classification, detection, and estimation of the position of the tracked objects [64].

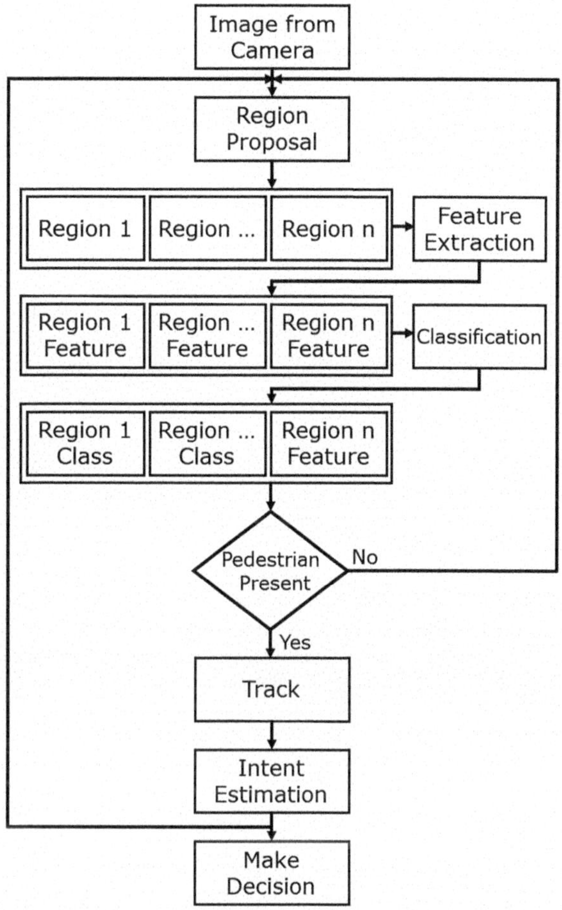

Fig. 2.43 Conditions for detecting and estimating VRU intentions

References

1. https://www.intechopen.com/books/electric-vehicles-modelling-and-simulations/electrical-vehicle-design-and-modeling. 6 Oct 2020
2. https://www.ti.com/solution/electric-drive?variantid=26121&subsystemid=28521. 21 Mar 2021
3. Iclodean, C., Cordos, N., Varga, B.O.: Autonomous shuttle bus for public transportation: a review. Energies **13**, 2917 (2020)
4. https://www.ti.com/solution/battery-management-system-bms?variantid=14008&subsystemid=17303. 21 Mar 2021
5. Lu, L., Han, X., Li, J., Hua, J., Ouyang, M.: A review on the key issues for lithium-ion battery management in electric vehicles. J. Power Sources **226**, 272–288. ISSN 0378-7753. https://doi.org/10.1016/j.jpowsour.2012.10.060. 17 May 2022
6. https://easymile.com/vehicle-solutions/ez10-passenger-shuttle. 24 July 2020
7. https://navya.tech/shuttle/. 11 Apr 2020
8. https://techcrunch.com/2019/08/31/come-along-take-a-ride/. 22 April 2022
9. https://www.ntsb.gov/investigations/AccidentReports/Reports/HAB1906.pdf. 12 Apr 2020

10. Iclodean, C., et al.: Comparison of different battery types for electric vehicles. IOP Conf. Ser.: Mater. Sci. Eng. **252**, 012058 (2017)
11. Skouras, T.A., Gkonis, P.K., Ilias, C.N., Trakadas, P.T., Tsampasis, E.G., Zahariadis, T.V.: Electrical vehicles: current state of the art, future challenges, and perspectives. Clean Technol. **2**, 1–16 (2020)
12. Sun, X., Li, Z., Wang, X., Li, C.: Technology development of electric vehicles: a review. Energies **13**, 90 (2020)
13. https://www.thierry-lequeu.fr/data/SZZA058C.pdf. 12 Apr 2020
14. https://www.emobilitysimplified.com/2019/10/ev-charging-levels-modes-types-explained.html. 12 Apr 2020
15. https://www.iecee.org/dyn/www/f?p=106:49:0::::FSP_STD_ID:6032. 12 Apr 2020
16. https://www.sae.org/standards/content/j1772_201710/. 12 Apr 2020
17. https://www.pngkey.com/maxpic/u2e6e6i1a9o0q8a9/. 12 Apr 2020
18. https://landtransportguru.net/easymile-ez10/. 18 Apr 2020
19. https://www.theverge.com/2016/6/17/11962776/local-motors-olli-3d-printed-autonomous-bus-photos. 18 Apr 2020
20. Ainsalu, J., Arffman, V., Bellone, M., Ellner, M., Haapamäki, T., Haavisto, N., Josefson, E., Ismailogullari, A., Lee, B., Madland, O., Madžulis, R., Müür, J., Mäkinen, S., Nousiainen, V., Pilli-Sihvola, E., Rutanen, E., Sahala, S., Schønfeldt, B., Smolnicki, P.M., Soe, R.-M., Sääski, J., Szymańska, M., Vaskinn, I., Åman, M.: State of the art of automated buses. Sustainability **10**, 3118 (2018)
21. Eskandarian, A.: Handbook of Intelligent Vehicles. Springer, Heidelberg, pp. 33–58 (2012)
22. https://www.ti.com/applications/automotive/adas/overview.html?HQS=sys-auto-adas-automotivevanity-vanity-lp-adas-wwe. 28 Aug 2020
23. https://www.ti.com/solution/hev-ev-on-board-obc-wireless-charger. 21 Mar 2021
24. https://eduscol.education.fr/sti/concours_examens/epreuve-de-sciences-de-lingenieur-septembre-2018-nouvelle-caledonie#fichiers-liens. 16 Apr 2020
25. Chen, Y., Yu, H., Graaf, R., Wang, X., Wan, J.: Robust vehicle longitudinal motion control subject to in-wheel-motor driving torque variations. In: Proceedings of the 2017 American Control Conference (ACC), Seattle, WA, USA, pp. 4316–4321, 24–26 May 2017
26. Zhang, Z., Pan, H., Salman, W., Rasim, Y., Liu, X., Wang, C., Yang, Y., Li, X.: A novel steering system for a space-saving 4ws4wd electric vehicle: design, modeling, and road tests. IEEE Trans. Intell. Transp. Syst. **18**, 114 (2017)
27. Isa, K.B., Jantan, A.B.: An autonomous vehicle driving control system. Int. J. Eng. Educ. **21**, 855–866 (2005)
28. Zheng, H., Yang, S.: A trajectory tracking control strategy of 4WIS/4WID electric vehicle with adaptation of driving conditions. Appl. Sci. **9**, 168 (2019)
29. https://www.ti.com/solution/electric-power-steering-eps. 22 Mar 2021
30. IPG Automotive, Reference Manual Version 9.1 CarMaker (2020)
31. https://e2e.ti.com/blogs_/b/behind_the_wheel/archive/2014/09/25/cars-are-becoming-rolling-sensor-platforms. 12 Apr 2020
32. https://uk.farnell.com/wcsstore/ExtendedSitesCatalogAssetStore/cms/asset/images/europe/common/applications/automotive/pdf/ti-adas-solution-guide.pdf. 12 Apr 2020
33. https://www.wevolver.com/article/a.review.of.autonomous.vehicle.safety.and.regulations. 12 Apr 2020
34. https://github.com/ApolloAuto/. 12 Apr 2020
35. https://velodynelidar.com/downloads/. 12 Apr 2020
36. https://www.hesaitech.com/en/download?product=Pandora. 12 Apr 2020
37. https://www.conti-engineering.com/en-US/Industrial-Sensors/Sensors-Overview. 12 Apr 2020
38. https://leopardimaging.com/product-category/usb30-cameras/. 12 Apr 2020
39. https://docs.novatel.com/oem7/Content/PDFs/PwrPak7_Installation_Operation_Manual.pdf. 12 Apr 2020

40. Rosique, F., Navarro, P.J., Fernández, C., Padilla, A.: A systematic review of perception system and simulators for autonomous vehicles research. Sensors **19**, 648 (2019)
41. Zheng, L., Li, B., Yang, B., Song, H., Lu, Z.: Lane-level road network generation techniques for lane-level maps of autonomous vehicles: a survey. Sustainability **11**, 4511 (2019)
42. https://www.itf-oecd.org/sites/default/files/docs/safer-roads-automated-vehicles.pdf. 12 Apr 2020
43. http://umich.edu/~umtriswt/PDF/SWT-2017-12.pdf. 12 Apr 2020
44. https://cleantechnica.com/2016/07/29/tesla-google-disagree-lidar-right/. 12 Apr 2020
45. https://www.vie-publique.fr/sites/default/files/rapport/pdf/184000287.pdf. 02 Oct 2020
46. https://www.semagarage.com/assets/pdf/advanced-driver-assistant-system-paper.pdf. 07 Sept 2020
47. Im, J.-H., Im, S.-H., Jee, G.-I.: Extended line map-based precise vehicle localization using 3D LIDAR. Sensors **18**, 3179 (2018)
48. Royo, S., Ballesta-Garcia, M.: An overview of lidar imaging systems for autonomous vehicles. Appl. Sci. **9**, 4093 (2019)
49. https://www.sick.com/kr/en/detection-and-ranging-solutions/2d-lidar-sensors/c/g91900. 12 Apr 2020
50. https://www.sick.com/kr/en/detection-and-ranging-solutions/3d-lidar-sensors/c/g282752. 12 Apr 2020
51. Sualeh, M., Kim, G.-W.: Dynamic multi-LiDAR based multiple object detection and tracking. Sensors **19**, 1474 (2019)
52. Yu, X., Marinov, M.: A study on recent developments and issues with obstacle detection systems for automated vehicles. Sustainability **12**, 3281 (2020)
53. https://www.ti.com/solution/automotive-mechanically-scanning-lidar. 22 Mar 2021
54. https://www.ti.com/lit/wp/slyy150a/slyy150a.pdf?ts=1606728761407. 12 Apr 2020
55. Steinhardt, N., Leinen, S.: Data fusion for precise localization. In: Winner, H., Hakuli, S., Lotz, F. (eds.) Handbook of Driver Assistance Systems, pp. 154–196. Springer International Publishing, Switzerland (2016)
56. Stateczny, A., Kazimierski, W., Gronska-Sledz, D., Motyl, W.: The empirical application of automotive 3D radar sensor for target detection for an autonomous surface vehicle's navigation. Remote Sens. **11**, 1156 (2019)
57. Bosch Automotive Electrics and Automotive Electronics Systems and Components, Networking and Hybrid Drive. https://www.springer.com/gp/book/9783658017835
58. Bishop, R.: Intelligent Vehicle Technology and Trends. Artech House, Norwood, MA, USA, pp. 7–24 (2005)
59. Robert Bosch GmbH: Bosch Automotive Electrics and Automotive Electronics. Springer, Heidelberg. https://doi.org/10.1007/978-3-658-01784-2_7 (2014)
60. https://www.bosch-mobility-solutions.com/en/products-and-services/passenger-cars-and-light-commercial-vehicles/driver-assistance-systems/automatic-emergency-braking/mid-range-radar-sensor-(mrr)/. 12 Apr 2020
61. https://www.ti.com/solution/automotive-long-range-radar?variantid=18307&subsystemid=18539. 22 Mar 2021
62. Yu, H., Shi, W., Alawieh, H.B., Yan, C., Zeng, X., Li, X., Yu, H.: Efficient statistical validation of autonomous driving systems. In: Yu, H., Li, X., Murray, R.M., Ramesh, S., Tomlin, C. (eds.) Safe, Autonomous and Intelligent Vehicles, pp. 5–32. Springer International Publishing, Cham (2019)
63. Yu, F.-M., Jwo, K.-W., Chang, R.-S., Tsai, C.-T.: Dispensing technology of 3D printing optical lens with its applications. Energies **12**, 3118 (2019)
64. Ahmed, S., Huda, M.N., Rajbhandari, S., Saha, C., Elshaw, M., Kanarachos, S.: Pedestrian and cyclist detection and intent estimation for autonomous vehicles: a survey. Appl. Sci. **9**, 2335 (2019)
65. https://www.ti.com/solution/automotive-front-camera?variantid=14995&subsystemid=15228. 22 Mar 2021

66. Wu, X., Cheng, C., Zhao, X.: Kinematic and dynamic vehicle model-assisted global positioning method for autonomous vehicles with low-cost GPS/Camera/In-Vehicle sensors. Sensors **19**, 5430 (2019)
67. Odijk, D.: Positioning model. In: Teunissen, P.J.G., Montenbruck, O. (eds.) Springer Handbook of Global Navigation Satellite Systems, pp. 606–635. Springer International Publishing, Switzerland (2017)
68. Onyekpe, U., Palade, V., Kanarachos, S.: Learning to localise automated vehicles in challenging environments using Inertial Navigation Systems (INS). Appl. Sci. **11**, 1270 (2021)
69. Özgüner, Ü., Acarman, T., Redmill, K.: Autonomous Ground Vehicles. Artech House, Norwood, MA, USA, pp. 193–216 (2011)
70. https://www.ti.com/solution/ultrasonic-park-assist-sensor. 22 Mar 2021
71. Wang, J., Shao, Y., Ge, Y., Yu, R.: A survey of vehicle to everything (V2X) testing. Sensors **19**, 334 (2019)
72. Eichberger, A., Markovic, G., Magosi, Z., Rogic, B., Lex, C., Samiee, S.: A Car2X sensor model for virtual development of automated driving. Int. J. Adv. Robot. Syst. (2017). https://doi.org/10.1177/1729881417725625. 17 May 2022
73. https://www.ti.com/solution/vehicle-to-everything-v2x. 22 Mar 2021
74. Shrestha, R., Nam, S.Y., Bajracharya, R., Kim, S.: Evolution of V2X communication, and integration of blockchain for security enhancements. Electronics **9**, 1338 (2020)
75. Herrtwich, R.G., Nöcker G.: Cooperative driving: taking telematics to the next level. In: Fukui, M., Sugiyama, Y., Schreckenberg, M., Wolf, D.E. (eds.) Traffic and Granular Flow'01. Springer, Heidelberg (2003)
76. https://www.sae.org/standards/content/j2735_202007/. 17 May 2022
77. https://www.mathworks.com/help/mpc/ref/adaptivecruisecontrolsystem.html (July 31, 2020).
78. https://www.nhtsa.gov/sites/nhtsa.dot.gov/files/documents/812327-electronicbrakelight.pdf. 31 July 2020
79. https://savari.net/wp-content/uploads/2017/01/Savari-V2VApps-DataSheet-FINALOct2016.pdf. 31 Aug 2020
80. Dimitrievski, M., Veelaert, P., Philips, W.: Behavioral pedestrian tracking using a camera and LiDAR sensors on a moving vehicle. Sensors **19**, 391 (2019)
81. Soni, A., Hu, H.: Formation control for a fleet of autonomous ground vehicles: a survey. Robotics **7**, 67 (2018)
82. Anaya, J.J., Merdrignac, P., Shagdar, O., Nashashibi, F., Naranjo, J.E.: Vehicle to pedestrian communications for protection of vulnerable road users. IEEE Intelli. Veh. Symp. Proc. **2014**, 1037–1042 (2014). https://doi.org/10.1109/IVS.2014.6856553
83. Sewalkar, P., Krug, S., Seitz, J.: Towards 802.11p-based vehicle-to-pedestrian communication for crash prevention systems. In 2017 9th International Congress on Ultra-Modern Telecommunications and Control Systems and Workshops (ICUMT), pp. 404–409 (2017). https://doi.org/10.1109/ICUMT.2017.8255154
84. Sewalkar, P., Seitz, J.: Vehicle-to-pedestrian communication for vulnerable road users: survey, design considerations, and challenges. Sensors **19**, 358 (2019)
85. Sugimoto, C., Nakamura, Y., Hashimoto, T.: Prototype of pedestrian-to-vehicle communication system for the prevention of pedestrian accidents using both 3G wireless and WLAN communication. In: 2008 3rd International Symposium on Wireless Pervasive Computing, pp. 764–767 (2008). https://doi.org/10.1109/ISWPC.2008.4556313
86. Vargas Rivero, J.R., Gerbich, T., Teiluf, V., Buschardt, B., Chen, J.: Weather classification using an automotive LIDAR sensor based on detections on asphalt and atmosphere. Sensors **20**, 4306 (2020)

Chapter 3
Autonomous Driving Systems

3.1 Autonomous Driving Algorithm

3.1.1 Overview

The autonomous driving ability is provided by the autonomous driving system based on blocks consisting of combinations of functions and properties that define the characteristics of the environment and the objects found in this environment. These functions and properties are follows [1]:

(1) Safety is integrated into the decision-making process for information technology.
(2) Behavior control uses information from the safety, planning, and control block to ensure a safe and comfortable driving behavior in traffic.
(3) Simulation eases the development and continuous improvement of the performance of the autonomous vehicle.
(4) Machine Learning—on the perception of the environment and the objects found in this environment.
(5) Localisation implies knowing the position of the autonomous vehicle at any time.
(6) The local map generator generates HD maps of the environment for the precise movement of the autonomous vehicle.
(7) Perception manages the position of the autonomous vehicle relative to the position of objects in the environment, based on information provided by the sensors and the HD maps.
(8) Navigation planning determines the dynamic behavior of the autonomous vehicle resulting from the prediction correlated with the travel rules.
(9) Motion control controls the powertrain, steering and braking systems based on data from the location, perception, prediction, and planning blocks.
(10) Remote assistance—monitoring the movement of the autonomous vehicle using the management platform.

C. Iclodean et al., *Autonomous Vehicles for Public Transportation*, Green Energy and Technology, https://doi.org/10.1007/978-3-031-14678-7_3

(11) Networking performs the transfer of large amounts of data between the command-and-control systems of the autonomous vehicle.

The autonomous driving system for public transportation is a hardware and software technological solution that has the ability to drive without the need for human intervention on the control mechanisms of the vehicle (depending on the level of automation according to SAE J3016™ specifications), with it being remotely monitored by a human operator using the management platform [2–4].

The degree of automation of an autonomous shuttle bus intended for public transport falling within the level 3 or 4 of automation, allows the human operator not to intervene in its operation (all the dynamic functions) while it is in operation. After stopping in the vicinity of an object, respectively after the object in question is removed from its vicinity, the autonomous vehicle will independently resume its function of autonomous driving.

For level 4, in the event of a problem—a system error, an incident, a deviation from the path etc.—the autonomous driving system will have control to overcome this situation, and the human operator will intervene only if he considers it necessary to change the perception and planning of the actions of the autonomous driving system. When a problem arises for a level 3 system, the human operator will be the one to intervene to reactivate the autonomous driving system after the problem is solved.

The autonomous driving system control algorithm is a substitute for a human driver who, by fully delegating control over driving and describing behavioral conditions, performs the following functions [2, 5]:

(1) Discovering and avoiding existing objects in the direction of travel, stopping safely, or changing lanes as appropriate.
(2) Moving the autonomous vehicle in the center of the direction of travel, at maximum velocity.
(3) Changing the direction of travel for reasons other than avoiding or overtaking existing objects, where possible.
(4) Overtaking over several lanes by securing over existing objects, changing direction, overtaking, and returning.
(5) Overtaking over a single lane by securing over existing objects, changing direction, overtaking, and returning.
(6) Cancelling the overtaking in the event of an unforeseen situation arising from a change of direction and returning to the lane.
(7) Complete overtaking when the maneuvering space allows it, after overtaking the autonomous vehicle returns to the lane.
(8) When another vehicle engages in overtaking, the autonomous vehicle avoids increasing the velocity and maintains a safe distance from the overtaking vehicle.
(9) Ensuring and maintaining a minimum safety distance in the vicinity of existing objects to avoid any collision.
(10) Ensuring and maintaining a low velocity in the vicinity of existing objects to avoid any collision.

(11) Avoiding interaction with existing objects by keeping its distance, adjusting its velocity, overtaking, or stopping as appropriate.
(12) Constant movement in the direction of travel when no object intervenes in the vicinity.

The autonomous driving system uses a decision-making algorithm based on the following principles [2, 3, 6, 7]:

(A) Perception of the position of the autonomous vehicle, respectively of the position of the objects in the environment in the vicinity of the vehicle.
(B) Prediction/ Planning of the resulting dynamic behaviors correlated with the rules of travel/operation.
(C) Motion control of the dynamic behaviors based on the decisions made in the prediction/planning stage.

The Machine Learning algorithms used by the autonomous driving system make the connections between a set of inputs and a set of outputs, based on a set of predefined instructions. These algorithms are based on Deep Learning methodologies, which are an integral part of the autonomous driving algorithm and can be of the following types: CNN (Convolutional Neural Networks), RNN (Recurrent Neural Networks) or DRL (Deep Reinforcement Learning) [8].

The CNN are used to process images and information provided by sensors and to extract features of interest and to identify static and dynamic objects in the environment. These neural networks consist of a convolution layer consisting of a set of filters that have the ability to distinguish the elements of an image and label them. The data provided at the output of this convolution layer is processed based on an algorithm that combines them to predict the best description of an image. The final software component is called an object classifier and has the ability to classify the objects into images.

The RNN are used when information provided by sensors is video captures, which are considered time sequences and allow the autonomous driving algorithm to adopt a dynamic time behavior. Recurrent neural networks use internal memory to process sequences of varying length and allow information and knowledge to be contextualized.

The DRL neural networks combine DL (Deep Learning) and DR (Deep Reinforcement). The methods of these neural networks allow the software algorithm to retain the best actions to achieve its objectives in the virtual environment. Goal-oriented algorithms learn how to achieve a goal along a specific dimension by following a number of steps.

The autonomous driving system, regardless of the technical solution adopted for the implementation of the functional algorithm, provides decisional support for achieving the perception and movement of the vehicle, respectively for stopping in conditions of maximum safety, based on data provided by the sensors. According to these data, the system establishes the characteristics that define the trajectory of movement and regulates the velocity so that depending on the circumstances it can operate the stop in complete safety, depending on the characteristics of the route, the objects on the route, the points defined as stopping points for embarkation/disembarkation. All these functions are characteristic of an autonomous driving solution for public transportation [2].

Autonomous shuttle buses (levels 3 and 4 of automation) intended for public transportation are constantly monitored through the management platform by a human operator, who, in particular situations, potentially dangerous, can order their stop.

3.1.2 Perception

Perception manages the position of the autonomous vehicle relative to the position of objects in the environment (Fig. 3.1) [9]. The images captured by the video cameras (1a and 1b) are analyzed by the block for detecting and classifying static and dynamic objects (detector) and in parallel by the locating block (localizer). Information about identified (1c) and located (2) objects is transmitted to the tracking block (tracker) to anticipate the future trajectory (3). This information is transmitted (4) to the motion planning block (motion planner) for the command and control of the dynamic behaviors (propulsion, steering, braking) of the autonomous vehicle.

All information on object detection is used to manage the priorities of the system and to integrate the autonomous vehicle in road traffic, through a programmed control area, in order to continue its movement in autonomous driving mode. The velocity of the autonomous vehicle is adjusted to safely manage the objects detected on the path by prediction. In most cases for public transportation the maximum speed limit is 25 km/h. The autonomous vehicle accelerates/decelerates depending on the trajectory followed and the velocity of the dynamic objects.

The perception module uses a machine learning platform based on visual detection on road sectors that do not have large turning angles, respectively unmarked lanes or intersections, public roads with heavy traffic [10]. The main characteristics of the perception module are the following:

(1) Detecting the travel trajectory and maintaining the safe distance to the vehicle in front (closest in-path vehicle).

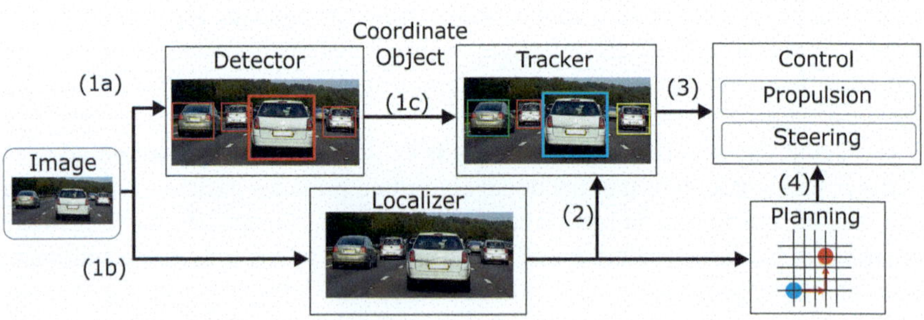

Fig. 3.1 Mechanism of the perception module

(2) Fusion of data from Lidar sensors, radar, video camera for a clear perception of the environment (asynchronous sensor fusion).
(3) Estimation of the position of the object pursued based on 3D location (online pose estimation).
(4) Use of ultrasonic sensors for automated emergency braking in the event of an AEB object appearing, or for park assist.
(5) Accurate detection of the whole lane line.
(6) Visual localisation through the video camera.
(7) Use of 16 beam Lidar sensors.

The block diagram of the perception module is shown in Fig. 3.2 [10]. The deep network takes over the images transmitted by the sensors (image) and based on the collected data classifies the detected objects, assigning them data labeling and separates the detected travel lanes. The network transmits at the output the labeled objects that will be tracked on the trajectory (object tracking), respectively the information about the path broken down into segments (lane line detection).

Objects detected/tracked in traffic are classified into two categories: static objects and dynamic objects. The dynamic objects will be classified and tracked in the order of importance established by their trajectory of travel (closest in-path vehicle). Depending on the position of the tracked object (online pose estimation), the object tracking block estimates its location in a 3D coordinate system by an association between labels and object, thus obtaining a correct identification for each tracked object.

The detected static objects will be tracked by the lane detection/tracking system. The framing limits of a lane ensure lateral control when the autonomous vehicle is moving, the vehicle being guided along the lane line by controlling its longitudinal movement. Each left–right turn made by the lane line is marked and stored in the

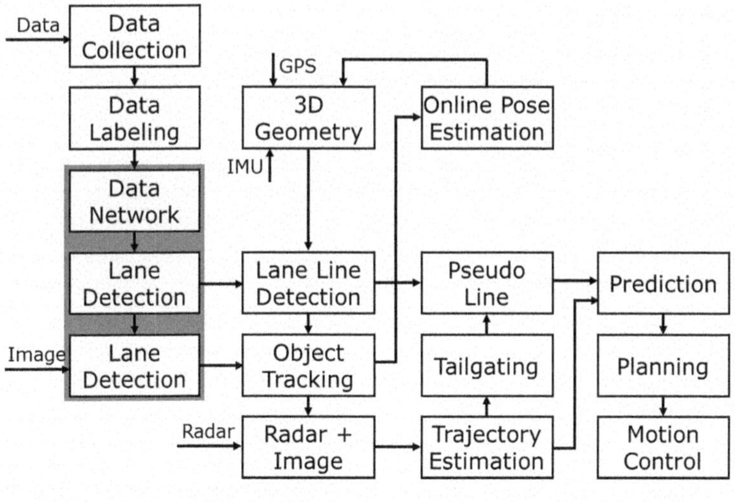

Fig. 3.2 Diagram of the perception module

autonomous driving system, which coordinates the maneuvers that the vehicle must perform while travelling based on the traffic markings.

Based on the signals transmitted by the radar sensor together with the data about the detected and tracked object, the system performs an estimate of the trajectory of this object (trajectory estimation) and determines the travel trajectory of the autonomous vehicle so as to maintain a safe distance to the tracked object (tail-gating). Tracking this object can also be correlated with maintaining the travel lane (pseudo line) and the travel history buffer.

The trajectory of the tracked object, together with the trajectory of the travel lane of the autonomous vehicle in the perception stage determines the autonomous driving system to issue a prediction on the trajectory that the vehicle will have to follow and based on this prediction to plan (planning) the control of the propulsion, steering and braking systems.

3D obstacle perception is based on the data provided by Lidar and radar sensors. The Lidar sensor scans the coverage area and delimits the limits of the route traveled by the autonomous vehicle ROI (Region Of Interest) and highlights the objects in the area using points that it marks on the HD map. The autonomous driving system transforms the sets of points provided by the Lidar sensors into a set of polygons that frame the contour of the objects in the area. There are three main components of 3D obstacle perception [10]:

(1) The perception of obstacles detected by Lidar sensors.
(2) The perception of obstacles detected by radar sensors.
(3) The fusion of the results of the obstacles detected by the sensors.

Perception of the obstacles detected by Lidar sensors includes the following steps that facilitate the generation of input data consisting of a cloud of 3D points:

(a) Filter for delimiting the route traveled by the autonomous vehicle in ROI on the HD map.
(b) Segmentation of the CNN.
(c) Reproduction of the 3D objects using the MinBox builder utility.
(d) Tracking the static and dynamic objects using the HM object tracker.
(e) Sequential data fusion (typefusion).

The filter for delimiting the route traveled by the autonomous vehicle is based on the ROI which is specific to the driving area of the autonomous vehicle, and it includes road surfaces and respectively the intersections on the HD map. The ROI filter applied to the HD map processes the point cloud generated by the Lidar sensors that are outside the ROI, removing background objects. This results in the concentration points within the ROI that will be used for further processing. Given the details of the HD map, the affiliation of each Lidar point indicates whether it is inside or outside the ROI. Each Lidar point can be identified from a 2D quantification LUT (Look-Up Table) of the region around the autonomous vehicle. The ROI filter applied to the HD map is achieved by following three successive steps:

(1) Coordinate transformation: for the ROI filter applied to the HD map, the data interface of the HD map is defined by a set of polygons, each of which being an

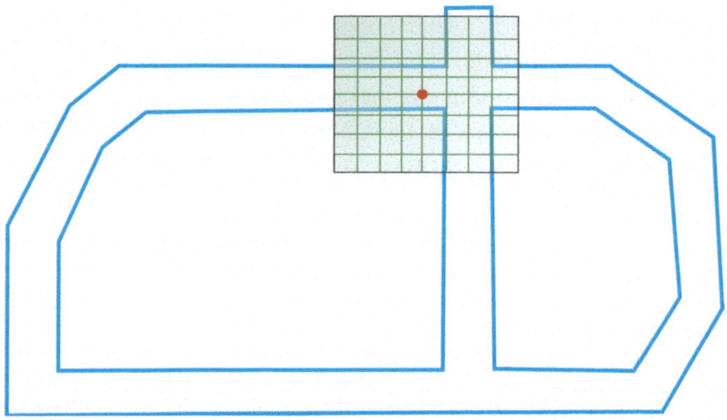

Fig. 3.3 ROI LUT construction

ordered set of points in the coordinate system. In order to perform an evaluation on the set of points with the ROI filter applied to the HD map, the point cloud and polygons must be represented in the same coordinate system. To this end, the coordinate transformation transforms the input point cloud points and the polygons of the HD map into a local coordinate system that indicates the position of the autonomous vehicle.

(2) ROI LUT construction: the determination of entry points, regardless of whether it is inside or outside the ROI, is performed according to the network that quantifies ROI in a 2D grid (Fig. 3.3). This LUT covers a rectangular region, delimited by a predefined spatial range around the general top view, at the edge of the HD map.

(3) Punctual investigation with ROI LUT involves querying each entry point using a two-step check. The following operations are performed for the punctual investigation process:

(a) The identification of the point inside or outside the rectangular region of the ROI LUT.

(b) The request for the cell corresponding to the point in the LUT for its affiliation in relation to ROI.

(c) The collection of all points that belong to ROI and the generation of clues in relation to the input point cloud.

The segmentation of the CNN is performed after identifying the environment based on the data obtained from the scan performed with Lidar sensors and using an ROI filter applied to the HD map, where the filtered point cloud includes only the points inside the ROI (roadway sector and junction areas). The segmentation performed by the CNN convolutional neural network is performed in the following successive steps:

(1) Channel feature extraction. After quantification, a number of eight statistical point measurements are calculated for each cell of the network, which will be the characteristics of the input channel in the next step.
(2) Prediction of obstacles based on the CNN.
(3) Grouping the obstacles into clusters.
(4) Post-processing.

The perception module is based on data from four sources of information: Lidar, video camera, GPS and V2X communication. The main features of the perception mode are the following [2, 11]:

(1) Detection of static and dynamic objects (object detection) performed with the help of video cameras and location of these objects in space through the Lidar sensor. The mechanisms for detecting and classifying static and dynamic objects are based on the SSD (Single Shot multibox Detector) model and the YOLO (You Only Look Once) model.
(2) Ground profile extraction by separating the points from which it is formed from the points from which the objects detected by the Lidar sensor are formed.
(3) Establishing communication with interconnected objects via the V2X connection that helps the autonomous vehicle make decisions about future actions.
(4) Establishing the location based on the characteristics of the area (region base mode selector) in which the parameters provided depend on the specifics of the area:
(a) General standard road conditions in which the area in the vicinity of the autonomous vehicle is scanned for the detection of all static and dynamic objects.
(b) Urban agglomeration (dense urban) extends the scanned region by increasing the scanning resolution.
(c) Obstacle on highway classification of static or dynamic objects on the highway.
(d) Give way changing the lane in the situation when an emergency vehicle is approaching.
(5) Clustering parameter selector selecting the cluster parameters from the cluster-grouping of the point cloud provided by the Lidar sensor for the local route planning.
(6) Object attribute allocator allocation of attributes for the detected, labeled, and classified static and dynamic objects, respectively measurement of the Euclidean distance to their location.

Pendleton et al. in [12] defined the environment perception as a fundamental function that enables the autonomous vehicles, by providing the information on the driving environment: free drivable areas, the locations of obstacles, velocities, and predictions of their future states. Based on the sensors implemented, the environment perception task can be tackled by using Lidars, cameras, or a fusion between these two kinds of devices.

The vision system in the autonomous vehicle environment perception involves road detection and on-road object detection. The road detection also includes lane marking line detection and road surface detection. Lane marking line detection is to identify the lane line markings on the road and estimate the vehicle's position reported to the detected lines. Lane line feature extraction is to identify the pixels that belong to the lane markings line and eliminate non-lane line marking pixels.

Road surface detection informs the autonomous vehicle on the locations of free space where it can drive without collision, which is why it is the prerequisite for any online path planning and control operations. The approaches can be divided into three categories: feature/cue-based detection, feature/cue based learning, and Deep Learning.

The localisation determines the position of the vehicle and measures its own motion. To estimate the location of the vehicle, a fusion of satellite-based navigation systems and inertial navigation systems is needed. Satellite navigation systems, such as GPS/GNSS can provide a regular fix on the global position of the vehicle. Their accuracy can vary from a few tens of meters to a few millimeters depending on the signal strength, and the quality of the equipment used. IMU, which use accelerometer, gyroscope, and signal processing techniques to estimate the attitude of the vehicle, do not require external infrastructure.

SLAM (Simultaneous Localisation And Mapping) algorithms use the features that have been observed by the autonomous vehicle's sensors to estimate its position on the map and to locate new features.

3.1.3 Prediction

The prediction of the dynamic behaviors of the autonomous vehicle results from its perception correlated with the travel/operation rules.

The prediction generates the shortest route on the HD map between two points (nodes) to be traveled by the autonomous vehicle, according to an algorithm implemented in two stages:

(A) Generating the route on a predefined map where information about static objects is available and marked on this map, and the route is generated by avoiding the marked static objects.
(B) Generating the route in a heuristic mode when there is no predefined map, using the information provided by the video camera and the Lidar sensor, which allow the autonomous vehicle to detect static and dynamic objects located in a limited field of view.

The improvement of the prediction mechanism is achieved through the V2X connection, a situation in which the autonomous vehicle receives from the infrastructure information about the static and dynamic objects on the route (Fig. 3.4) [11].

V2X transmitters installed in the road infrastructure transmit traffic jam information to all autonomous vehicles within their range. V2X receivers on vehicles extract this information (GPS coordinates) and perform the following operations [11]:

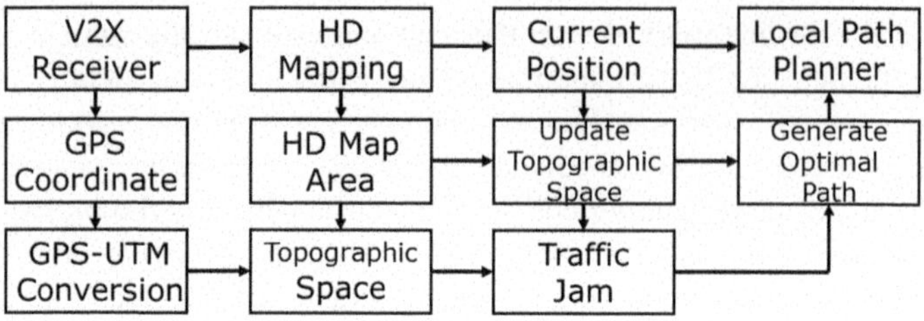

Fig. 3.4 Diagram of the prediction/planning module

(1) GPS to UTM (Universal Transverse Mercator) coordinate conversion.
(2) Location and delimitation of the topographic space in the HD map area in which the traffic jams were identified.
(3) Identification and marking of traffic jams.
(4) The topographic space is updated, and an optimal path is generated.
(5) Moving the autonomous vehicle on the optimal route (local path planner) to the destination point.

Traffic prediction using the gray prediction model is based on the total number of vehicles in traffic (N) between which the autonomous vehicle (i) is located, where $1 \leqslant i \leqslant N$ and moving at different velocity values $\{v_i(1), v_i(2) \ldots, v_i(m)\}$ measured at discrete time moments $\{1, 2, \ldots, m\}$, where m is the current moment of time. The algorithm used for road traffic prediction considers that the succession of future velocities of the autonomous vehicle $\{v_i(m+1), v_i(m+2) \ldots, v_i(n)\}$ for discrete time moments $\{m+1, m+2 \ldots, n\}$ can be obtained by calculating the sum of the velocity intervals (v_i) [13]:

$$X_i(k) = \sum_{j=1}^{k} v_i(j),\ 1 \le k \le m \tag{3.1}$$

where X_i is an exponentially increasing sequence that satisfies the differential Eq. (3.2), and a and u are constant scalar parameters.

$$\begin{aligned} &\Delta X_i(k) + aX_i(k) = u \\ &\Delta X_i(k) = X_i(k) - X_i(k-1) = v_i(k) \\ &X_i(k+1) = \left[X_i(1) - \frac{u}{a}\right]e^{-ak} + \frac{u}{a} \end{aligned} \tag{3.2}$$

By combining the relations in Eq. (3.2), the following matrix results:

$$\begin{aligned} &v_i(k) = \left[-\tfrac{1}{2}(X_i(k-1) + X_i(k))\ 1\right]\begin{bmatrix} a \\ u \end{bmatrix} \\ &Y_i = [v_i(2) \ldots v_i(m)]^T \end{aligned}$$

$$Y_i = \begin{bmatrix} X_i(1) + X_i(2) \; -2 \\ X_i(2) + X_i(3) \; -2 \\ \cdots \\ X_i(m-1) + X_i(m) \; -2 \end{bmatrix}$$

$$(a_i, u_i)^T = \left(B_i^T B_i\right)^{-1} B_i^T Y_i \tag{3.3}$$

Velocity prediction for the discrete time intervals {m + 1…,n} is calculated:

$$v_i(k) = X_i(k) - X_i(k-1), m+1 \leq k \leq n \tag{3.4}$$

where the prediction of $X_i(k)$ is obtained according to the exponential increase:

$$X_i(k+1) = \left(X_i(1) - \frac{u_i}{a_i}\right)e^{-a_i k(k-1)} + \frac{u_i}{a_i} \tag{3.5}$$

Due to the random nature of the road traffic, the prediction accuracy for the considered model may change as the prediction limits are extended.

3.1.4 Planning

Dynamic behavior planning allows the autonomous vehicle to interact with other traffic participants, preventing accidents and automatically adjusting the travel velocity. The characteristics of dynamic behavior planning are synchronized with the characteristics of motion control and result in the activation of the dynamic functions of the propulsion, steering or braking systems to determine the best behavior of the autonomous vehicle depending on the presence of some static or dynamic objects on the road [2, 3]. The determination of the trajectory of the autonomous vehicle is based on the decision regarding the following factors:

(1) The position of static and dynamic objects in traffic (traffic decider).
(2) The decision on the direction of travel (path decider).
(3) The decision on the travel velocity (speed decider).

As the position of the objects participating in the traffic cannot be changed by the autonomous driving system, in order to avoid any incident, it must intervene in optimizing the direction of travel (path optimizer) and the velocity (speed optimizer) of the autonomous vehicle (Fig. 3.5) [14].

Optimization of the direction of travel and travel velocity is done using two approaches: dynamic programming and quadratic programming:

(A) Dynamic programming involves solving a complex problem by breaking it down into simple sub-problems, in a recursive manner. The autonomous driving system cannot decide with maximum precision which must be the exact direction that the autonomous vehicle must follow in order to avoid any incident

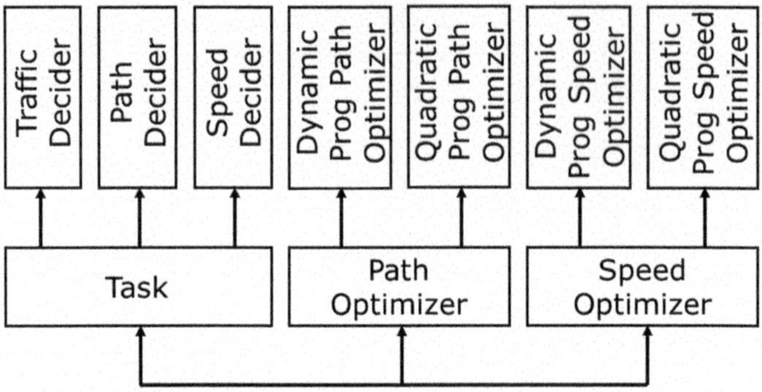

Fig. 3.5 Dynamic behavior planning

with the static and dynamic objects existing on the route. But, if this route will be broken down into several segments, correlated with the position and velocity of the tracked objects, it will be easier to make a prediction of the direction that the autonomous vehicle must follow. The same process will be followed when determining the travel velocity.

(B) Quadratic programming is an optimization solution associated with the dynamic programming method based on the Bellman equation, which demonstrates that the solution of a complex problem is based on a partial solution and the part of the problem left to be solved, and so on, until the final solution is found based on multiple partial solutions [15].

After optimizing the travel direction and velocity of the autonomous vehicle, the decision regarding these data is transmitted to the motion control system for the execution of the commands.

Generating the trajectory of an autonomous vehicle based on movement tracking of the static and dynamic objects detected in traffic is done using an approach based on the recordings of the sensors that equip the autonomous vehicle, respectively on the dynamics of its movement.

The generation of the trajectory for an autonomous vehicle is based on the dynamics of the vehicle and uses the application of genetic algorithms and the gradient reduction method.

The geometric path tracking algorithm uses the tracking of a point along the roadway that is located at a certain distance from the current position of the autonomous vehicle. The virtual trajectory is tangential to the velocity vector and passes through the tracked point along the path. According to the geometric path shown in Fig. 3.6 the radius of curvature of the arc R can be calculated geometrically with the following relations [12]:

$$\frac{L_{fw}}{\sin(2\eta)} = \frac{R}{\sin\left(\frac{\pi}{2} - \eta\right)}$$

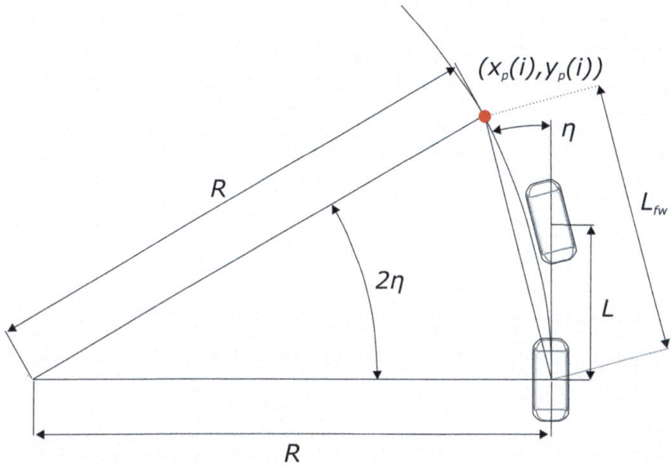

Fig. 3.6 Geometric path tracking algorithm

$$\frac{L_{f_w}}{2\sin(\eta)\cos(\eta)} = \frac{R}{\cos(\eta)}$$
$$\frac{L_{f_w}}{\sin(\eta)} = 2R \tag{3.6}$$

where R is the radius of rotation, and η is the direction of travel. The curvature k of the arc of a circle is defined by the relation:

$$k = \frac{2\sin(\eta)}{L_{f_w}} \tag{3.7}$$

Therefore, the angle of the direction of travel of the autonomous vehicle (δ) can be calculated by applying the kinematic model:

$$\delta = \tan^{-1}(kL)$$
$$\delta(t) = \tan^{-1}\left(\frac{2L\sin(\eta(t))}{L_{f_w}}\right) \tag{3.8}$$

The distance traveled by the autonomous vehicle is formulated as a function of the longitudinal velocity and is given by the equation:

$$\delta(t) = \tan^{-1}\left(\frac{2L\sin(\eta(t))}{kv(t)}\right) \tag{3.9}$$

The tracking algorithm calculates a circular arc based on the geometry of the autonomous vehicle model. The Stanley algorithm is a geometric steering controller (Fig. 3.7) that calculates the steering command based on a nonlinear control law that

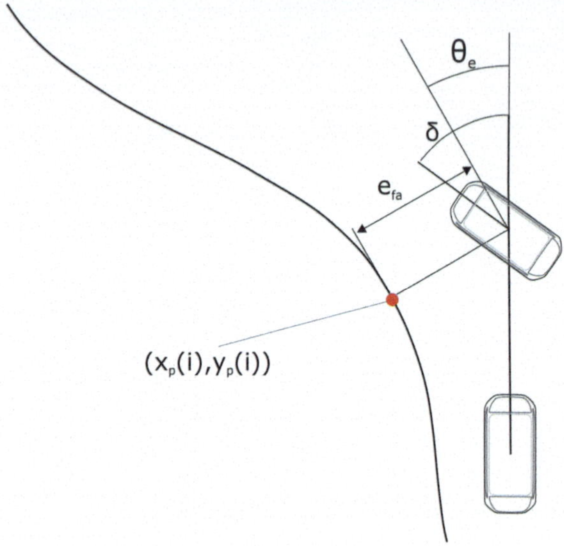

Fig. 3.7 Stanley method of geometric path tracking

takes into account the transverse error (e_{fa}) measured from the center of the front axle of the vehicle to the roadway, as well as the vehicle steering error (θ_e). The resulting control law for the steering angle δ can be written so that the trajectory of the autonomous vehicle intersects tangent to the roadway, where k is the gain parameter, and $v_x(t)$ is the instantaneous longitudinal velocity:

$$\delta(t) = \theta_e(t) + \tan^{-1}\left(\frac{ke_{fa}(t)}{v_x(t)}\right) \tag{3.10}$$

Compared to the geometric path tracking algorithm, the Stanley algorithm has better tracking results and does not cut corners because it uses information about the cross-track error and the autonomous vehicle steering error. The vehicle position can oscillate between the inside and outside of the curvature, and the calculated steering angle can oscillate between positive and negative values (left or right). The Stanley method also requires a continuous curvature path made of discrete waypoints [12].

The kinematic model in the coordinates of the geometric path traveled by the autonomous vehicle is defined by the relations [12]:

$$\begin{bmatrix} \dot{s} \\ \dot{e}_{ra} \\ \dot{\theta}_e \\ \dot{\delta} \end{bmatrix} = \begin{bmatrix} \frac{\cos(\theta_e)}{1-\dot{e}_{ra}k(s)} \\ \sin(\theta_e) \\ \left(\frac{\tan\delta}{L}\right) - \frac{k(s)\cos(\theta_e)}{1-\dot{e}_{ra}k(s)} \\ 0 \end{bmatrix} v \begin{bmatrix} 0 \\ 0 \\ 0 \\ 1 \end{bmatrix} \dot{\delta} \tag{3.11}$$

$x_1 = s$

$$
\begin{aligned}
x_2 &= k^l(s)e_{ra}\tan(\theta_e) - k(s)(1 - e_{ra}k(s))\frac{1+\sin^2(\theta_e)}{\cos^2(\theta_e)} + \frac{(1 - e_{ra}k(s))\tan(\delta)}{L\cos^3(\theta_e)} \\
x_3 &= (1 - e_{ra}k(s))\tan(\theta_e) \\
x_4 &= e_{ra} \\
v &= \frac{1 - e_{ra}k(s)}{\cos(\theta_e)}u_1 \\
\dot{\delta} &= \alpha_2(u_2 - \alpha_1 u_1) \\
\alpha_1 &= \frac{\partial x_2}{\partial s} + \frac{\partial x_2}{\partial e_{ra}}(1 - e_{ra}k(s))\tan(\theta_e) + \frac{\partial x_2}{\partial \theta_e}\left(\frac{\tan(\delta)(1 - e_{ra}k(s))}{L\cos(\theta_e)} - k(s)\right) \\
\alpha_2 &= \frac{L\cos^3(\theta_e)\cos^2(\delta)}{(1 - e_{ra}k(s))^2}
\end{aligned}
\tag{3.12}
$$

where e_{ra} is the transverse error, θ_e is the steering error, δ is the steering angle, k is the curvature of the arc of a circle.

The velocity of the autonomous vehicle is considered to be constant along the geometric path traveled by the vehicle throughout the prediction horizon. The prediction of movement for the vehicles found in the traffic in which the autonomous vehicle is travelling is based on data sets recorded in real traffic, with the help of which the density functions are calculated. The side position of the autonomous vehicle is taken into account by means of the Gamma density function [16]. The prediction of longitudinal motion is based on the Gaussian density functions. The planning of the movement of the autonomous vehicle will be done considering the maximum value of the lateral acceleration for each node considered along the entire length of the geometric path traveled. In order to plan a collision-free trajectory in the movement of the autonomous vehicle, the planning algorithm must receive data sets on the static position or movement trajectory of all the objects in the environment so that it can calculate the prediction for the movement trajectory and the travel velocity, respectively. For a collision-free trajectory, the probabilities of possible collisions must be kept to a minimum along the reference trajectory. The lateral movement of the autonomous vehicle P_{lat} can be estimated using the following equation [16]:

$$
P_{lat}\left(y_j\right) = \int_{a_{min}}^{a_{max}} f_{lat}(x, \alpha, \beta)dx\, P_w(\sigma, \mu) \tag{3.13}
$$

where f_{lat} is the function of gamma density and (a_{min}, a_{max}) represents the ensemble of possible lateral accelerations. P_w is the function that expresses the width of the vehicle, using variables y_w for vehicle width $\left(\sigma = \frac{\sqrt{y_w}}{4}\right)$, respectively μ for the position of the center of gravity.

The longitudinal movement of the autonomous vehicle P_{long} can be estimated using the following equation [16]:

$$
P_{long}\left(x_j, a_{d,i}\right) = \int_0^{t_i v_x} f_{j,i}(x)dx \tag{3.14}
$$

where v_X is the longitudinal velocity of the autonomous vehicle, x_j is the longitudinal position of the node on the considered trajectory, j-th the moment the vehicle passes through the th node, $a_{d,i}$ is the current acceleration of the vehicle.

Considering the Eqs. (3.8) and (3.9), the probability of a possible collision along the geometric path traveled by the autonomous vehicle can be reduced to a single node and calculated based on the equation [16]:

$$P(x_j, y_j, a_{d,i}) = P_{long}(x_j, a_{d,i}) P_{lat}(y_j) = \int_0^{t_i v_x} f_{j,i}(x)dx \int_{a_{min}}^{a_{max}} f_{lat}(x, \alpha, \beta)dx\, P_w(\sigma, \mu) \tag{3.15}$$

And the maximum probability of a possible collision along the geometric path is calculated based on the equation [16]:

$$P(x_m, y_m) = \max(P(x_j, y_j, a_{d,i})), y \in \left[y_m - \frac{y_{b,p}}{2}, y_m + \frac{y_{b,p}}{2}\right] \tag{3.16}$$

where y_b, is used to apply the trajectory planning algorithm (Fig. 3.8).

The planning algorithm of the geometric trajectory travelled by the autonomous vehicle uses the data generated and provided by the following functional blocks [16]:

(A) Section 1 that generates the travel graph based on the data regarding the maximum lateral acceleration value calculated.

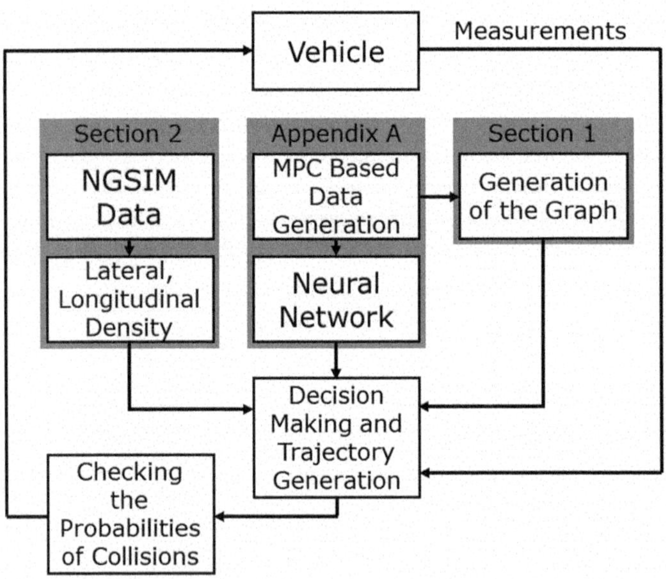

Fig. 3.8 Trajectory planning algorithm

(B) Appendix A which generates the MPC (Model Predictive Control) used to evaluate all feasible trajectories required during the neural network training process, which throughout the optimization process uses Machine Learning algorithms to identify an optimal trajectory.
(C) Section 2 which uses data sets recorded in real traffic NGSIM (Next Generation SIMulation), respectively calculates, and generates the density functions, which are used in the predictions of the longitudinal and lateral movement of the autonomous vehicle.

After calculating the trajectory, the probabilities of collisions are checked. If the probability values are higher than the defined maximum, the longitudinal velocity of the vehicle will be adjusted. The training of the neural network is made using a previously generated data set, and the results obtained based on the MPC will be validated using the test data set.

3.1.5 Motion Control

The command and control of the operation of the propulsion, steering and braking systems are taken based on the decisions established in the prediction/planning stage. The main command and control parameters are collected in a non-volatile internal memory in the form of data from all the sensors that equip the autonomous vehicle. These parameters are:

(1) The permanent position in RT (Real Time), respectively a history of the movement.
(2) Data on the main functional parameters of the propulsion, steering and braking systems and a history of the operation of these systems.
(3) Data on the main functional parameters of the comfort systems in the passenger compartment of the autonomous vehicle.
(4) Data on the main events that took place during operation (abnormal operation, emergency stop etc.).
(5) Data regarding the video camera data streams.

The main functions of the command-and-control unit are [2]:

(6) The logic and command, monitoring, and diagnosis of the autonomous operation of the main systems: propulsion, steering and braking.
(7) Logic and control, monitoring, and diagnosis of the autonomous operation of auxiliary lighting systems, door/ramp actuation, comfort in the passenger compartment of the vehicle, multimedia systems.
(8) Logic and command of blockages to external access to main/auxiliary systems for a safe operation of the autonomous vehicle.
(9) Interconnection with the management platform, recording and transmission of data, events, and operating errors.

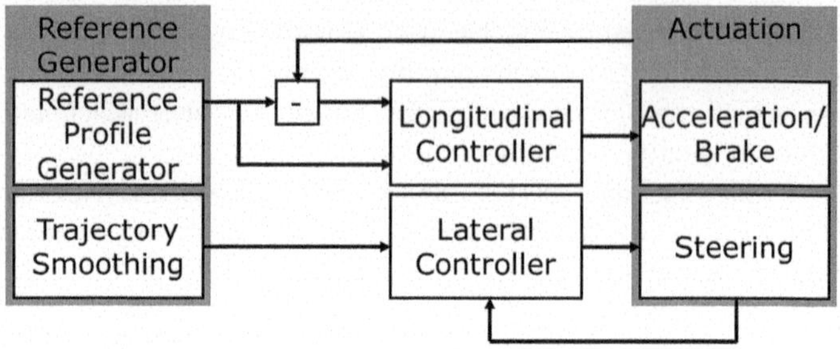

Fig. 3.9 Diagram of the motion control module

The control component of the autonomous vehicle will ensure the local path tracking received from the planning module which consists in the graph of the waypoints and of the maximum velocity intervals for these points (Fig. 3.9) [2, 11].

The reference generator defines the velocity profile of the autonomous vehicle on the established path. It correlates with the longitudinal and lateral coordinates of the movement and is associated with each intermediate point of the trajectory and is dependent on the structure of the road, the presence of static and dynamic objects, traffic signs etc.

The longitudinal controller ensures the local path tracking and works on the basis of the PI (Proportional Integral) control method controlling the powertrain, respectively the braking system.

The lateral controller ensures the tracking of the steering angle for the route followed by the autonomous vehicle controlling the steering system.

Longitudinal control or motion control of the autonomous vehicle is performed by the autonomous driving system in an acceleration stage by the command transmitted to the powertrain regarding the desired value of the velocity and the time in which the vehicle must reach this velocity. Based on these data, the acceleration value is automatically set so that the vehicle falls within these values. In the braking stage, the motion control is achieved by transmitting the distance to which the autonomous vehicle must stop, respectively a velocity value (reduced compared to the travel velocity) and the time in which the vehicle must reach this velocity. As in the first stage, the value of the deceleration is automatically set on the basis of these data so that the autonomous vehicle falls within these values (Fig. 3.10) [15].

The lateral control or the steering control of the autonomous vehicle is based on the signals received from the sensors, correlated with the information communicated by V2X transmissions, superimposed over the trajectory implemented in the memory of the autonomous driving system. These signals are interpreted by the autonomous driving system that intervenes on the steering control system so that the autonomous vehicle can follow the established path (Fig. 3.11) [17].

The general logic diagram for longitudinal and lateral control is shown in Fig. 3.12 [17]. Functionally, the autonomous vehicle is accelerated by the command from the acceleration controller on the throttle device until it reaches the established velocity,

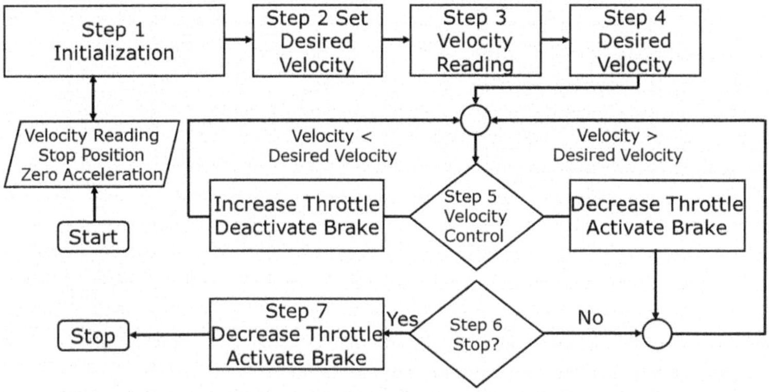

Fig. 3.10 Control of the autonomous vehicle velocity

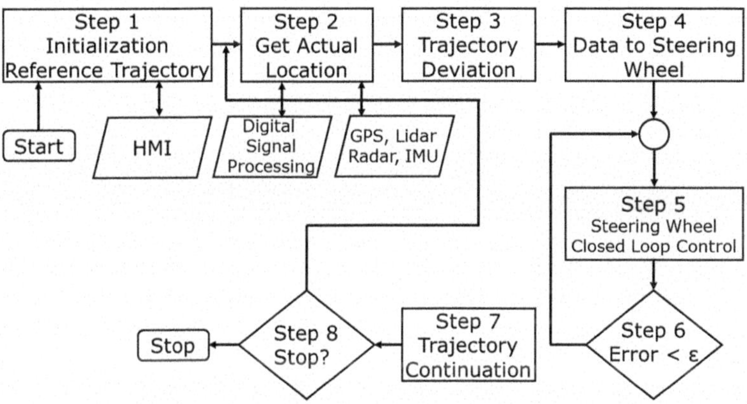

Fig. 3.11 Steering control of the autonomous vehicle

at which point the acceleration limiter stops the acceleration and transmits the brake request to the brake controller, which acts through the brake device. The brake controller has minimum and maximum limits set by means of the brake limiter.

The velocity of the autonomous vehicle is determined by the speed sensor, which transmits the measured value to the acceleration controller. The braking of the vehicle is also determined by the information transmitted by the Lidar sensor in case of the appearance of static or dynamic objects on the path. Depending on the route profile, it also intervenes in the steering controller command. The steering device is controlled by the steering controller so that the autonomous vehicle follows the established trajectory. The angle of rotation of the wheels and the limits of rotation are set by the steering position controller who transmits these limits to the steering actuator.

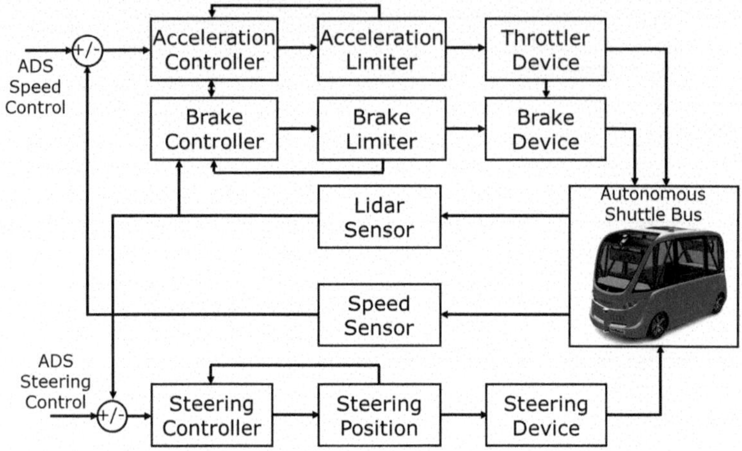

Fig. 3.12 Logical diagram of the longitudinal and lateral control

3.1.6 Object Classification Mechanism

SAE proposed six-level classification criteria for autonomous vehicles [4]. The level 1 autonomous vehicles can travel to the ACC, which is fulfilled by the LRR. The level 1 autonomous vehicles can always detect the location and velocity of its preceding vehicle in the same lane. From level 2 to level 5, autonomous vehicles gradually take control from human drivers. To achieve that, autonomous vehicles need to continuously monitor the surrounding traffic conditions. From the perspective of traffic sensing, level 2 to level 5 vehicles can detect or track the vehicles in their surrounding areas. Detection is the localisation of a vehicle when it is in the detection area of an autonomous vehicle and tracking means that the autonomous vehicle can keep track of a vehicle if it is within the detection area. The task of detection does not require one to memorize the detected vehicles in each time frame, while tracking requires the autonomous vehicle to keep track of the detected vehicles when they are within the detection range. If the autonomous vehicle processes 30 frames per second, tracking requires one to accurately detect all the vehicles in each frame and match them correspondingly, while detection does not require one to match the vehicles in different frames. Detection only provides the location of each vehicle, but tracking can provide additional velocity information. The precise descriptions of the three perception levels are as follows (Fig. 3.13) [18]:

(S1) Tracking the preceding vehicle the primary task for S1 is to track the vehicle within the same lane, and this is originally used for the ACC. The velocity and location of the preceding vehicle are obtained for TSE (Traffic State Estimation).

(S2) Locating surrounding vehicles in addition to S1, the primary task for S2 is to detect and locate vehicles surrounding an autonomous vehicle.

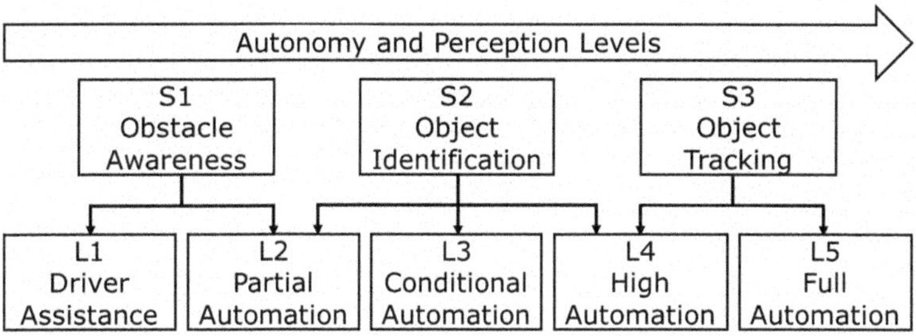

Fig. 3.13 Overview of the perception levels

(S3) Tracking surrounding vehicles in addition to S2, the primary task for S3 is to track every uniquely identified vehicle in the detection area, hence the location and velocity of each vehicle are monitored over time by the autonomous vehicles in S3.

In the perception level S1, an autonomous vehicle can only detect and track its preceding vehicle, and its detection area for density and velocity is O_l^1 [18]:

$$\widetilde{k}_l(h,s) = \begin{cases} \frac{\sum_{i\in IA} t^i_{lh}}{\sum_{i\in IA} \mu(a^i_l(h,s))} if(h,s) \in O_l^1 \\ * \ else \end{cases}$$

$$\widetilde{v}_l(h,s) = \begin{cases} \frac{\sum_{i\in IA} d^i_{lh}}{\sum_{i\in IA} t^i_{lh}} if(h,s) \in O_l^1 \\ * \ else \end{cases} \tag{3.17}$$

For autonomous vehicles that have a LRR equipped to track the vehicle following behind the autonomous vehicles, ($I^A = (I^A)$ U following (I^A)), where following (I^A) represents all the vehicles that follow the autonomous vehicles in (I^A). The set of time–space indices O_l^1 covers a small fraction of all cells in the time–space region (multi-lane highways).

In the perception level S2, both D_1 and D_2 are enabled by the LRR, Lidar and cameras, while D_2 can only detect the location of the surrounding vehicles. The density can be observed in both D_1 and D_2, and the velocity is only observed in D_1. D_2 provides a snapshot of the traffic density at a certain time point, and we can compute the density of time interval h by taking the average of all snapshots.

$$\widetilde{k}_l(h,s) = \begin{cases} \frac{\sum_{i\in A} t^i_{lh}}{\sum_{i\in A} \mu(a^i_l(h,s))} & if(h,s) \in O_l^1 \\ \frac{1}{\mu(T_h^0(l,s))} \int_{t\in T_h^0(l,s)} \frac{|I_l^0(t,s)|}{\Delta X} dt & elseif(h,s) \in O_l^2 \\ * & else \end{cases}$$

$$\widetilde{v}_l(h,s) = \begin{cases} \frac{\sum_{i\in A} d^i_{ls}}{\sum_{i\in IA} t^i_{lh}} if(h,s) \in O_l^1 \\ * \ else \end{cases} \tag{3.18}$$

where $T_h^0(l, s)$ represents the set of time indices when X_{ls} is covered by the D_2 in T_h, and $I_l^0(t, s)$ represents the set of vehicles detected by D_2 on X_{ls} at the time t.

In the perception level S3, both localisation and tracking are enabled by the LRR, Lidar and cameras. In addition, velocity information of surrounding vehicles is also available in D_2. The instantaneous speed of a cell at certain time points is calculated by taking the harmonic mean of all detected vehicles, and then the average speed of a cell is calculated by taking the average of all time points [18]:

$$\widetilde{k}_l(h,s) = \begin{cases} \dfrac{\sum_{i\in A} t_{lh}^i}{\sum_{i\in A} \mu\left(a_l^i(h,s)\right)} & if(h,s) \in O_l^1 \\ \dfrac{1}{\mu\left(T_h^0(l,s)\right)} \displaystyle\int_{t\in T_h^0(l,s)} \dfrac{\left|I_l^0(t,s)\right|}{\Delta X} dt & else\ if(h,s) \in O_l^2 \\ * & else \end{cases}$$

$$\widetilde{v}_l(h,s) = \begin{cases} \dfrac{\sum_{i\in A} d_{ls}^i}{\sum_{i\in I A} t_{lh}^i} & if(h,s) \in O_l^1 \\ \dfrac{1}{\mu\left(T_h^0(l,s)\right)} \displaystyle\int_{t\in T_h^0(l,s)} hmean(\{v^i(t)|x^i(t) \in X_{ls}, i \in I\})dt & else\ if(h,s) \in O_l^2 \\ * & else \end{cases} \tag{3.19}$$

where $hmean(\{v^i(t)|x^i(t) \in X_{ls}, i \in I\})dt$ represents the harmonic mean.

The mechanism for detecting and classifying static and dynamic objects is based on the detection and classification models: SSD and YOLO.

SSD and YOLO are designed in a way that a single network predicts region proposals as well as the class of those region proposals. This design saves a significant amount of computational time, allowing it to perform faster than the state-of-the-art fast R-CNN while achieving higher accuracy [19].

The SSD model is an algorithm for detecting dynamic objects that move in the front of the autonomous vehicle. SSD is based on the Deep Learning model and uses a CNN to analyze visual images and extract the characteristics of the object being tracked and its classification. The SSD model consists of the following neural networks:

(1) Basic network.
(2) Feature extraction network.
(3) Object detection network.

CNNs are based on the feed-forward neural network (where the output of a neuron would be the input of another set of neurons in the preceding layer). CNNs use convolutional operations to extract features from the input data with each layer using a kernel (filter). Feature maps are then generated after the convolutional operation is performed across the entire input data. The feature map is a representation of the activation of different parts in the image [19].

A CNN consists of an input layer, an output layer, and several convolutional layers. They apply a convolution operation at the input, passing the result to the next layer, so that each time it passes from one layer to another it reduces the size of the image [20].

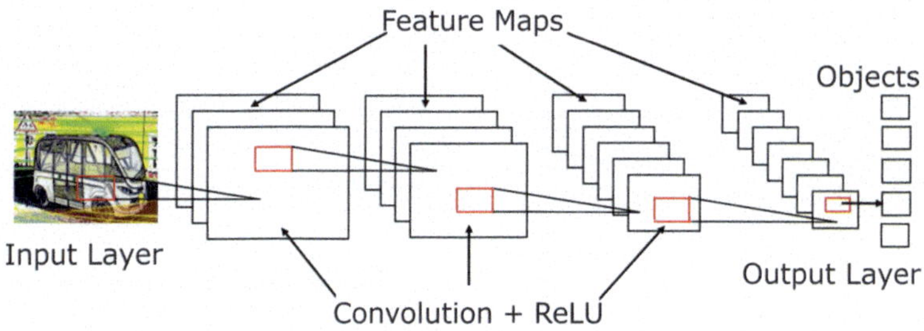

Fig. 3.14 SSD model based on CNN network

Convolution is the first layer that extracts certain features from an image taken at the input and by applying a filter it maintains a relationship between the number of pixels by learning the features of the image. Thus, for an image with the dimensions h(height)*w(width)*d(depth), by applying a filter of the (fh*fw*d) type, a volume with the dimensions (h-fh + 1)*(w − fw + 1)*1 results at the output. The Deep Learning process is based on a non-linear function of the ReLU (Rectified Linear Unit) type, such as f(x) = max(0,x). With each convolutional step, the image taken at the input is transformed by applying different filters, resulting in an image reduced in size, but which has improved feature maps (Fig. 3.14).

The convolutional operator that is used is dependent on the type of input data. 2D kernels are used for 2D temporal sequences and 1D kernels are used for 1D temporal sequences. With several layers, CNNs can represent data in a hierarchical mode. The network can automatically learn to extract useful information from images/frames. The deep layers will attempt to combine the features to identify objects in the frame. The classifying layer typically consists of several neurons. The number of neurons is dependent on the number of desired outputs [19].

In order to increase the extraction capacity of the tracked object features, using the SSD model the extension of the convolutional layers is necessary. To this end, Cao et al. [20] proposed the introduction of an inception block which, using convolutional nuclei of different sizes, it extracts features of the same input layer, thus extending the total width of the network. The SSD model creates a multi-dimensional extraction network of convolutional layers. Each layer extracts the features of the model independently of the other layers in the inception block, then the information merges and a new function layer is formed which is transmitted to the following layers (Fig. 3.15).

The YOLO model is a RT (Real Time) object detection system. YOLO uses a neural network to detect the complete image and divides the image into grid units by estimating the delimitation probabilities for this image. Images are marked and framed in bounding boxes that have the dimensions (height and width) estimated by the model and a location predicted though localisation. YOLO detects marked images and retains specific information resulting from detection (prediction and objectivity of box coordinates), as well as how to classify common objects, for which it uses labels grouped by object classes for categories of images [21].

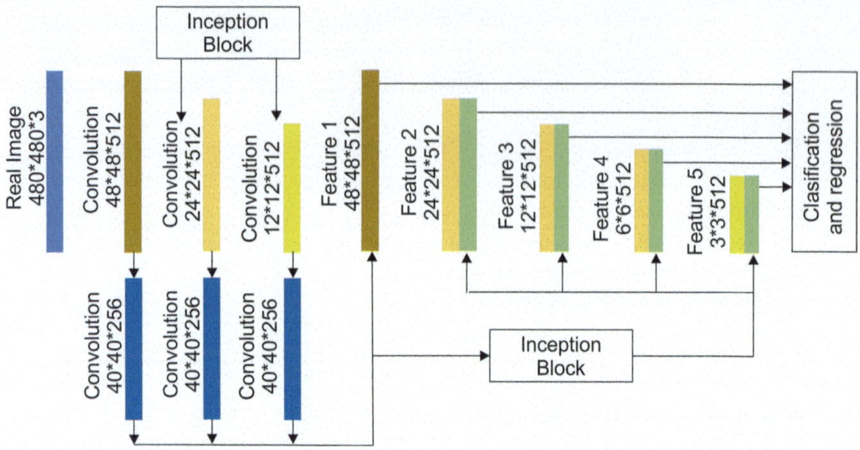

Fig. 3.15 Base structure of the SSD model

The YOLO model is a RT object detection system. YOLO uses a neural network to detect the complete image and divides the image into grid units by estimating the delimitation probabilities for this image. The images are marked and framed in bounding boxes that have the dimensions (height and width) estimated by the model and a location predicted through localisation. YOLO detects the marked images and retains specific information resulting from detection (prediction and objectivity of box coordinates), as well as the way to classify common objects. By using tags grouped by object classes for image categories, the number of detectable categories can be expanded. The RT object detection and classification mechanism must have a high detection accuracy and a fast response of the vehicle controller for road safety.

The uncertainty of the coordinate variation (Fig. 3.16) is modeled on the Gaussian model, the mean (μ) and the variation of the parameters (Σ) [22].

$$\mu_{t_x} = \sigma\left(\hat{\mu}_{t_x}\right), \mu_{t_x} = \sigma\left(\hat{\mu}_{t_y}\right), \mu_{t_w} = \hat{\mu}_{t_w}, \mu_{t_h} = \hat{\mu}_{t_h} \tag{3.20}$$

$$\Sigma_{t_x} = \sigma\left(\hat{\Sigma}_{t_x}\right), \Sigma_{t_x} = \sigma\left(\hat{\Sigma}_{t_y}\right), \Sigma_{t_w} = \hat{\Sigma}_{t_w}, \Sigma_{t_h} = \hat{\Sigma}_{t_h} \tag{3.21}$$

$$\sigma(x) = \frac{1}{(1 + exp(-x))} \tag{3.22}$$

The YOLO model assumes a convolutional network with Deep Learning consisting of convolutional layers (Fig. 3.17) that allows the detection and classification of objects.

The loss function (L) for the considered parameters is redesigned as an NLL (Negative Log Loss) probability [22]:

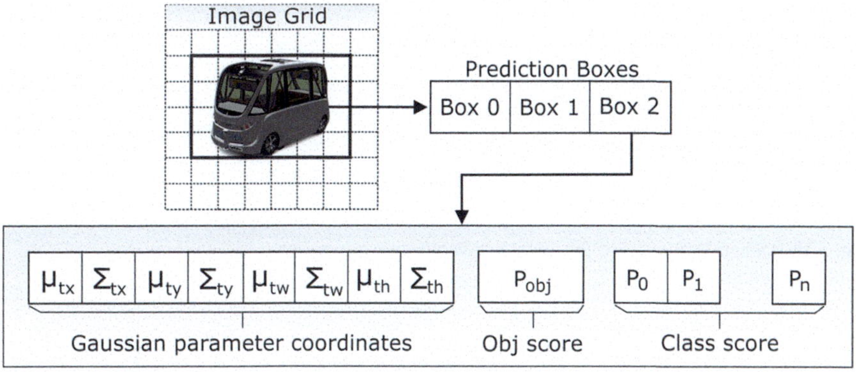

Fig. 3.16 Variation of coordinate values (the Gaussian model)

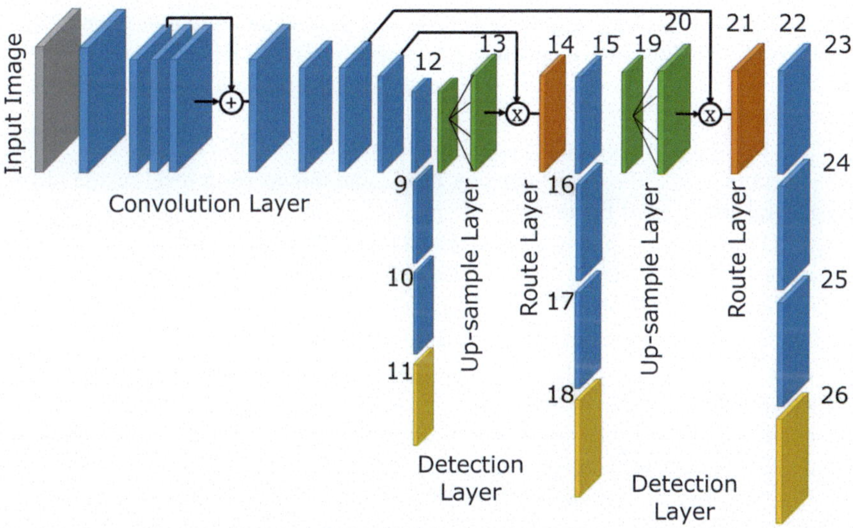

Fig. 3.17 Convolutional network with Deep Learning (YOLO)

$$L_x = -\sum_{i=1}^{W}\sum_{j=1}^{H}\sum_{k=1}^{K}\gamma_{ijk} log\left(N\left(x_{ijk}^{G}|\mu_{t_x}(x_{ijk}), \sum_{t_x}(x_{ijk})\right) + \varepsilon\right) \tag{3.23}$$

$$L_y = -\sum_{i=1}^{W}\sum_{j=1}^{H}\sum_{k=1}^{K}\gamma_{ijk} log\left(N\left(y_{ijk}^{G}|\mu_{t_y}(y_{ijk}), \sum_{t_y}(y_{ijk})\right) + \varepsilon\right) \tag{3.24}$$

$$L_w = -\sum_{i=1}^{W}\sum_{j=1}^{H}\sum_{k=1}^{K}\gamma_{ijk} log\left(N\left(w_{ijk}^{G}|\mu_{t_w}(w_{ijk}), \sum_{t_w}(w_{ijk})\right) + \varepsilon\right) \tag{3.25}$$

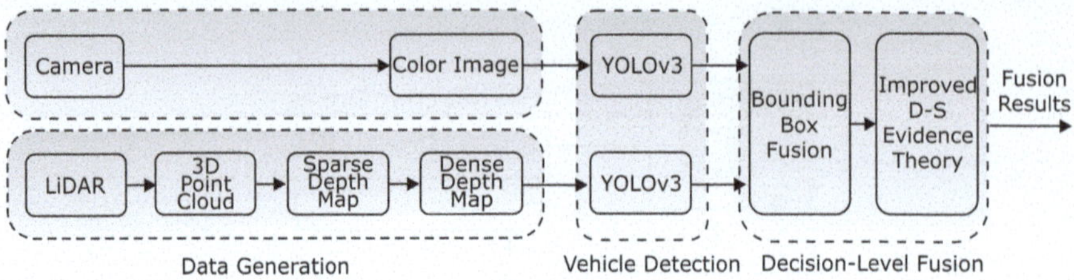

Fig. 3.18 General structure of the detection framework

$$L_h = -\sum_{i=1}^{W}\sum_{j=1}^{H}\sum_{k=1}^{K}\gamma_{ijk}log\left(N\left(h_{ijk}^{G}|\mu_{t_h}(h_{ijk}),\sum_{t_h}(h_{ijk})\right)+\varepsilon\right) \quad (3.26)$$

where L_x (respectively L_y, L_w, L_h) is the NLL loss function for the t_x coordinates (respectively t_y, t_w, t_h), μ_{tx} (respectively μ_{ty}, μ_{tw}, μ_{th}) is the output of the convolutional layer for the proposed algorithm, x_{ijk}^{G} (respectively y_{ijk}^{G}, w_{ijk}^{G}, h_{ijk}^{G}) is a term that refers to data resulting from direct GT (Ground Truth) observations of the parameters of the image.

The static and dynamic object detection framework consists of three blocks: the data generator block, the object detector block, and the decision-level fusion block. The general structure of the detection framework is presented in Fig. 3.18 [23].

The 3D Lidar point cloud was transformed into a 2D sparse depth map by the joint calibration of camera and Lidar, and then it was converted into a dense depth map by depth completion so that the laser data and image have the same resolution and are aligned with each other in space and time. Then the color image and dense depth map were input into the YOLOv3 (version 3) detection network and the bounding box and confidence score of each detected vehicle were obtained. The bounding box fusion and the improved D–S (Dempster–Shafer) evidence theory was proposed to obtain the final detection results.

Before the depth completion, a pre-processing operation is required to convert the 3D Lidar point cloud into a 2D sparse depth map. In pre-processing, the precise calibration, joint calibration, and synchronization of the Lidar and camera are needed so that each 3D Lidar point cloud can be projected accurately onto the 2D image plane to form the sparse depth map. The coordinate conversion relationship between the sensors is shown in Fig. 3.19 [23].

After the pre-processing, the sparse depth map is transformed into a dense depth map through the depth completion framework so that the resolution of Lidar data and image are the same. The depth-completion method can be divided into two types, guided depth completion and non-guided depth completion.

The 3D point cloud generated by the Lidar sensor will be transformed into a 2D map by merging the data from the video camera and the Lidar and it will later be transformed into a high-resolution map. The fusion of the sensors overlaps the data from the video camera with the cloud of laser points by entering the image obtained

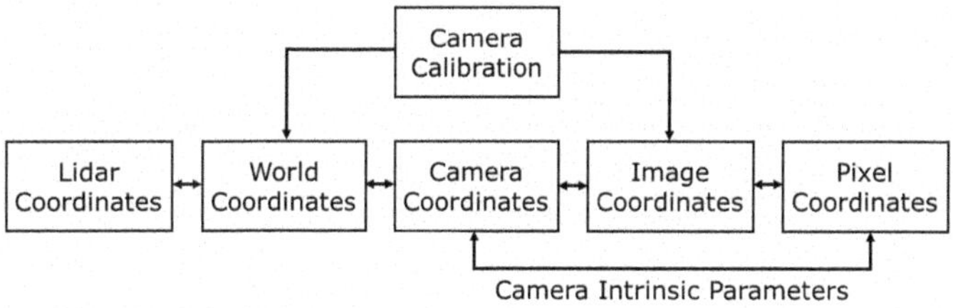

Fig. 3.19 The coordinate conversion between camera and Lidar

in a YOLOv3 detection network resulting in a bounding box and a reliable score for each detected object.

For an objective assessment of the quality of the high-resolution map, the following is introduced: the RMSE (Root Means Square Error), the MAE (Mean Absolute Error), the iRMSE (inverse Root Means Square Error) and the iMAE (inverse Mean Absolute Error) [23]:

$$RMSE = \sqrt{\frac{\sum_{m=1}^{M}\sum_{n=1}^{N}[R(m,n) - I(m,n)]^2}{MN}} \tag{3.27}$$

$$MAE = \frac{1}{MN}\sum_{m=1}^{M}\sum_{n=1}^{N}|R(m,n) - I(m,n)| \tag{3.28}$$

$$iRMSE = \sqrt{\frac{\sum_{m=1}^{M}\sum_{n=1}^{N}\left[\frac{1}{R(m,n)} - \frac{1}{I(m,n)}\right]^2}{MN}} \tag{3.29}$$

$$iMAE = \frac{1}{MN}\sum_{m=1}^{M}\sum_{n=1}^{N}\left|\frac{1}{R(m,n)} - \frac{1}{I(m,n)}\right| \tag{3.30}$$

where R(m,n) and I(m,n) represent the reference image and the target image.

For the regression method in YOLO, the loss function F (loss) can be calculated using the relation [24]:

$$F(loss) = \lambda coord \sum_{i=0}^{S^2}\sum_{j=0}^{B} 1_{ij}^{obj}\left[(x_i - \hat{x}_i)^2 + (y_i - \hat{y}_i)^2\right]$$
$$+ \lambda coord \sum_{i=0}^{S^2}\sum_{j=0}^{B} 1_{ij}^{obj}\left[\left(\sqrt{\omega_i} - \sqrt{\widehat{\omega_i}}\right)^2 + \left(\sqrt{h_i} - \sqrt{\widehat{h_i}}\right)^2\right]$$
$$+ \sum_{i=0}^{S^2}\sum_{j=0}^{B} 1_{ij}^{obj}\left(C_i - \hat{C}_i\right)^2 + \lambda noobj \sum_{i=0}^{S^2}\sum_{j=0}^{B} 1_{ij}^{noobj}\left(C_i - \hat{C}_i\right)^2$$

$$+\sum_{i=0}^{S^2} 1_{ij}^{obj} \sum_{c \in classes} \left(p_i(c) - \hat{p}_i(c)\right)^2 \tag{3.31}$$

where 1_i^{obj} is the grid cell that contains information about objects in traffic, 1_{ij}^{obj} is the bounding box j in the network cell i. The inverse term, 1_i^{noobj} is the bounding box j in the network cell i that does not contain information about traffic objects.

3.2 Electronic Control Unit of Autonomous Driving System

3.2.1 Overview

The ECU is a complex of electronic modules used to command and control some parameters that determine the operation of the autonomous vehicle based on the autonomous driving algorithm and in direct connection with the operating mode of the vehicle. The principle according to which the command-and-control unit operates is called IPO (Input Process Output) data input, data processing, data output [25].

A modern vehicle is equipped with a large number of electronic command and control units, or embedded systems that are intelligent systems, designed to perform various functions in real time, consisting of hardware elements (controllers) and software elements (code).

The control unit is the central component of the autonomous driving system that receives information from sensors and based on the specifications of the autonomous driving system control algorithm and the operating mode of the vehicle, it generates control signals for the main drive systems, the powertrain, steering and braking systems. As the central communication node, the control unit acts as a router for communication between the control units in a vehicle on the data buses LIN (Local Interconnect Network), CAN (Controller Area Network), FlexRay, MOST (Media Oriented Systems Transport), Ethernet (Fig. 3.20).

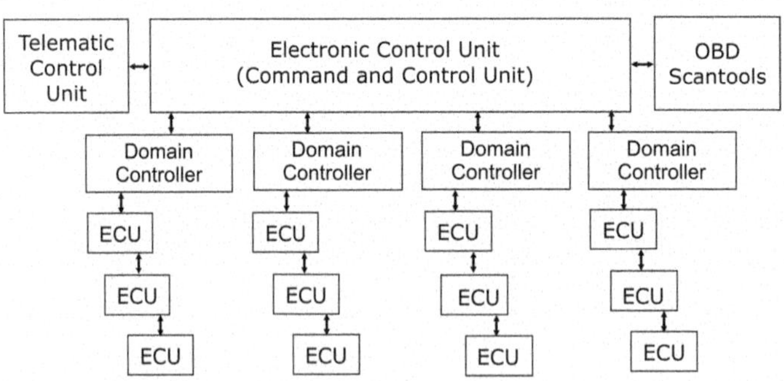

Fig. 3.20 The command-and-control unit

The sensors that equip the autonomous vehicle perform a complete scan of the environment in which it travels to distinguish all types of static and dynamic objects (pedestrians, vehicles), the environment, the particularities of the road etc., both during the day and during the night (low visibility). The information from the sensors is collected by the autonomous driving system, which will analyze this data and will control the movement in maximum safety conditions. The information is provided by the following categories of users/sensors:

(A) Users who will provide data about the selected destination via the control panel in on-demand travel mode.
(B) Location sensors (Lidar, radar, camera etc.) that will allow the vehicle to be framed on a map defined with the data provided.
(C) Position sensors (GPS/GNSS, IMU, odometers) that will provide data on the position of the autonomous vehicle.

3.2.2 Hardware Architecture

The operation of the command-and-control unit as a central component of the autonomous control system is shown in the functional block diagram in Fig. 3.21 [2, 26–29].

Autonomous driving is based on the RT position of the autonomous vehicle, which is generated by the localisation module, based on data from the sensor block, data that is taken and managed by the data integrator block. These data are interpreted and compared with the reference data on the geographical coordinates (latitude, longitude, and altitude) of the selected and implemented route (local map). These

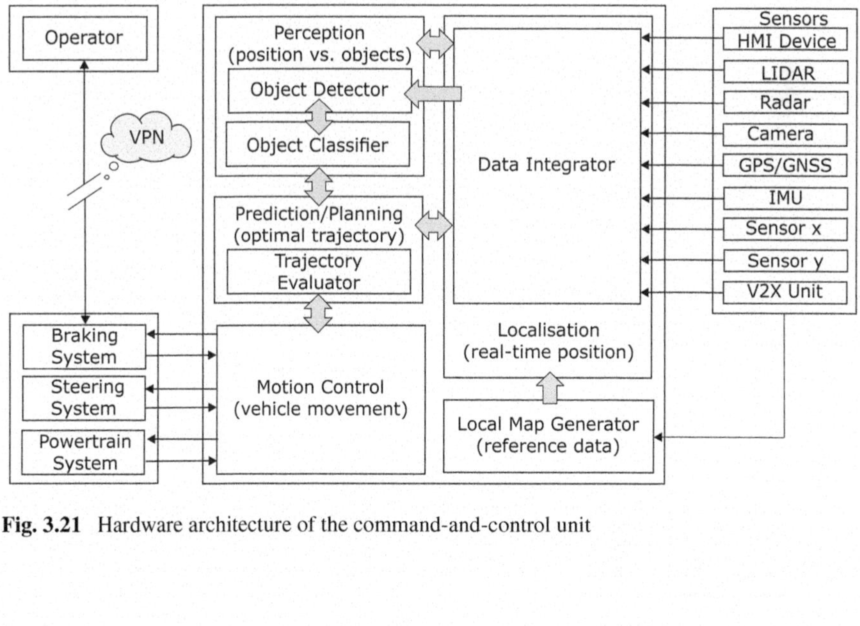

Fig. 3.21 Hardware architecture of the command-and-control unit

data are constantly updated with the data from the sensor block and are stored in a memory block (local map generator) [2, 30, 31].

The perception module receives the location data of the autonomous vehicle from the localisation block and the data from the sensor block for the objects in proximity. This module has the role of detecting objects (object detector) and of classifying the detected objects (object classifier) in the vicinity of the autonomous vehicle into static objects and dynamic objects. At the same time, this module has the role of estimating whether the dynamic objects in proximity can intersect their trajectory with the route selected for movement (of the autonomous vehicle). Based on the velocity of these objects (velocity determined by the radar sensor) relative to the velocity of the autonomous vehicle, the perception module will determine if there is a risk of a collision with the monitored object.

The prediction/planning module receives the perception data that contain the map of static and dynamic objects existing in the proximity, superimposed over the selected and implemented trajectory (local map). Based on these data, the prediction and planning module will select an optimal trajectory based on certain predetermined criteria (trajectory evaluator) in order to direct the movement of the autonomous vehicle in safe conditions without collisions.

Motion control transmits commands to the powertrain, steering and braking systems based on data received from the localisation, perception, and prediction and planning modules (Fig. 3.22). The braking system can be operated in case of emergency by the monitoring staff on the management platform through a secure VPN (Virtual Private Network).

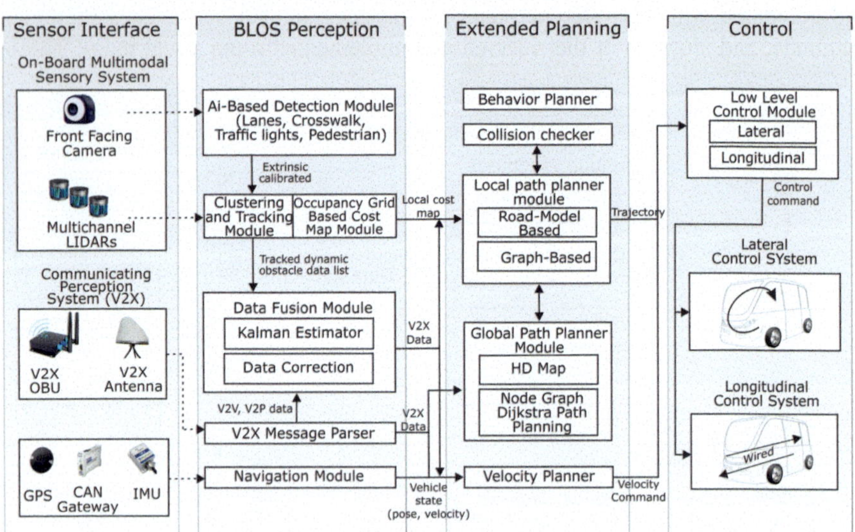

Fig. 3.22 Autonomous driving system architecture

V2X communication is the information exchange between a vehicle and various elements of an intelligent transportation system, other vehicles, pedestrians, and transport infrastructure. V2X communication is seen as an alternative that overcomes the drawback of the LOS (Limited of On-board Sensors), and it is proven that V2X can improve road safety [32].

3.2.3 *Software Architecture*

The software architecture of the command-and-control unit of the autonomous control system consists of the following modules [2]:

(A) Perception identifies the position of the autonomous vehicle in the environment, respectively the position of static and dynamic objects relative to its position.
(B) Prediction anticipates future trajectories of object movement relative to the position of the autonomous vehicle.
(C) Routing is the road traveled by the autonomous vehicle consisting of traffic lanes.
(D) Planning establishes the future dynamic behavior of the autonomous vehicle resulting from the prediction correlated with the rules of the road.
(E) Motion control controls the operation of the propulsion, steering and braking systems based on the decisions established in the prediction/planning stage.
(F) CAN bus is the connection interface between the software architecture and the hardware architecture of the autonomous vehicle.
(G) HD map is a support of information about the geography, communication routes, watercourses, human settlements, which provides structured information about the route.
(H) Localisation determines the current position of the autonomous vehicle based on the signals provided by the sensors (GPS, Lidar, IMU).
(I) HMI (Human Machine Interface) establishes the physical connection between the user and the autonomous driving system that helps with data input.
(J) Monitoring is the remote supervision of the autonomous vehicle by the human operator, performed through a software application called management platform.
(K) Guardian is the monitoring of the operation of the autonomous driving system that intervenes in case of a failure.
(L) Storytelling is the isolation and management of complex scenarios related to the operation of the autonomous vehicle.

The software architecture of the command-and-control unit of the autonomous driving system can be organized as an open and standardized AUTOSAR (AUTomotive Open System ARchitecture) platform (Fig. 3.23), being organized on layers with the possibility to be integrated and managed in applications such as application layer, consisting of software components that perform different functions and come from

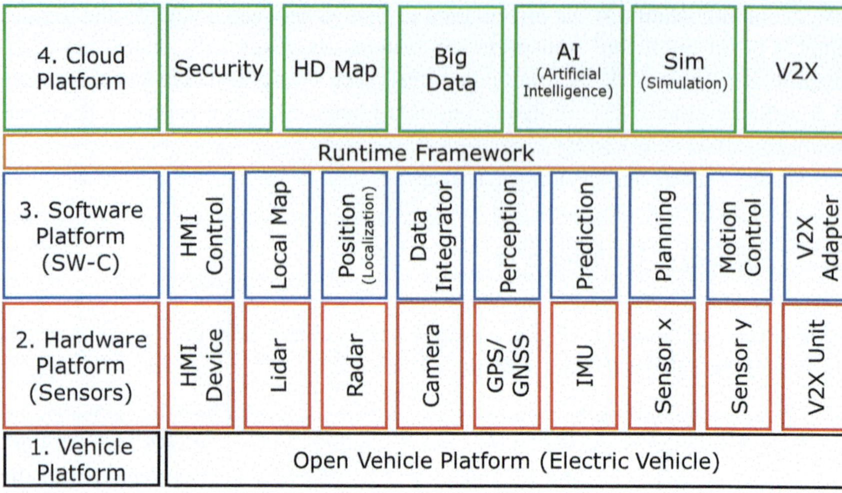

Fig. 3.23 AUTOSAR for autonomous driving

different manufacturers, so that it can use microcontroller resources and allow periodic software updates (Firmware) [2, 33, 34]. AUTOSAR is a software architecture used in the design of a command-and-control unit, on the levels of which different components can be integrated into the application layer so that the resources of the microcontroller can be used.

(1) Vehicle platform is the open vehicle platform used for the development of the autonomous driving system. Vehicle platform contains internal drivers, which are software modules with direct access to the microcontroller.
(2) Hardware platform contains sensors, regardless of their internal/external location (Lidar, radar, camera, GPS/GNSS), HMI devices, with installation drivers for each device. This layer provides an API (Application Programming Interface) for accessing and connecting to the Open Vehicle Platform.
(3) Software platform contains the RTOS (Real Time Operating System) and the runtime framework, respectively the layer of functions specific to the application software between the software components and the sensor components. The components of the software platform consist of:
(a) Local map generation and localisation on these maps.
(b) Collection and multiplexing of information (data integrator) provided by sensors.
(c) Perception of the position of the autonomous vehicle in relation to objects.
(d) Prediction of the behavior of objects in proximity.
(e) Planning the optimal trajectory of the autonomous vehicle.
(f) Motion control of the autonomous vehicle.
(g) Control of the human machine interaction (HMI control).
(h) Connection with the V2X units via a V2X adapter.

(4) Cloud platform offers a suite of Cloud Computing services that run a series of modular services that require a large storage space and high data traffic on the same infrastructure:
(a) Local HD maps.
(b) Large volumes of reference data (big data).
(c) Security protocols.
(d) AI (Artificial Intelligence) algorithms.
(e) Virtual model for prediction & planning (simulation).
(f) Transmission of information from intelligent entities (V2X).

The AUTOSAR software architecture is useful when there are software modules that can be used to switch to different levels of automation (from level 3 to level 4) or when there are software modules that can be used on different models of autonomous vehicles. In these situations, modules from various OEM (Original Equipment Manufacturer) developers can be integrated into the software architecture, which can be adapted to work on the existing platform through an update to the software version.

De Miguel et al. in [35] presented the platform of an autonomous vehicle that contains a customized software architecture. This software architecture is based on a ROS (Robotic Operating System), which provides a framework for the development and integration of applications used in robotics. The main modules that constitute the ROS software architecture are presented in Fig. 3.24. Each software module has the responsibility to perform a specific task, and the connections between the modules follow predefined formats based on ROS interfaces. This property allows flexibility in replacing similar components which are compatible with each other as functions.

The main modules of the ROS are following [35]:

(I) Data acquisition consists of a module composed of drivers that read raw data from sensors, ensures the preprocessing of this data by applying a filtering algorithm to clean the noise from the sensor and an algorithm for selecting and generating complex information extracted from the received data. At the same time, the drivers must convert the data provided by the sensors to a ROS compatible format in order to share them with the other software modules in the system.
(II) Localisation consists of a module whose architecture exploits different sensors and algorithms that ensure local and global localisation. Local localisation or vehicle odometry determines the movement of the vehicle relative to its position of origin based on two methods of odometry:
(i) Wheel odometry uses the velocity and steering angle provided by the vehicle to calculate the position of the autonomous vehicle based on the odometry algorithm.
(ii) Lidar odometry is a method of exteroceptive odometry that uses within the odometry algorithm the point cloud data generated by the Lidar 3D sensor in order to generate the position of the autonomous vehicle in a 3D map.

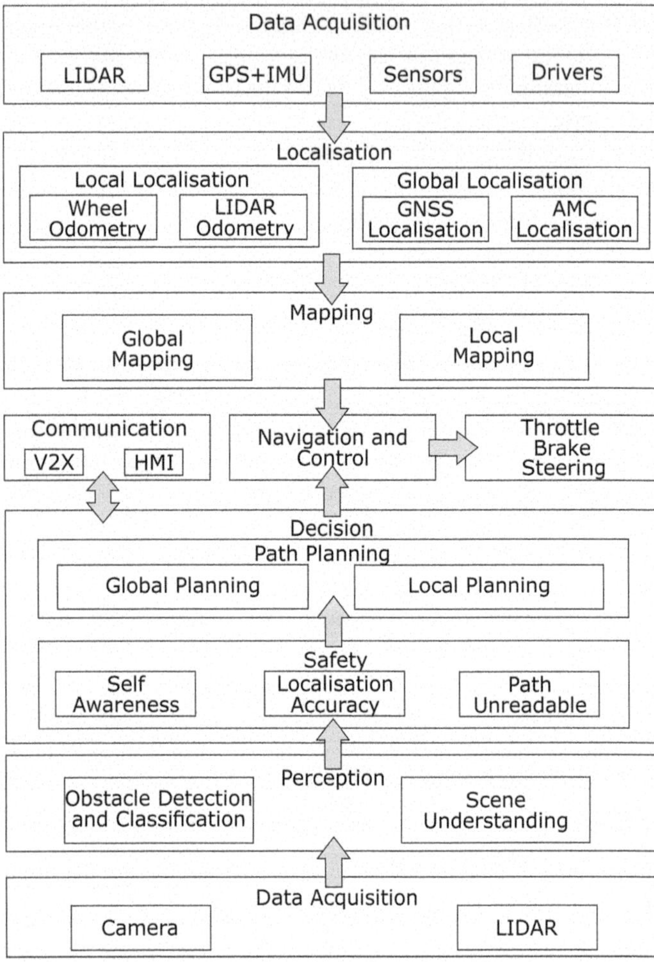

Fig. 3.24 Robotic Software Architecture (ROS)

(III) Global localisation determines the position of the autonomous vehicle relative to the reference coordinates in a digital map that identifies with the real world and uses two odometry methods for localisation:

(i) GNSS localization uses the data provided by the GNSS sensor, combined with the data provided by the IMU inertial sensor to estimate the position and orientation of the vehicle.

(ii) AMCL (Adaptive Monte Carlo Localisation) calculates the global position of the autonomous vehicle in a reference map based on data provided by Lidar sensors, which together with the odometry data of the vehicle are superimposed over the digital map of the area where the movement takes place.

(IV) Mapping consists of a module that has the role of interpreting and modeling the geography of the environment, generating digital maps. These maps are used by the vehicle to generate new routes and to facilitate the navigation of the autonomous vehicle through the environment. Mapping techniques can be classified into global and local mapping:

(i) The role of global mapping is to plan and generate routes on a global map of the area of movement of the autonomous vehicle. The global map of an area is generated offline by integrating the coordinates of previously recorded measurements (manually) around the area to be mapped. Global maps can be uploaded in real time from digital map providers (Google Maps, Bing Maps, OpenStreetMap etc.). Global maps are static maps that can be used to plan global routes, but they do not provide information about dynamic obstacles (vehicles, pedestrians etc.).

(ii) The role of local mapping is to generate location maps online, in RT and to cover the areas around the autonomous vehicle, integrating the information from the sensors. This method will generate an OGM (Occupancy Grid Map), which will be used by the planning and navigation modules to detect the obstacles around the autonomous vehicle in real time.

(V) Perception consists of a module whose function is to analyze the raw data provided by the sensors of the autonomous vehicle and to extract significant information about the static and dynamic obstacles that are in its immediate vicinity. This module is classified into two groups:

(i) Obstacle detection and classification is the property of the software module to identify and classify all road traffic participants (static and/or dynamic objects). This task will be performed either by analyzing the images recorded by the video camera that equips the autonomous vehicle, or by analyzing the 3D information provided by the Lidar sensors.

(ii) Scene understanding is the property of the software module to recognize various elements in the scene of the environment. The scene recognition algorithm performs a complete semantic segmentation of the images provided by the video camera that equips the autonomous vehicle, providing information about the various elements and objects captured in the scene, respectively how meteorological and climatic conditions influence this landscape.

(VI) Safety is one of the most critical issues of autonomous vehicles. The software module responsible for safety monitoring must be able to detect any anomaly or problem that may occur during the entire operation of the autonomous vehicle and in that situation to hand over the driving commands to the human operator. Consequently, the software architecture includes an algorithm for continuous introspective analysis of the system to assess the safety conditions that determine with maximum precision when the autonomous vehicle is able to drive in autonomous driving mode. This module is classified into three groups:

(i) Self-awareness is an indicator that checks the status of the sensor inputs and the error-free operation of each software process running in the system.

(ii) Localisation accuracy is an indicator that represents the SOH of the self-locating system of the autonomous vehicle, warning the human operator and asking him to take command of the vehicle if the situation requires it.

(iii) Unreachable path is an indicator that monitors the correct planning and navigation of the autonomous vehicle on the travel path. In the event that the autonomous vehicle does not recognize this route, the system shall hand over control of the vehicle to the human operator.

(VII) Path planning is a module whose function is to plan the route by generating a feasible trajectory that the autonomous vehicle can follow to reach the destination. This module is classified into the following groups:

(i) Global planning that takes into account all the characteristics of the route (number of lanes, road width, connections between different road sectors etc.) to find the shortest path between the current position of the autonomous vehicle and the destination. These features are obtained automatically by accessing the information available from digital map providers (Google Maps, Bing Maps, OpenStreetMap etc.) and by generating the shortest path using planning algorithms. The global planning module sets out information on the maximum velocity for each route, respectively the lane on which the vehicle will have to travel.

(ii) Local planning that takes into account all local obstacles that are not found on the global map but are detected with perception algorithms based on information transmitted by sensors. All this information determines the optimal trajectory that will be transformed into geographical coordinates, so that the control module of the autonomous driving system can follow it. Thus, the autonomous vehicle will have the ability to overtake slower vehicles or will accelerate or brake, accordingly, always maintaining the global path as a reference and the velocity set as a goal.

(VIII) Navigation and control are a module whose function is to coordinate the movement of the autonomous vehicle on the route established in accordance with the localisation information, respectively with the route planning decisions. The route tracking algorithm permanently controls the steering angle of the vehicle and the travel velocity (by controlling acceleration and braking) determining its movement within the limits of the established route to the destination.

(IX) Communication software architecture includes V2X communication modules, which provide information capabilities in smart cities, respectively cooperation between autonomous vehicles and other intelligent vehicles. Together with the V2X communication architecture, the autonomous vehicle can be equipped with HMI, consisting of a touch screen used by passengers to enter data (destinations, stop at the station etc.).

(X) Simulation is a software module used in the development of autonomous vehicles, respectively in the planning and prediction part of the travel in order to verify the scenarios to be implemented. Local applications integrated in the autonomous driving system, or cloud platforms specialized in simulating the operation of autonomous vehicles (Baidu Apollo, Carla Simulator etc.) are used for simulation.

3.2.3.1 Network Architecture

The complexity and large number of embedded command and control systems that are part of the E/E Architecture (Electrical/Electronic Architecture) of an autonomous vehicle, more precisely in the composition of the autonomous driving system have led to the need to implement and use different constructive types of communication networks.

The characteristics of the most important communication networks that are found implemented in the constructive architecture of an autonomous vehicle are presented in Table 3.1.

The communication networks described are approached from the point of view of the constructive architecture (hardware), respectively the structure of nodes and the interconnections between them and of the communication protocol (software), respectively the algorithm according to which the messages are transmitted between nodes, the priorities are established, and the errors are managed [36].

The vehicle-centric, field-oriented E/E Architecture maximizes the possibilities of reusing hardware systems and software applications, making the complexity of future in-vehicle systems easier to manage. Multi-domain centralized E/E Architecture uses a smaller number of computers instead of a larger number of individual ECUs. These vehicle computers will be connected to the other built-in ECUs, as well as to the sensors and actuators that equip the autonomous vehicle via communication networks (Fig. 3.25) [36].

Table 3.1 Comparison between communication networks

Name	LIN	CAN	FlexRay	MOST
Physical layer	Cable	Cable	Cable, optical	Optical
Applications	Comfort	Drivetrain	Drivetrain	Infotainment
Control of the nodes	Master/slave	Autonomous	Autonomous	Master/slave
Topology	Linear	Linear	Star	Ring
Bandwidth	20 kbps	1 Mbps	10 Mbps	150 Mbps
Nodes	16	10	22	64
Control	Time	Event	Event	Event
Timing	Time	Priority	Time	Time
Retransmission	Limited	Possible	Possible	–
Errors	CRC, ID	CRC, ERR	Bus guardian	–
Bytes/frame	8	8	254	64
Standard	LIN	ISO 1198	FlexRay	MOST
SAE Class	A	B/C	Drive-by-wire	Mobile media

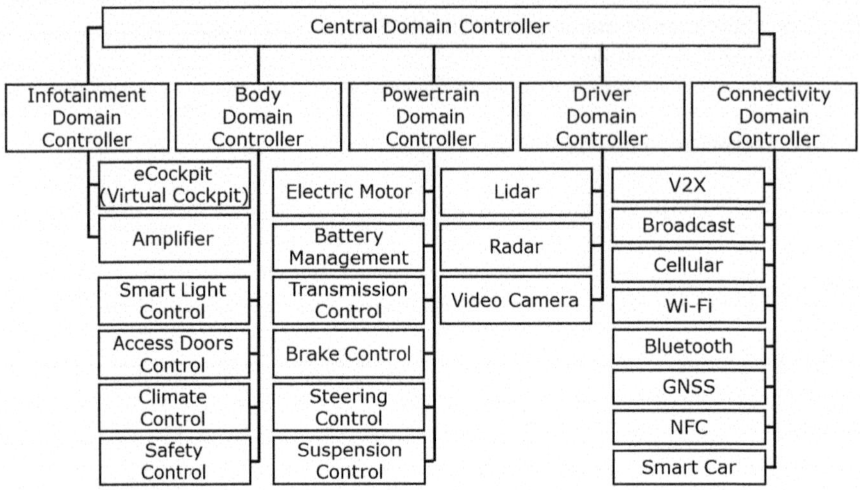

Fig. 3.25 Domain-oriented E/E architecture

A constructive in-vehicle architecture based on a small number of electronic command and control units reduces the complexity of the autonomous driving system alongside other systems and ensures an increase in cybersecurity.

The E/E Architecture is structured on the following domains [36]:

(1) Infotainment domain controller:

  (a) eCockpit (virtual cockpit) is the cluster of digital instruments on LED (Light Emitting Diode) display, which presents a wide range of information (functional parameters, navigation, multimedia, or assisted driving systems).
  (b) Amplifier is an electronic amplifier device that increases the power of a signal by increasing the amplitude of the signal applied to the input, generating a proportionally higher amplitude signal at the output.

(2) Body domain controller:

  (a) Smart light control of the internal and external lighting, respectively monitoring of the diagnosis of the entire lighting system.
  (b) Access, doors control sensors for controlling the motors for operating the mirrors, windscreen wipers, locks, respectively the command and control of the climate control system.
  (c) Climate control is the climate control system.
  (d) Safety control security control for RKE (Remote Keyless Entry), anti-theft locking device, passive starting system, air pressure monitoring system inside the tires, electricity management system.

(3) Powertrain domain controller:

(a) Electric motor is the electric traction motor (asynchronous/synchronous) powered by a DC/AC inverter from the electric battery.
(b) Battery management system is the management and monitoring system of the electric battery.
(c) Transmission is the transmission system of the kinematic chain of motion from the propulsion source to the wheels.
(d) Brake is the braking system equipped with the ABS and it has the role of preventing the locking of the wheels the autonomous vehicle is braking and it allows maintaining control of the steering during braking and shortening the braking distance.
(e) Steering the steering system that ensures the current position of the autonomous vehicle on the planned trajectory.
(f) Suspension is the suspension system that uniformly transmits the forces acting on the autonomous vehicle to the running surface and isolates the vehicle from the forces appearing from the road.

(4) Driver replacement domain controller:

(a) Lidar are sensors that use laser beam distance measurement technology to generate a virtual image of the environment, respectively to determine the position of the autonomous vehicle and to detect static or dynamic objects on the road.
(b) Radar are sensors that measure the distance between the transmitter and the target object by calculating the flight time of the transmitted signal and the time of receiving the echo reflected by the target object.
(c) Video cameras are perception systems that provide information about the surroundings: BSD (Blind Spot Detection), side view control, accident recording, object identification, LCA and sign detection.

(5) Connectivity domain controller:

(a) V2X is a communication link between the autonomous vehicle and the intelligent environment (vehicles, infrastructure, pedestrians etc.), designed to increase traffic safety and efficiency.
(b) Broadcast radio is the telecommunication carried out by means of guided radio waves between a transmitter and a receiver.
(c) Cellular is a mobile data communications network distributed in terrestrial areas called via fixed or mobile cells.
(d) Wi-Fi is a communication standard used to make local wireless communication networks.
(e) Bluetooth is a communication standard for a wireless PAN (Personal Area Network), through radio waves.
(f) GNSS operates on the principle of detecting the position of the autonomous vehicle against a fixed reference consisting of a group of geostationary satellites that emit signals containing information about their position, the

orbital parameters etc. Based on these signals, the reception systems extract information about geographical position, velocity etc.

(g) NFC (Near Field Communication) is a technology based on RF (Radio Frequency) identification that allows mobile terminals to communicate with each other by touch or nearness, at a distance of less than 10 cm.

(h) Smart car access to the dashboard of the autonomous vehicles based on accessing smart applications (Android or iOS).

The multi-domain centralized E/E Architecture offers an optimal level of autonomy, while ensuring a maximum level of security and cyber security. The main advantages of this architecture are [36]:

(I) Modularity consists in dividing the functionality of the autonomous vehicle into separate domains to highlight the functional requirements of safety and cybersecurity for each subsystem and to facilitate the development and implementation of the autonomous driving algorithm.

(II) Optimization consists in grouping and isolating similar functions within the multi-domain-structured E/E Architecture, based on the common requirements characteristic to each domain.

(III) Scalability consists in the modular approach of the characteristics and constructive performances specific to the systems included in each domain based on the production requirements of each autonomous vehicle model (from the specific performances of the entry-level equipment to the specific performances of high-end equipment).

(IV) Reusability is the use of the same basic elements for each domain, with the possibility to perform software updates or to implement new software versions on the same basic software structure.

(V) Connectivity consists of all the ways to connect the autonomous vehicle with the outside world to ensure the exchange of information between the vehicle and all smart devices (V2X).

3.2.3.2 LIN (Local Interconnect Network)

LIN (Local Interconnect Network) is a local communication network designed for small networks of the vehicles, which provides the connection between the command-and-control units and certain sensors and actuators within the composition of a vehicle. In the constructive architecture of autonomous vehicles, LIN in the format of local networks interconnected through the CAN network (where the master node of each LIN network becomes master/slave node in the CAN network) manages the following systems:

(1) Light blocks (headlights, positions, signaling, interior etc.).

(2) Closing, opening, locking, unlocking doors, or operating the automatic platform for wheelchairs.

(3) Interior climate control: air conditioning, heating, ventilation.

(4) The operating system of the steering system engines.

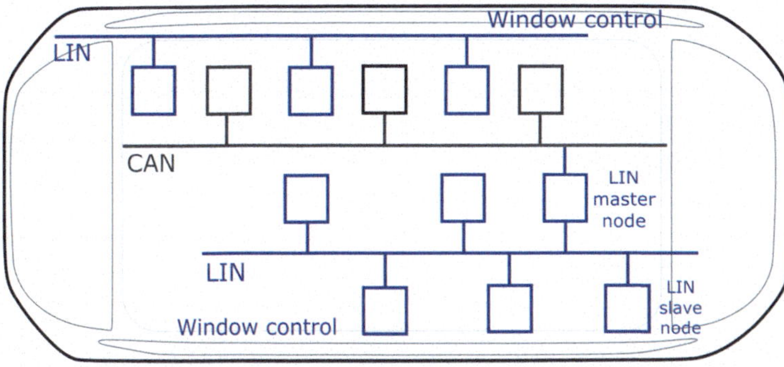

Fig. 3.26 Structure of the LIN communication network

The LIN communication protocol is developed based on the SCI (Serial Communication Interface) standard and works with the UART (Universal Asynchronous Receiver Transmitter) data format, on a single–master multiple–slave type architecture. The data transfer is done on a single 12 VDC, the data transmission speed is up to 20 kbps, and the synchronization of the internal clocks for nodes takes place without a time base [37].

The access environment in the LIN network is controlled by a master node, so that no arbitration or collision management is required in the slave nodes, providing a latency time in the transmission of the data message [38]. The messages are transmitted in one direction only, from the main master command equipment to the next sensor or actuator considered slave. The main equipment can transmit commands to a maximum of 15 slave nodes, each slave node having a distinct address in the network (Fig. 3.26) [37].

The internal structure of the master/slave communication nodes is presented in Fig. 3.27 and it consists of Transmission (Tx) and Reception (Rx) buffers, respectively of the task buffers of the two nodes: master task and slave task.

The addresses of the slave nodes can be configured by the master node when the network is put into operation through a SNPD (Slave Node Position Detection) sequence.

Communication in a LIN network is initiated by the master node (Fig. 3.28), which transmits the header of the message consisting of [39, 40]:

(1) The 14-bit break from synchronization field (Break) acts as a frame start notification for all LIN nodes on the bus.
(2) The 8-bit synchronization field (Sync) identifies the beginning of the message transmitted by the master node and allows the synchronization of Slave nodes.
(3) The 6-bit identifier field (ID) is followed by 2-bits of parity and defines an action to be performed by a slave node.
(4) 2-bit parity field (Parity). A slave node is activated upon reception, which filters the identifier and starts receiving the response message that contains:

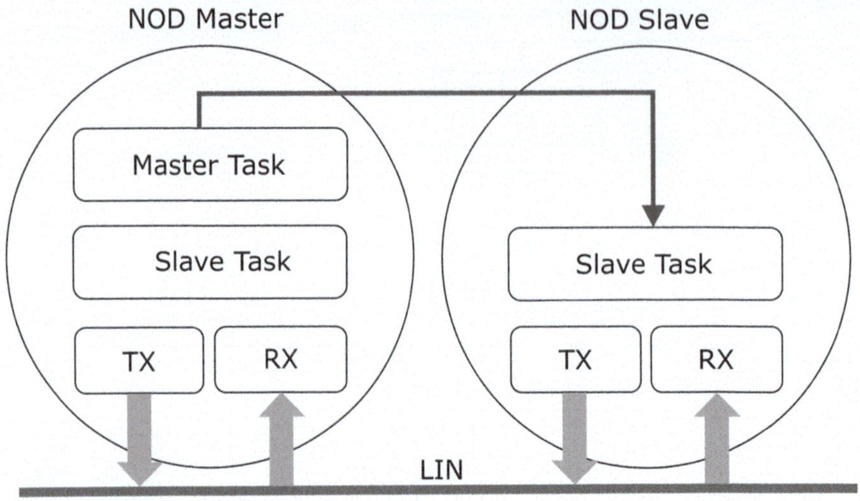

Fig. 3.27 Internal structure of nodes in the LIN network

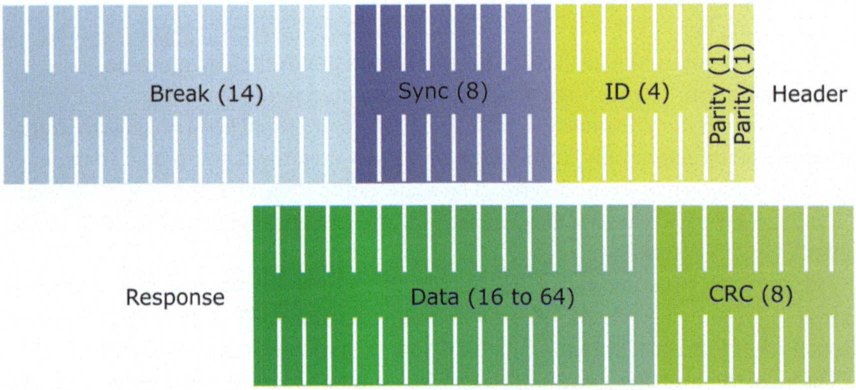

Fig. 3.28 Communication in a LIN network

(5) 16–64-bit data field (Data), consisting of the start bit marking the beginning of an 8-bit data frame and a stop bit marking the end of a data frame.
(6) The 8-bit data field CRC (Cyclic Redundancy Check) has the checksum calculated by adding the bytes in the data field and ignoring the synchronization and identification bytes.

The identifier of a message defines the content of the message, but does not contain information about its destination, a concept that allows the exchange of data from the master node to one or more slave nodes and from the slave node to the master node and/or other slave nodes.

The data transmitted in the LIN network is divided into recessive data (logical high) and dominant data (logical low), and the two states are the sleep mode and the active mode. When the vehicle is not in use, the network enters sleep mode. The master node sends the request (sleep CMD) three times in order to switch to sleep mode.

When there is data in the network all LIN nodes are required to be activated by a wake up CMD signal, a signal transmitted three times by any node requesting network activity. After all nodes have been activated, the master node will program the next identifier.

Data transmission starts with a start bit (level 0) and ends with one or two stop bits. There is no error diagnosis on the LIN bus, errors that may occur are not transmitted on this bus, they are managed by the ECU via the master node [39]:

(I) Bit error—the node that transmits the first bit also monitors the bus. The bit error occurs when the value of the monitored bit is different from that of the transmitted bit.
(II) Checksum error occurs if inverted modulo—256 sum of all received bytes and the checksum is not 0xFF.
(III) Parity identifier error occurs for transmitted messages that cannot distinguish between a valid identifier and a corrupt identifier. Consequently, it is mandatory for all slave nodes for a known identifier to evaluate all 8-bits in the identification field and to distinguish between known and corrupt identifiers.
(IV) The slave node response error is detected if the message is not transmitted in full for the maximum length allowed after transmitting a synchronization field and an identification field.
(V) Sync field error is detected if a slave node detects the edges of a sync field outside the allowed tolerances.
(VI) Bus idle error is detected if no sync field or byte was received within the interval from the reception of the last valid message.

3.2.3.3 CAN (Controller Area Network)

CAN (Controller Area Network) is a serial communication protocol that supports RT operation and has a high level of security in the transmission of data packets. The CAN network is a bus that ensures a digital transfer of data between the command and the execution equipment through a master/slave node system (Fig. 3.29).

In the composition of autonomous vehicles, CAN, interconnects the network of master/slave nodes, respectively makes the connections between other communication networks in the structure of the autonomous vehicle managing the command and control of the following systems [39, 40]:

(I) Motion control system which is part of the autonomous driving system:
(i) Powertrain system.
(ii) Steering system.
(iii) Braking system.

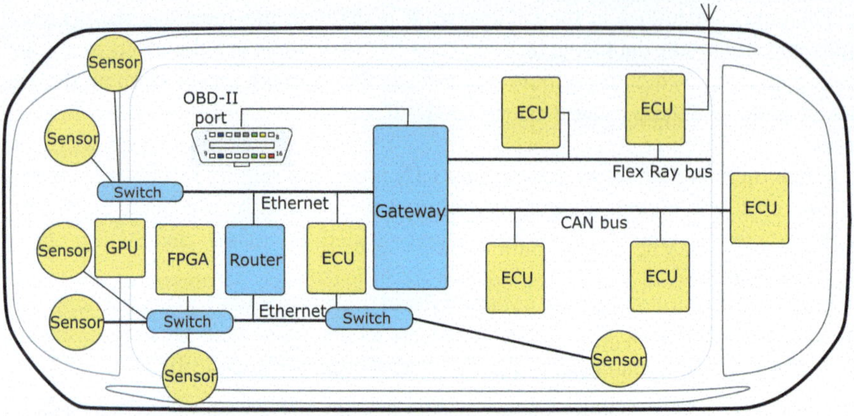

Fig. 3.29 Structure of the CAN communication network

(II) BMS (Battery Management System).
(III) ESP (Electronic Stability Program).
(IV) ACC (Adaptive Cruise Control).
(V). TCU* (Telematic Control Unit).
(VI) Interconnects to local LIN networks.
(VII) OBD (On-Board Diagnostics).

The transmission of data messages in the CAN network is performed on two twisted wires, with a maximum transmission speed of 1 Mbps, allowing the connection of 10 nodes at most. In the CAN network each node can be both a master node and a slave node. Each node attached to the CAN network can send and receive messages and has a unique address in the network. The internal structure of the master/slave communication nodes (Fig. 3.30) consists of the data transmission (Tx) and reception (Rx) buffers, respectively of the instruction buffers (tasks) of the two nodes: master task and slave task [39, 41].

The identifier sets the priorities of the messages through binary values, so that the message with the identifier consisting of the lowest binary number has the highest priority in the transmission. When the network is free, any node can start the transmission of a message. If two or more nodes start transmitting simultaneously, the network access conflict will be resolved through bit-by-bit arbitration, using the identifier. Since the data transmission line is unique for all nodes, the level that has the logical value 0 is called dominant, and the level that has the logical value 1 is called recessive [39, 42]. The CAN network does not involve a hierarchy of nodes when the network is free any node can start transmitting a message. The node that transmits the highest priority message will gain access to the bus.

Data messages transmitted in the CAN frame consist of the fields shown below [39, 43]. Standard CAN message (CAN Standard) (Fig. 3.31):

(A) SOF* (Start Of Frame) 1-bit informs the nodes in the network that a node is about to transmit a message.

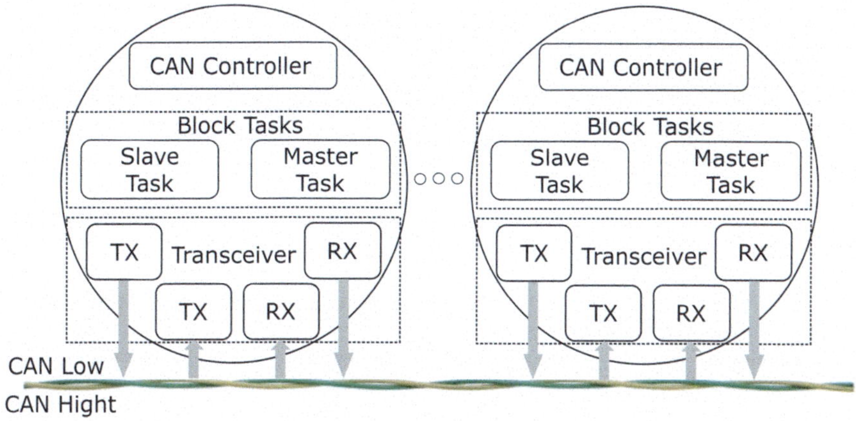

Fig. 3.30 Internal structure of the nodes in the CAN network

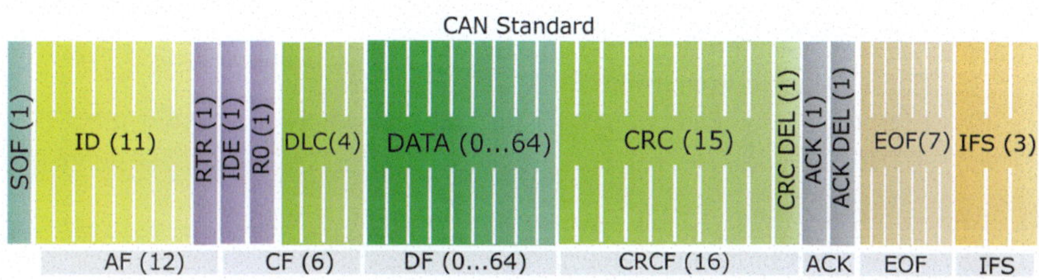

Fig. 3.31 Standard CAN message (CAN Standard)

(B) AF (Arbitration Field) 12-bits consists of:

(1) ID (IDentifier) 11-bits, message identifier.
(2) RTR (Remote Transmission Request) 1-bit, indicates whether a node is transmitting or requesting data from another node.

(C) CF (Control Field) 6-bits consists of:

(1) IDE (IDentifier Extension) 1-bit.
(2) R0 (Reserved) 1-bit, reserved frame.
(3) DLC (Data Length Code) 4-bits.

(D) Data (Data Field) 0–64-bits.

(E) CRCF (Cyclic Redundancy Check Field) 16-bits, the check field is:

(1) CRC (Cyclic Redundancy Check) 15-bits, the sequence of checking the integrity of the data packet.
(2) CRC DEL (CRC Delimiter) 1-bit, check delimiter.

(F) ACK (ACKnowledgement field) 2-bits is:

(1) ACK (ACKnowledgement) 1-bit, acknowledgement sequence.
(2) ACK DEL (ACKnowledgement Delimiter) 1-bit, acknowledgement delimiter.

(G) EOF (End Of Frame) 7-bits—the end frame of the message.

(H) IFS (Inter Frame Space) 3-bits, the pause sequence.

Standard CAN FD message (Flexible Data rate) (Fig. 3.32):

(A) SOF* (Start Of Frame) 1-bit informs the nodes in the network that a node is about to transmit a message.

(B) AF (Arbitration Field) 12-bits consists of:

(1) ID (IDentifier) 11-bits, message identifier.
(2) R1 (Reserved) 1-bit, reserved frame.

(C) CF (Control Field) 9-bits consists of:

(1) IDE (IDentifier Extension) 1-bit, extended identifier.
(2) EDL (Extended Data Length) 1-bit, difference between CAN-CAN FD.
(3) R0 (Reserved) 1-bit, reserved frame.

(4) BRS (Bit Rate Switch) 1-bit, data transfer speed (1–5 Mbps).
(5) ESI (Error State Indicator) 1-bit.
(6) DLC (Data Length Code) 4-bits, length of the data frame.

(D) Data (Data Field) 0–64 bytes.

(E) CRCF (Cyclic Redundancy Check Field) 18-bits, the check field is:

(1) CRC (Cyclic Redundancy Check) 17-bits, the sequence of checking the integrity of the data packet.
(2) CRC DEL (CRC Delimiter) 1-bit, check delimiter.

(E) ACK (ACKnowledgement Field) 2-bits is:

(1) ACK (ACKnowledgement) 1-bit, acknowledgement sequence.

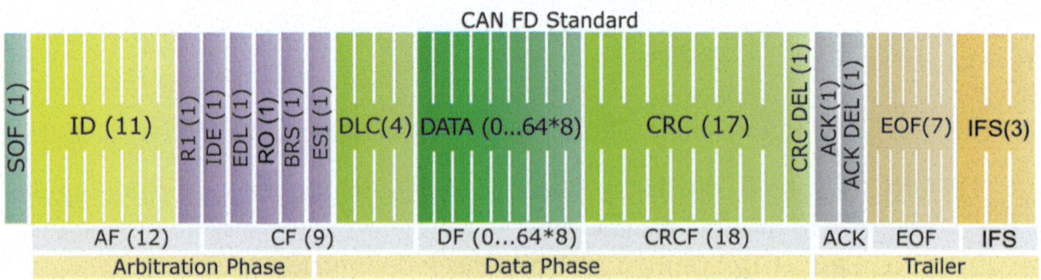

Fig. 3.32 Standard CAN FD message (Standard CAN FD)

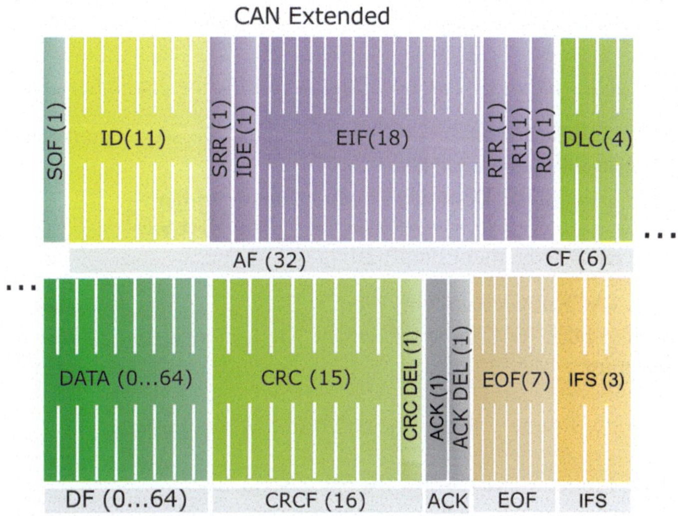

Fig. 3.33 Extended CAN message (Extended CAN)

(2) ACK DEL (ACKnowledgement Delimiter) 1-bit, acknowledgement delimiter.

(G) EOF (End Of Frame) 7-bits—the end frame of the message.

(H) IFS (Inter Frame Space) 3-bits, the pause sequence.

Extended CAN message (Extended CAN) (Fig. 3.33):

(A) SOF* (Start Of Frame) 1-bit informs the nodes in the network that a node is about to transmit a message.

(B) AF (Arbitration Field) 32-bits consists of:

(1) ID (IDentifier) 11-bits, message identifier.
(2) SRR* (Substitute Remote Request) 1-bit, the request for remote transmission of the replaced bits within the RTR frame.
(3) IDE (IDentifier Extension) 1-bit, extended identifier.
(4) EIF (Extended Identifier Field) 18-bits.
(5) RTR (Remote Transmission Request) 1-bit, the request for remote transmission indicates whether a node is transmitting or requesting data from another node.

(C) CF (Control Field) 6-bits consists of:

(1) R1 (Reserved) 1-bit, reserved frame.
(2) R0 (Reserved) 1-bit, reserved frame.
(3) DLC (Data Length Code) 4-bits, length of the data frame.

(D) Data (Data Field) 0–64-bits.

(E) CRCF (Cyclic Redundancy Check Field) 16-bits, the check field is:

(1) CRC (Cyclic Redundancy Check) 15-bits, the sequence of checking the integrity of the data packet.
(2) CRC DEL (CRC Delimiter) 1-bit, check delimiter.

(F) ACK (ACKnowledgement Field) 2-bits is:

(1) ACK (ACKnowledgement) 1-bit, acknowledgement sequence.
(2) ACK DEL (ACKnowledgement Delimiter) 1-bit, acknowledgement delimiter.

(G) EOF (End Of Frame) 7-bits—the end frame of the message.

(H) IFS (Inter Frame Space) 3-bits, the pause sequence.

Extended CAN FD message (Extended CAN FD) (Fig. 3.34):

(A) SOF* (Start Of Frame) 1-bit informs the nodes in the network that a node is about to transmit a message.

(B) AF (Arbitration Field) 32-bits consists of:

(1) ID (IDentifier) 11-bits, message identifier.

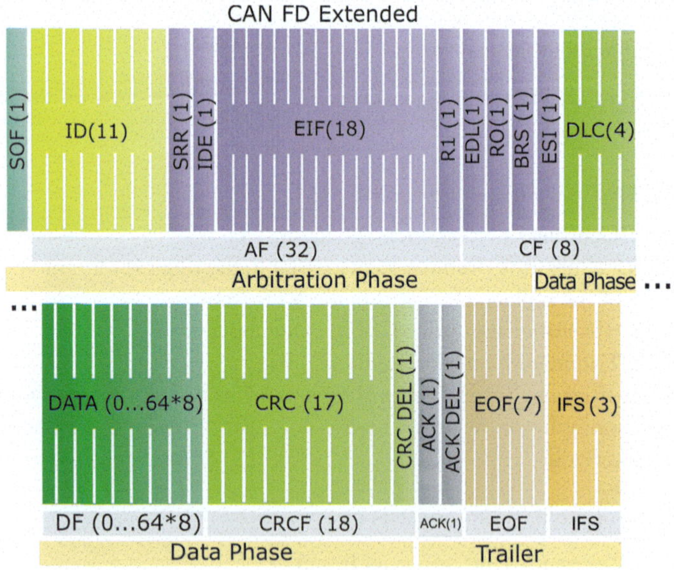

Fig. 3.34 Extended CAN FD message (Extended CAN FD)

   (2) SRR* (Substitute Remote Request) 1-bit, the request for remote transmission of the substituted bits.
   (3) IDE (IDentifier Extension) 1-bit, extended identifier.

(4) EIF (Extended Identifier Field) 18-bits.
(5) R1 (Reserved) 1-bit, reserved frame.

(C) CF (Control Field) 8-bits consists of:

   (1) EDL (Extended Data Length) 1-bit, the difference between CAN-CAN FD.
   (2) R0 (Reserved) 1-bit, reserved frame.
   (3) BRS (Bit Rate Switch) 1-bit, data transfer speed (1–5 Mbps).

(4) ESI (Error State Indicator) 1-bit.
(5) DLC (Data Length Code) 4-bits, length of the data frame.

(D) Data (Data Field) 0–64 bytes.

(E) CRCF (Cyclic Redundancy Check Field) 18-bits, the check field is:

   (1) CRC (Cyclic Redundancy Check) 17-bits, the sequence of checking the integrity of the data packet.
   (2) CRC DEL (CRC Delimiter) 1-bit, check delimiter.

(F) ACK (ACKnowledgement field) 2-bits is:

   (1) ACK (ACKnowledgement) 1-bit, acknowledgement sequence.
   (2) ACK DEL (ACKnowledgement Delimiter) 1-bit, acknowledgement delimiter.

(G) EOF (End Of Frame) 7-bits—the end frame of the message.

(H) IFS (Inter Frame Space) 3-bits, the pause sequence.

The CRC are performed on the CAN bus at the message level and the following five types of errors may occur [39]:

(I) Bit error—occurs when a node does not receive the same transmitted bit from the line.
(II) Bit stuffing error—is generated when there are six equal bits in the frame start, arbitrary, control, data, and CRC sequence fields. Bit stuffing consists of a process directly performed by the transmitter, or vice versa performed by the receiver according to the algorithm: sequences of type 00000 will be followed by a 1 added bit, respectively sequences of type 11111 will be followed by a 0 added bit.

(III) CRC error—is reported if the rest of the calculated CRC polynomial is not identical to the received code. The CRC field contains the CRC sequence followed by the CRC delimiter. The CRC sequence is obtained from the data stream between SOF* and LSB (Least Significant Bits) which constitute the polynomial to be divided and the divisor polynomial. The rest of this division represents the CRC sequence transmitted on the CAN. The CRC delimiter contains a recessive bit.

(IV) Form error—is detected if the presence of a dominant bit is found in the space reserved for delimiters.

(V) ACK error—is generated by a transmitter that does not monitor a dominant bit on the position of the ACK slot. The ACK field is two bits long and contains the ACK slot and the ACK delimiter. In the ACK field, the transmitting node transmits two recessive bits, and the receiving node responds with a dominant bit for the ACK slot if it received the message correctly. In this situation all receivers overwrite the response, the ACK delimiter contains a recessive bit, and the ACK slot is bordered by two recessive bits: the CRC delimiter and the ACK delimiter.

3.2.3.4 FlexRay

FlexRay is a deterministic, error-tolerant communication system that works at a speed of 10 Mbps and uses two channels to transmit data synchronously or asynchronously. In the autonomous vehicle configuration, the FlexRay network is present through X-by-Wire technology, which has an increased tolerance to possible faults (fault tolerance) and increased operational safety, managing the command and control for the active safety controller due to response time [42].

The control system of the prediction and planning mechanism manages:

(1) Maintaining the separation line of the lanes (lane driving).
(2) Traffic and intersection situations (traffic, intersection).
(3) Emergency stops.
(4) Real time diagnostics.

The maximum number of nodes connected to a FlexRay network is 22 in a linear connection, 64 nodes in a star connection, over a maximum distance of 24 m between nodes. Each node is a command-and-control unit, which has in its internal structure (Fig. 3.35) a host controller, a communication controller, two bus guards and two bus drivers.

X-by-Wire technology replaces traditional mechanical control systems with electronic control systems that use electro-magnetic actuators and HMI (Human–Machine Interfaces), such as pedal and steering emulators. The constructive architecture of the FlexRay network is based on the use of two communication channels at most (channel A and channel B). To minimize the risk of error in transmitting the message, FlexRay ensures the redundant operation on both channels. Each of the two communication channels can be used with a data transmission rate of up

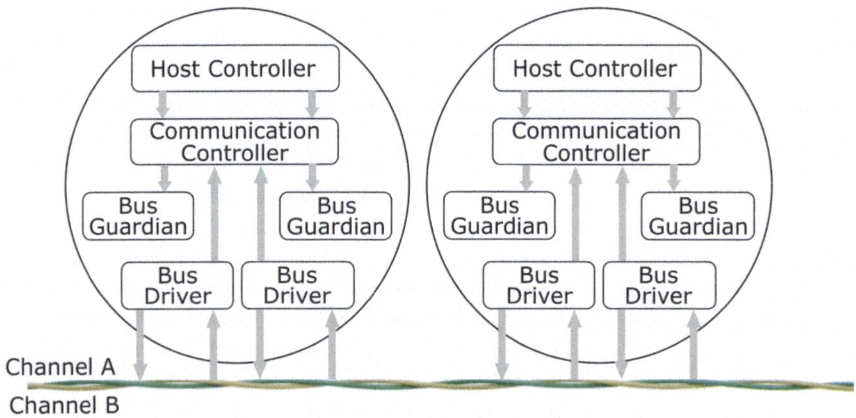

Fig. 3.35 Internal structure of the nodes in the FlexRay network

to 10 Mbps. Alternatively, the redundant channel can be used to increase the data transmission rate to 20 Mbps. A FlexRay node can be connected to a group structure (FlexRay cluster) via a two-channel constructive architecture. A FlexRay node at cluster-level can be assigned to an ECU and a logical command application of an action depending on the purpose served (sensor or actuator).

The FlexRay node consists of the following blocks [39, 44]:

(A) The host controller ensures the transmission and reception of data between the command-and-control unit and the FlexRay communication bus and configures the communication controller.
(B) The communication controller is a FlexRay interface that provides node access to the FlexRay bus.
(C) The bus guardian manages the transmission of data messages only to the nodes connected to the network that are authorized to perform the transmission of messages.
(D) The bus controller ensures the transmission-reception of signals to the FlexRay node.

The host controller transmits Tx Data (Transmit Data) to the communication controller and configures its TxEN (Transmit Data Enable Not) status indicating whether there is data on the communication line. The communication controller transmits to the host controller the received data Rx and the working statuses, and to the bus controller it transmits information about the time of the communication cycle, about its alteration, as well as commands for inhibiting TxEN transmissions. The bus driver transmits information about the status of the program to the host controller and configures it according to the current application, controlling the STBN (STandBy Not) operating mode and indicating the detected errors ERRN (ERRor Not) [38, 39].

The structure of a FlexRay data frame is shown in Fig. 3.36 [45] and consists of the header, the payload segment, and the trailer segment.

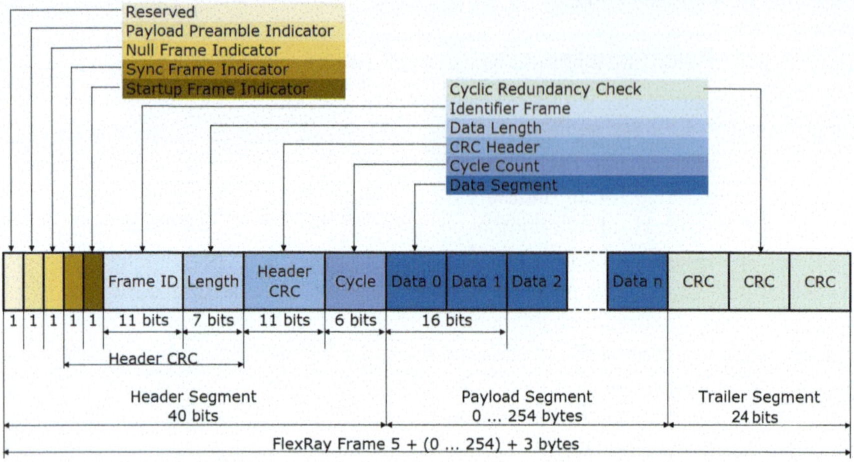

Fig. 3.36 FlexRay data frame

(A) The Header Segment, 40-bits, consists of the data structure [37]:
(1) Reserved, 1-bit, implementation of new applications.
(2) Payload Preamble Indicator, 1-bit, indicates a task in the static segment, or an ID identifier in the dynamic segment.
(3) Null Frame Indicator, 1-bit, indicates whether the frame contains useful data in the load segment.
(4) Synchronized Frame Indicator, 1-bit, indicates whether or not it is a synchronization frame.
(5) Startup Frame Indicator, 1-bit, indicates whether or not it is a startup frame.
(6) Identifier Frame, 11-bits, is used only once on each channel in a communication cycle.
(7) Payload Length, 7-bits, indicates half of the total number of bytes in the payload through encoding.
(8) CRC Header, 11- bits, calculated based on the sync indicator, start indicator, ID value, and the payload length value.
(9) Cycle count, 6-bits, indicates the status of the nodes at a certain time.
(B) Payload Segment contains up to 254 bytes and consists of static or dynamic segments [37, 39].
(C) Trailer Segment contains a 24-bit CRC frame and consists of a cyclic redundancy check code, calculated on the header and payload segment of the frame [37].

The smallest unit of time in a FlexRay network is called the microtick (μT) and has 1 μs, and 1000 microtick units form a macrotick unit (1 ms). The microtick unit is obtained by dividing the frequency of the system's internal clock generator. The succession hierarchy of FlexRay frames is structured on physical levels: microtick, macrotick, arbitration grid level, respectively communication cycle level.

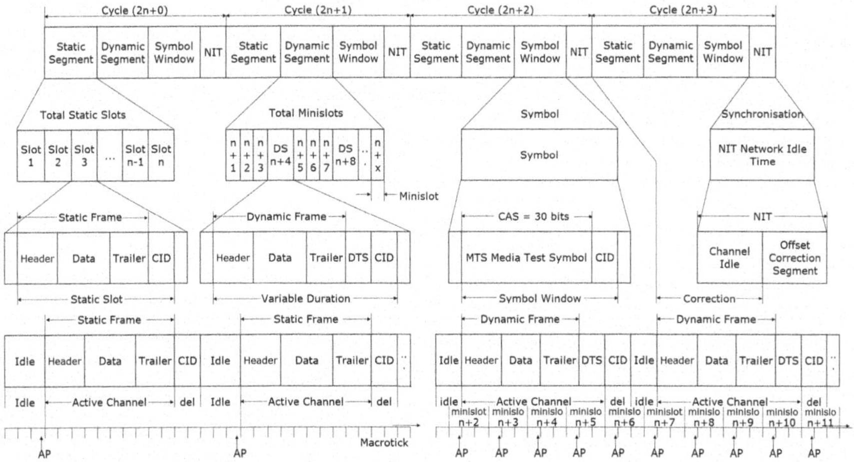

Fig. 3.37 FlexRay communication cycle

The FlexRay communication cycle (Fig. 3.37) is divided into segments, which are repeated cyclically at constant time intervals [46]:

(1) Static segments are dedicated to deterministic applications, with a known content and a well-established bandwidth.
(2) Dynamic segments are triggered by an event, with probabilistic content and a variable bandwidth.
(3) Symbol windows are used to avoid network collisions and to indicate the beginning of the first communication cycle.
(4) NIT (Network Idle Time) occurs when the bus is idle and the nodes in the network are synchronized.

The static segment consists of data frames that have a fixed length and are transmitted in fixed-duration slots, each static slot containing a slot frame reserved for data. A FlexRay network can contain dozens of static slots. The description of a static segment is presented in Fig. 3.38 [39, 46].

(1) A static segment is divided into 16 slots for each node, which according to the FlexRay protocol meet the characteristics [47]:
(a) All slots are characterized by the same duration.
(b) All communication cycles include the same number of slots.
(c) Each slot is identified by a unique number (identifier).
(d) When using the two-channel FlexRay network, the formats for the two-channel slots are identical.
e) The start time of the data transmission for each slot is synchronized based on the frequency of the bus's internal clock signal.

The data transmission duration for each slot is synchronized based on the frequency of the bus's internal clock signal.

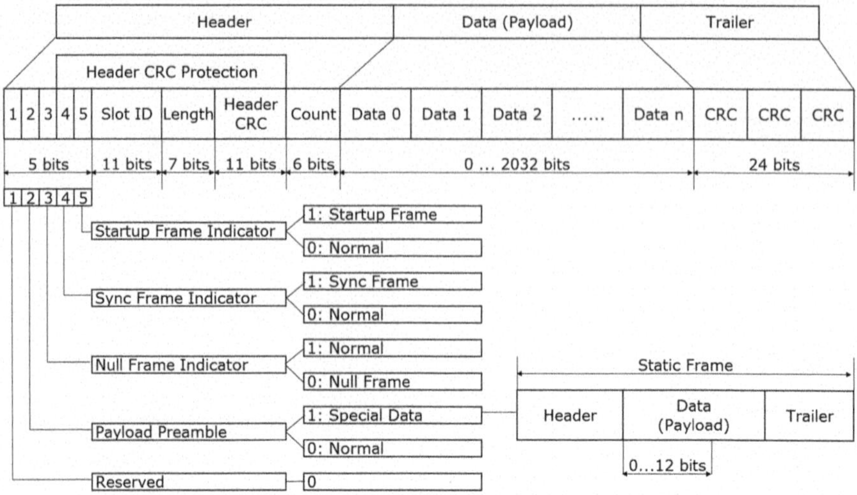

Fig. 3.38 FlexRay static segment

The dynamic segment is a part of the communication cycle in which access to the FlexRay bus is controlled according to the nodes with the highest priority in data transmission, according to a mechanism called FTDMA (Flexible Time Division Multiple Access). The dynamic communication slot is a time frame in which the access to a communication channel is granted exclusively for a node that transmits a frame with an ID identifier corresponding to a priority data frame. The length of a dynamic segment varies depending on the length of the data frame. The description of a dynamic segment is presented in Fig. 3.39 [37, 39].

Unlike the static segment which has a certain length, the length of a dynamic segment varies depending on the length of the data frame [38]. A dynamic segment is divided into mini slots that have an identical duration, much less than the duration of a slot that is part of a static segment.

(2) Dynamic segment mini slots meet the following characteristics according to FlexRay protocol specifications [37, 39, 46, 47]:

a) Dynamic segment cycles contain the same mini slots.
b) The duration of the mini slots is the same for all channels.
c) The bus access sequence during the progress of the dynamic segment may be different for the two channels.
d) The start procedure of a dynamic segment begins:
i. If during the communication cycle there are static and dynamic segments, the start is carried out according to the protocol.
ii. If there are only dynamic segments during the communication cycle, the start is initialized by the SOC* (Start Of Cycle).

The dynamic segment is intended for the transmission of messages that are triggered by events and or have no RT constraints, or for the flexibility of operation, so

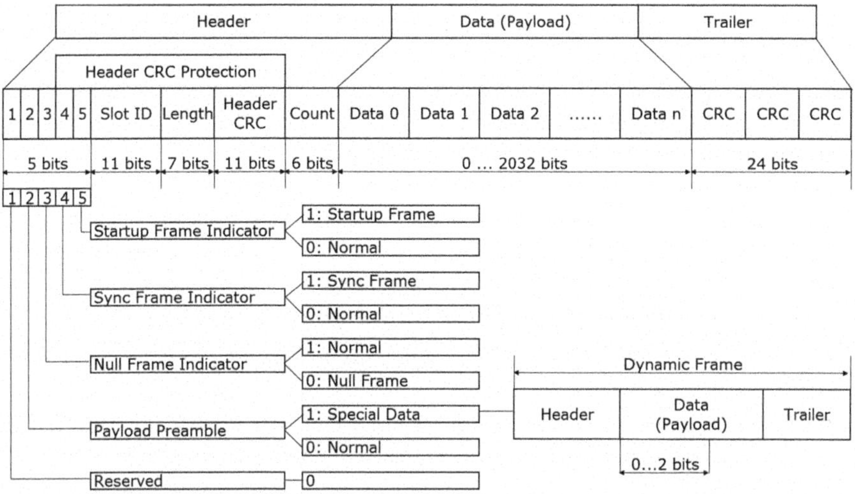

Fig. 3.39 FlexRay dynamic segment

as not to load the bus with static segments with a higher bandwidth. The following data messages are transmitted using dynamic segments [46]:

(1) Communication network management messages.
(2) Messages used in diagnostics.
(3) Messages used to unload the permanent memory.
(4) Calibration messages.
(5) Messages triggered by unforeseen events.

MOST (Media-Oriented Systems Transport)

MOST (Media-Oriented Systems Transport) is a fiber optic network, which transports large volumes of data in the Infotainment system of vehicles. The MOST network has been developed and optimized for multimedia applications and manages existing multimedia equipment, wireless internet connections, Bluetooth and 4G/5G mobile data in the vehicle equipment.

In the autonomous vehicle configuration, the MOST network manages the command and control of the following systems [48]:

(1) Infotainment and multimedia.
(2) Telematics and communication systems.
(3) Signals from data sources: video images taken from traffic, Lidar, radar, other signals received from communication networks.
(4) Diagnosis of embedded electronic components via OBD connector.

The MOST network achieves synchronous data communication via fiber optics, which supports up to 64 nodes [49]. The network topology is of the single master/multi-slave ring type. The synchronization of the nodes in the network is performed by the master node that continuously supplies the network with data [50]. The structure of the MOST network consists of HMI (Human Machine Interface) which is the physical connection with the user and the controllers for applications that manage the functional part of the system, part consisting of FBlock function [50]. A FBlock is used to control application-specific functions existing in the MOST network and defines the link between the HMI and the user-controlled application.

In the particular case of an autonomous shuttle bus for public transportation, the MOST network can manage the control panel materialized in a digital touch screen that is connected to the autonomous driving system and realizes the interface between the autonomous vehicle and the passengers. The control panel has the following functions:

(1) Passenger function for selecting the route and destination on the map and which will show the position of the autonomous vehicle in real time.
(2) Operator function for access to information on the operating status of the autonomous vehicle and the values of the functional parameters.
(3) Maintenance function that allows authorized personnel access to the command-and-control functions of the autonomous driving system.

The constructive architecture of a connection between two MOST nodes is presented in Fig. 3.40. The data packets provided by the MOST node are transmitted Tx through the NIC (Network Interface Controller) by the EOC (Electrical Optical Converter) which converts the electrical signal into a light signal and transmits it to the fiber optic network. From the network the light signal is converted into an electrical signal by the OEC (Optical Electrical Converter) receiver and the NIC interface, and the data packets are distributed to the destination node [39, 50].

The MOST data frame (Fig. 3.41) contains a segment for the synchronous transmission of data streaming, a segment used for the asynchronous transmission of data packets, as well as a segment for the transmission of the control data.

(A) Preamble is used to synchronize the slave nodes of the data stream. The master node generates the preamble frame based on the frequency of the input signal. After the data stream passes through all nodes of the MOST network, it reaches the master node again with a delay due to the propagation time [50].
(B) Boundary Descriptor adjusts the bandwidth of the data stream in the synchronous and asynchronous domains.
(C) The Synchronous area is used for RT transmission of audio and video data packets [48]. This requires a connection made via a master node, which will use the control channel for this purpose.
(D) The Asynchronous area is used to transmit high-bandwidth data packets. The protocol used handles two different lengths of the data field: 48-byte packets and 1014-byte packets. The data structure on the asynchronous domain consists of a header containing [6, 39]:
(1) Arbitration Field (AF)—1 byte.
(2) Destination Address (DA)—2 bytes.

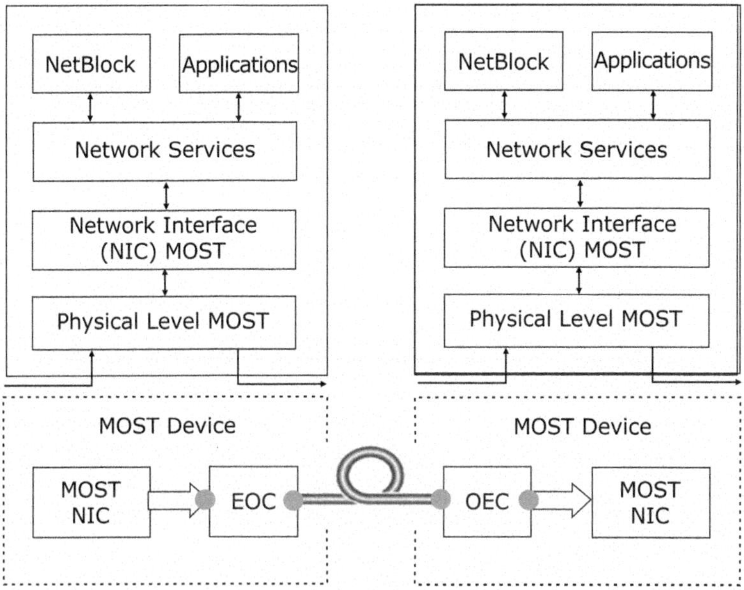

Fig. 3.40 The connection between two nodes in the MOST network

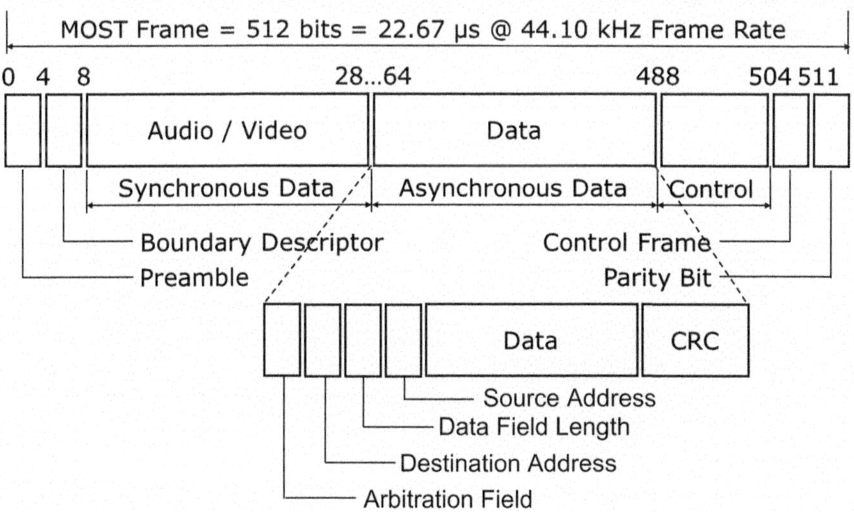

Fig. 3.41 MOST data frame

(3) Data Field Length (DFL)—2 bytes.
(4) Source Address (SA)—2 bytes.
(5) Cyclic Redundancy Check (CRC)—4 bytes.
(E) Control Channel is a communication channel for the administration and diagnosis of the MOST network. To prevent the allocation of a large bandwidth for the control channel, it manages a number of 16 data frames combined in a single block.

3.2.3.5 Ethernet

Ethernet is a high-speed communication network (1000 Mbps) used to connect certain categories of electronic systems that equip a vehicle. In a study conducted by Bosch [50], it was shown that by switching to a single communication standard, all different signals could be transmitted through a single communication network, which would reduce connectivity costs by up to at 80%, and the wiring weight by up to 30%. Ethernet has been used since the 1980s, but due to limitations it has not been widely used in the automotive industry [51]:

(1) EMI (ElectroMagnetic Interference) and RFI (Radio Frequency Interference) were far too high for the automotive industry.
(2) Network connections were affected by interferences from other adjacent devices in the vehicle construction.
(3) The latency of messages transmitted over the Ethernet network was too high for RT operating system applications used in automotive.
(4) Ethernet does not control bandwidth allocation for different streams, no shared data is transmitted from multiple sources.
(5) In the Ethernet network there is no possibility to synchronize the clock signal between the devices connected in the network and to have several data sampling devices at the same time.

The P802.3ch IEEE Draft 100BASE-T1 standard (the standard for Ethernet physical layer specifications and management parameters for greater than 1 Gbps Automotive Ethernet) is an automotive Ethernet standard that ensures reduced EMI, weight, cost, and wiring dimensions. 100BASE-T1 uses cable with shielded wires and can transmit and receive data at speeds of up to 100 Mbps for communication distances of up to 15 m [52].

The internal structure of the nodes in master/slave connection is presented in Fig. 3.42 and consists of the interface for access to the communication line (hybrid), the data transmission buffers (Tx) and receiver buffers (Rx), and the noise attenuators (echo canceller) [53].

In the autonomous vehicle configuration, Ethernet manages the control of the large volume of data from the sensor systems [39]: video cameras, radar, Lidar, respectively V2X connections.

Ethernet communication is managed by the TCP, which makes a virtual connection between two terminals, each point being defined by a unique IP address and a TCP

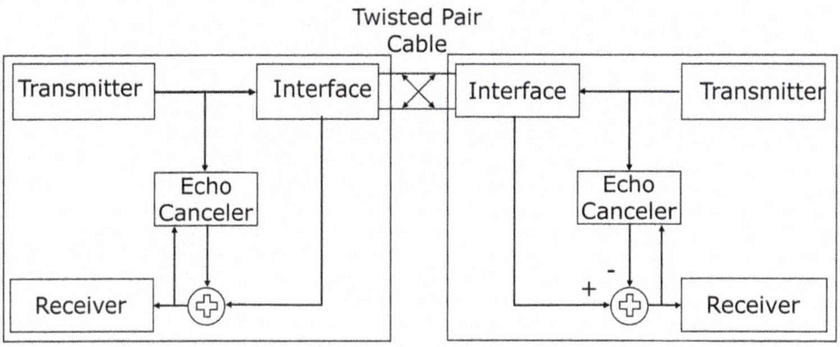

Fig. 3.42 Internal structure of master/slave nodes

port. The TCP protocol ensures the orderly delivery of a data stream between two networked systems. Each system has a unique IP address through which it is identified in the network. The data sent is divided into packets. Each data packet contains the address of the data sending system and the destination system. The Ethernet data frame has the structure as seen in Fig. 3.43 and consists of the following data:

(A) Preamble, 7-bits, the information sequence of the Rx receiver regarding a data transmission.
(B) SOF* (Start Of Frame), 1-bit, indicates the beginning of the data frame.
(C) DA (Destination Address), 6-bits, indicates the destination of the data frame based on the MAC (Media Access Control) address 6-bits.
(D) SA (Source Address) 6-bits, indicates the initial address of data frame.
(E) Length and Type, 2-bits, indicates the type of data frame depending on the length of the data field.
(F) Data Field, up to 1500-bits.
(G) CRC (Cyclic Redundancy Check), 4-bits, the frame checksum.

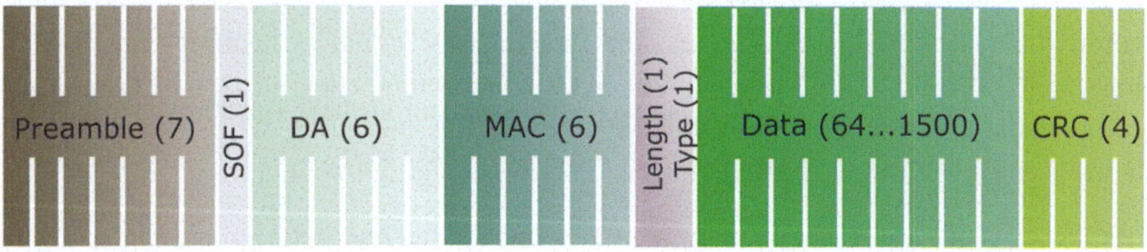

Fig. 3.43 Ethernet data frame

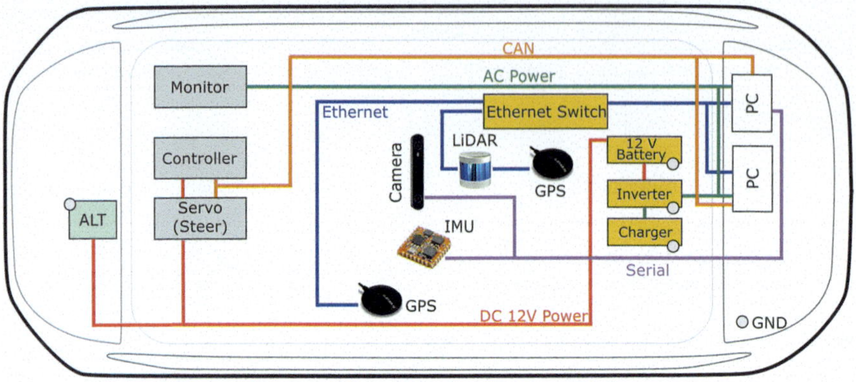

Fig. 3.44 Vehicle platform

3.3 Autonomous Drive Platform

3.3.1 Vehicle Platform

The autonomous driving system, regardless of the technical solution adopted for the implementation of the functional algorithm, provides decision support for the realization of the perception and movement of the vehicle, respectively for stopping in conditions of maximum safety, based on data provided by sensors. According to these data, the system establishes the characteristics that define the trajectory of the autonomous vehicle and regulates its velocity so that depending on the circumstances it can operate the stop in complete safety, depending on the characteristics of the route, on the objects (pedestrians, vehicles etc.) on the route, respectively on the points defined as stopping points for the embarkation/disembarkation of passengers [2].

The vehicle platform has 2WD (2-Wheel Drive) or 4WD (4-Wheel Drive). All the propulsion systems (throttle, brakes, and steering) are actuated through the vehicle's chassis CAN bus. The DC motor attached to the steering is designed to be controlled by electrical signals. The vital vehicle information including wheel velocity and steering angle is directly pulled via the CAN bus (Fig. 3.44) [11].

3.3.2 Hardware Platform

The hardware platform of an autonomous shuttle bus is developed on an E/E Architecture centralized on the vehicle, structured on the areas where the systems that equip the vehicle are located and equipped with computer systems customized on the functional areas (powertrain, body/comfort, chassis, telematics, diagnosis, infotainment) (Fig. 3.45).

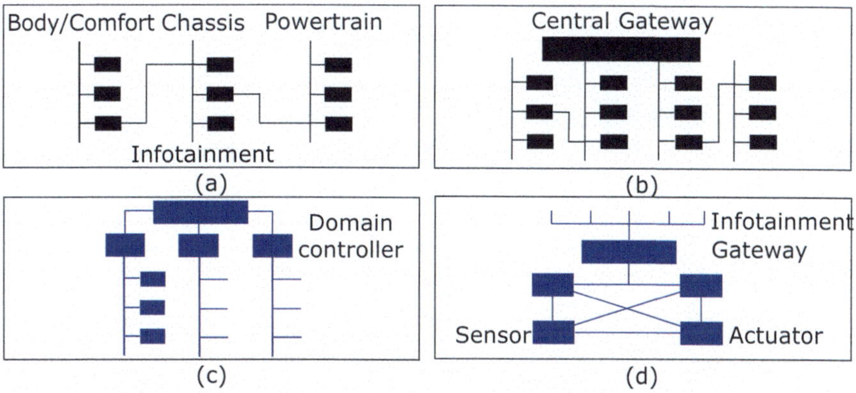

Fig. 3.45 Automotive architecture

Autonomous vehicles make the transition from the classical architecture based on (a) independent domains individually controlled by an electronic command and control unit (independent domains) and (b) domains centralized and interconnected by a central device (central gateway) to the E/E Architecture centralized on (c) functional domains (domain centralized), respectively on (d) the vehicle (vehicle centralized), where the last two centralized structures use vehicle computers to manage the functions and operations specific to each considered domain.

Vehicle computers are connected to the built-in command and control units, sensors and actuators that equip the autonomous vehicle. This way the number of command-and-control units is reduced, which leads to a reduction in the complexity of the vehicle's E/E Architecture, to a centralization of the data streams through the high-speed Ethernet network, and to an increase in cybersecurity and computing power [54].

The central communication gateway CGW (Central GateWay) acts as a router for communication between the electronic command and control units and the ECU command and control unit of the autonomous control system being a connection element for data exchange to all communication buses (LIN, CAN, FlexRay, MOST, Ethernet). CGW allows the exchange of data between all ECUs in the autonomous vehicle, sorting the data packets and distributing them on the communication buses according to the priorities and domains.

In the hardware platform structure of an autonomous shuttle bus developed on an E/E Architecture centralized on the vehicle, the BCU (Body Control Unit) is the electronic unit responsible for monitoring and controlling various electronic accessories in the vehicle equipment. BCU communicates with other electronic units through the CAN bus. The main functions of the BCU are (Fig. 3.46):

(1) Control of internal and external lighting: incandescent lamps, HID (High-Intensity Discharge lamps), LED lamps, respectively monitoring the diagnosis of the entire lighting system (overload and overtemperature protection, light bulb failure etc.).

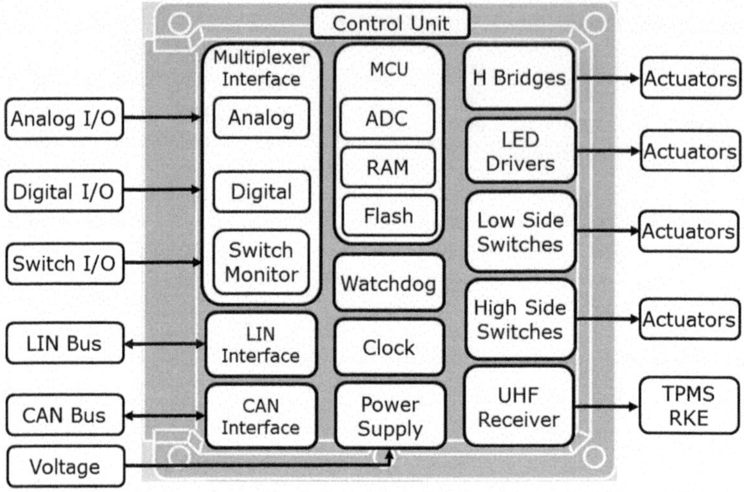

Fig. 3.46 Body control unit

(2) Sensors for controlling the motors step-by-step in order to operate the moving parts, and for controlling electromagnetic locks.
(3) Command and control of the climate control system.
(4) Security control for: RKE, anti-theft locking device, tire pressure monitoring system and power management system, respectively.

3.3.3 Software Platform

Autonomous vehicles collect large volumes of data on the driving environment using various sensor systems (Lidar, radar, video camera, GPS/GNSS etc.). According to a survey conducted by Hyundai Mobis Research Institute, an autonomous vehicle can generate a volume of data of up to 4 TB/h [55]. Due to the fact that the number of autonomous vehicles integrated in traffic will increase exponentially in the coming years, it is necessary that all these vehicles be connected to software, intermediate platforms, which manage this data, filter it, and send to the storage/processing servers from Cloud Computing only the data essential for the safe conduct of road traffic.

Lee et al. presented in [55] a software platform based on edge nodes, fog servers and Cloud Computing servers, which has the role of managing local/zonal large volumes of data generated by the autonomous vehicles (Fig. 3.47).

The edge node is a computer that collects data transmitted by the sensor systems of autonomous vehicles operating locally in the coverage area of that node. In general, the data collected by the edge nodes are RT location data of the autonomous vehicle (GPS/GNSS coordinates and Lidar point cloud, respectively).

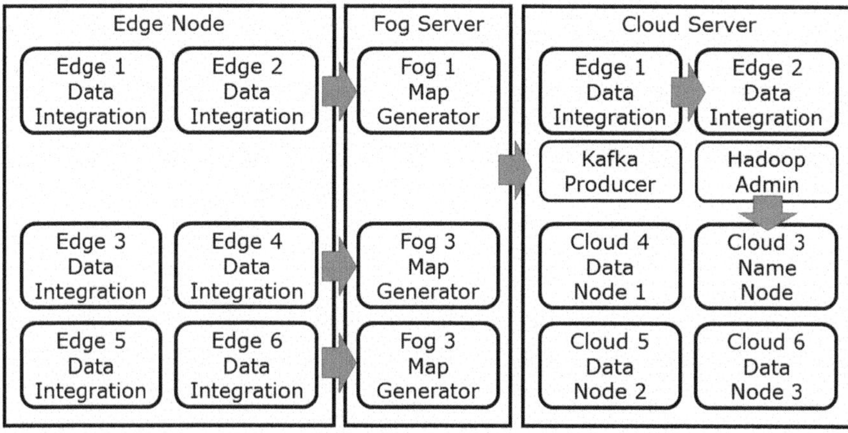

Fig. 3.47 Software architecture (Lee et al.)

The fog server receives data from the edge nodes in the server coverage area, analyzes the received data to filter and separate static objects from the dynamic objects detected by each autonomous vehicle, and based on this data it generates HD maps of the covered area. This HD map is constantly updated with information from all autonomous vehicles connected to the software platform.

The Cloud Computing server facilitates the transformation of data from all connected fog servers and the grouping of this data into a single coordinate plane based on which it will generate an extended HD map that will contain information about all tracked static and dynamic objects.

For optimal processing of this data, in the cloud server is implemented a HDFS (Hadoop Distributed File System) type software architecture, consisting of data nodes where the distributed data, information about nodes (name nodes) that manage the location of the distributed data through metadata and the Hadoop architecture are stored, which is responsible for the overall management of the big data software platform (Fig. 3.48) [56].

3.3.4 Cloud Platform

The constructive architecture of the Cloud Computing platform consists of the modules presented in Fig. 3.49. The platform constantly monitors the connected autonomous vehicles and based on the data received from their sensors it ensures the following functions [11]:

(i) Permanent maintenance of the safety limits established by the autonomous driving system algorithm.
(ii) Detection and avoidance of static and dynamic obstacles.

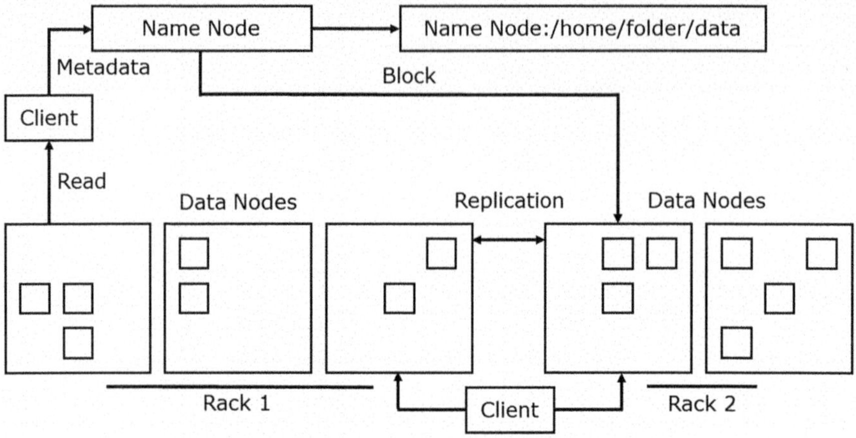

Fig. 3.48 HDFS software architecture

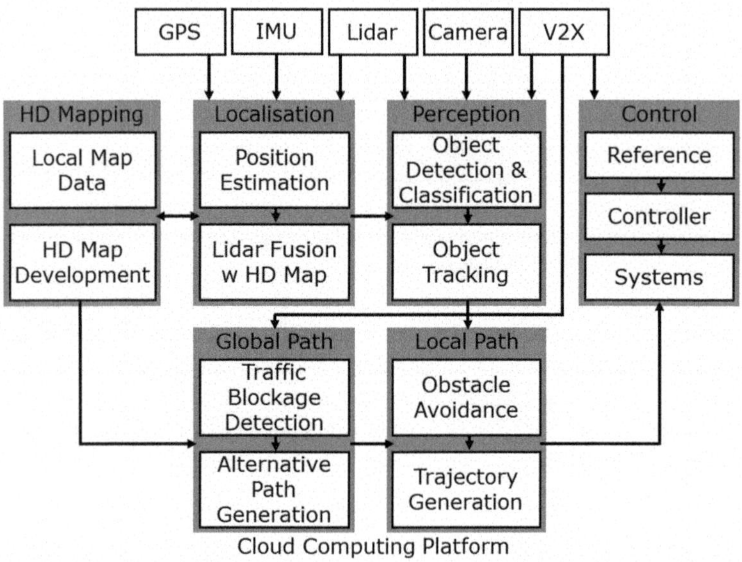

Fig. 3.49 Cloud computing platform

(iii) Detection of traffic jams located on the path and establishment of an alternative path.
(iv) Keeping the lane according to the plan established by the autonomous driving system.
(v) Maintaining the direction of travel even in the absence of the GPS/GNSS signal.

The constructive architecture of the Cloud Computing platform consists of the following modules [11]:

(A) HD mapping module develops the local HD map based on data stored in the cloud.
(B) Localisation module uses sensor data (GPS, IMU, Lidar) to locate the vehicle on the HD map.
(C) Perception module overlays the information from the video camera and the cloud platform via the V2X connection over the virtual environment created based on the HD map and the point cloud recorded by the Lidar sensor. The perception module locates the static and dynamic objects and based on the estimation of their trajectory it plans the future trajectory of the autonomous vehicle.
(D) The control module commands the drive systems (propulsion, steering, braking) based on the reference parameters in the planned trajectory correlated with the correction data received via the V2X connection.

The cloud platform used by Apollo Baidu contains a simulation platform based on Apollo Dreamland technology. This platform uses a large amount of data collected from the actual operation of Baidu Apollo autonomous vehicles, data on the basis of which various driving scenarios can be adopted. This data, together with the superior computing power of the cloud platform, allows the simulation platform to create a powerful test environment for the development of a virtual autonomous driving system [57]. The Apollo simulation platform creates a powerful virtual loop for the development of an autonomous driving system using algorithms. It allows developers and other users to run millions of kilometers of simulation daily for various models of autonomous vehicles, which allows the calibration and development of these vehicles.

The main items in the Apollo Dreamland simulation platform menu are as follows (Fig. 3.50) [58]:

(1) Array of scenarios that allow users to use elements of existing scenarios in the database to define their own scenario:
(a) Types of roads: intersections, turning lanes, junctions, obstacles, driving plans, traffic light sectors etc.
(b) Obstacles, static and dynamic objects (vehicles, pedestrians etc.).
(c) Driving plans: lane tracking, U-turn, lane change, left turn, right turn, lane fusion etc.
(d) Traffic light status: red, yellow, green.
(2) Execution modes the simulation platform provides users with a complete configuration to be able to run in the cloud, several parallel scenarios and to verify and develop the modules of autonomous vehicles simulated in the test environment.
(3) Automatic grading system which offers the possibility to perform tests on the following scenarios:
(a) Collision detection.

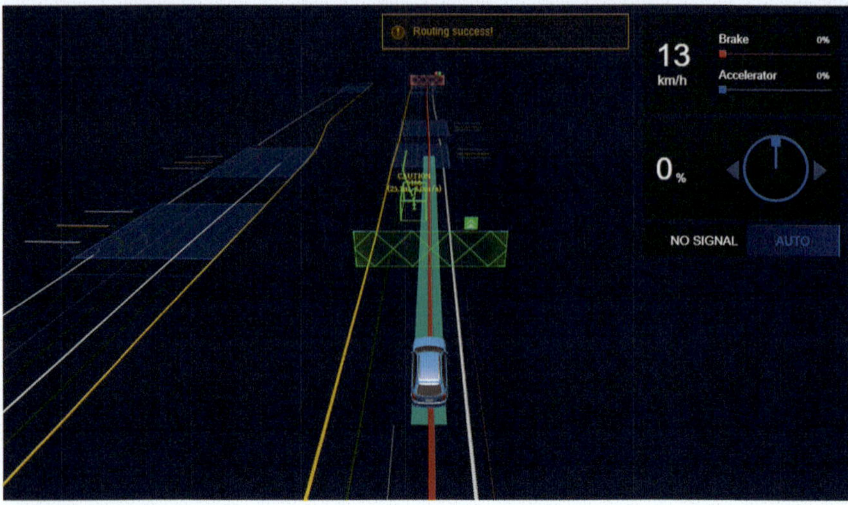

Fig. 3.50 Apollo Dreamland cloud computing simulation platform

(b) Red-light violation detection.
(c) Speeding detection.
(d) Off-road sector detection.
(e) Arrival test.
(f) Hard braking detection.
(g) Acceleration test.
(h) Routing test.
(i) Lane-change in junction detection.
(j) Yield to pedestrians at crosswalks.
(k) Brake taps.
(l) Stop at stop-signs.
(4) 3D visualization - ensures conditions of movement in a 3D environment (path, obstacles, static and dynamic objects etc.) in real time.
The Dreamland simulation platform run the following types of scenarios:
Worldsim user-created scenarios from the elements of existing scenarios in the database for testing autonomous vehicles in a well-defined environment, but which lacks the complexity of real traffic conditions.
Logsim scenarios, scenarios extracted from real situations using the data provided by the sensors on autonomous vehicles. These scenarios are realistic, but not deterministic. Existing obstacles can be vaguely defined, and traffic conditions are complex.

The main features of the Apollo Dreamland simulation platform are as follows:

(1) Web solution is a simulation application that runs online and does not require large resources and the installation of additional software, it can be accessed from any device connected to the internet.

(2) Customizable scenarios that include traffic elements and that adapt to all user requirements.
(3) Rigorous grading metrics as defined above.
(4) Verification and validation of the virtual model in the cloud before simulation.

Scenario management manages the scenarios and allows the definition of specific tasks for each scenario (task management). The main steps to be followed to perform a simulation using the Apollo Dreamland simulation platform are as follows:

(1) Development of the simulation process algorithm according to the interfaces of the Apollo Dreamland simulation platform.
(2) Management of the group of simulation scenarios.
(3) Running simulations based on different scenarios.
(4) Analysis and interpretation of simulation results.

Development of the simulation process algorithm according to the interfaces of the Apollo Dreamland simulation platform allows users to connect their own algorithms to define certain modules of the simulation model (vehicle dynamics).

API interfaces have a standard format for each module [57]:

(1) The localisation API mode (topic:/apollo/localisation/pose) displays the position and direction of travel of the autonomous vehicle, and after the simulation it provides information about the position, orientation, direction, linear/longitudinal velocity, longitudinal acceleration, angular velocity of the autonomous vehicle.
(2) The perception API mode (topic:/apollo/perception/obstacles) displays the position, direction of travel and velocity of the autonomous vehicle, respectively the shape of the obstacles around the vehicle, the colors of the traffic lights on the route, and after the simulation it provides information about the autonomous vehicle: position, direction, velocity, acceleration, and about the obstacles: type, dimensions (length, width, height), respectively about the traffic lights: the colors they indicate, the signaling time etc.
(3) The API mode on the CAN bus (topic:/apollo/canbus/chassis) displays the travel velocity of the autonomous vehicle, respectively the driving mode of the autonomous driving system, and after the simulation it provides information about the travel velocity of the autonomous vehicle.
(4) The routing API mode (topic:/apollo/routing_response) displays the results of the autonomous vehicle movement, and after the simulation it provides information about the autonomous vehicle movement.
(5) The prediction API mode (topic:/apollo/prediction) estimates the predicted trajectories for different obstacles (dynamic and/or static), and after the simulation it provides information about the prediction of the trajectory of these obstacles.
(6) The planning API mode (topic:/apollo/planning) generates the trajectory to be followed by the autonomous vehicle, and after the simulation it provides information about the timestamps along the trajectory.

The management of the simulation scenario group is done from the scenario management page, where users can group the test scenarios based on the test requirements. The group can handle the following elements:

(1) Type of scenario: Worldsim or Logsim.
(2) Road structure: Straight Road, Curve, Intersection, Four-Way Intersection, Multi-Way Intersection, T Junction, Y Junction.
(3) Ego vehicle behavior: traveling straight in a through lane, lane change in a through lane, traveling straight through an intersection, left turn at an intersection.

Running simulations based on different scenarios allows users to load their own algorithm to test scenarios based on existing elements in the database of the Apollo Dreamland simulation platform, respectively based on real situations using data provided by sensors equipping autonomous vehicles.

References

1. https://www.gm.com/content/dam/company/docs/us/en/gmcom/gmsafetyreport.pdf. 26 Sept 2020
2. Iclodean, C., Cordos, N., Varga, B.O.: Autonomous shuttle bus for public transportation: a review. Energies **13**, 2917 (2020)
3. Wolf, I.: The interaction between humans and autonomous agents. In: Maurer, M., Gerdes, J.C., Lenz, B. (eds.) Autonomous driving technical, legal and social aspects, pp. 103–124. Springer, Heidelberg (2016)
4. https://www.sae.org/standards/content/j3016_202104/. 19 May 2022
5. Kala, R.: On-Road Intelligent Vehicles Motion Planning for Intelligent Transportation Systems. Butterworth-Heinemann, Cambridge, MA, USA, pp. 63 (2016)
6. https://navya.tech/shuttle/. 11 Apr 2020
7. Jahromi, B.S., Tulabandhula, T., Cetin, S.: Real-time hybrid multi-sensor fusion framework for perception in autonomous vehicles. Sensors **19**, 4357 (2019)
8. https://www.wevolver.com/article/2020.autonomous.vehicle.technology.report. 29 Sept 2020
9. https://ro.scribd.com/document/453837110/AutonomousCar-ASPLOS18. 23 Apr 2022
10. https://github.com/ApolloAuto/apollo/blob/master/docs/specs/perception_apollo_5.0.md. 3 Aug 2020
11. Arshad, S., Sualeh, M., Kim, D., Nam, D.V., Kim, G.-W.: Clothoid: an integrated hierarchical framework for autonomous driving in a dynamic urban environment. Sensors **20**, 5053 (2020)
12. Pendleton, S.D., Andersen, H., Du, X., Shen, X., Meghjani, M., Eng, Y.H., Rus, D., Ang, M.H.: Perception, planning, control, and coordination for autonomous vehicles. Machines **5**, 6 (2017)
13. Wu, X., Qiao, B., Su, C.: Trajectory planning with time-variant safety margin for autonomous vehicle lane change. Appl. Sci. **10**, 1626 (2020)
14. https://github.com/ApolloAuto/apollo/blob/master/docs/specs/Class_Architecture_Planning.md. 4 Aug 2020
15. Dreyfus, S.: Richard Bellman on the birth of dynamic programming. Oper. Res. **50**(1), 48–51 (2002)
16. Hegedűs, T., Németh, B., Gáspár, P.: Design of a low-complexity graph-based motion-planning algorithm for autonomous vehicles. Appl. Sci. **10**, 7716 (2020)
17. Montes, H., Salinas, C., Fernández, R., Armada, M.: An experimental platform for autonomous bus development. Appl. Sci. **7**, 1131 (2017)

18. Ma, W., Qian, S.: High-resolution traffic sensing with probe autonomous vehicles: a data-driven approach. Sensors **21**, 464 (2021)
19. Ahmed, S., Huda, M.N., Rajbhandari, S., Saha, C., Elshaw, M., Kanarachos, S.: Pedestrian and cyclist detection and intent estimation for autonomous vehicles: a survey. Appl. Sci. **9**, 2335 (2019)
20. Cao, J., Song, C., Song, S., Peng, S., Wang, D., Shao, Y., Xiao, F.: Front vehicle detection algorithm for smart car based on improved SSD model. Sensors **20**, 4646 (2020)
21. https://arxiv.org/pdf/1612.08242.pdf. 24 Aug 2020
22. https://openaccess.thecvf.com/content_ICCV_2019/papers/Choi_Gaussian_YOLOv3_An_Accurate_and_Fast_Object_Detector_Using_Localisation_ICCV_2019_paper.pdf. 28 Aug 2020
23. Guan, L., Chen, Y., Wang, G., Lei, X.: Real-time vehicle detection framework based on the fusion of LiDAR and camera. Electronics **9**, 451 (2020)
24. Han, J., Liao, Y., Zhang, J., Wang, S., Li, S.: Target fusion detection of LiDAR and camera based on the improved YOLO algorithm. Mathematics **6**, 213 (2018)
25. Iclodean, C.: Interconectarea sistemelor virtuale de comandă și control pentru autovehicule, Volumul I, Software-in-the-Loop 2018, Risoprint Ed., Cluj-Napoca, Romania. ISBN: 978-973-53-2252-6
26. https://patents.google.com/patent/US7979173B2/en. 13 Apr 2020
27. https://patents.google.com/patent/US10334050B2/en. 20 Apr 2020
28. https://patents.google.com/patent/US10401852/en. 20 Apr 2020
29. https://patents.google.com/patent/US10446037/en. 20 Apr 2020
30. Rosique, F., Navarro, P.J., Fernández, C., Padilla, A.: A systematic review of perception system and simulators for autonomous vehicles research. Sensors **19**, 648 (2019)
31. Betz, J., Heilmeier, A., Wischnewski, A., Stahl, T., Lienkamp, M.: Autonomous driving—a crash explained in detail. Appl. Sci. **9**, 5126 (2019)
32. Jung, C., Lee, D., Lee, S., Shim, D.H.: V2X-communication-aided autonomous driving: system design and experimental validation. Sensors **20**, 2903 (2020)
33. https://github.com/ApolloAuto/. 12 Apr 2020
34. Herrmann, A., Brenner, W., Stadler, R.: Autonomous Driving: How the Driverless Revolution will Change the World. Emerald Publishing Limited, Bingley, UK, pp. 141 (2018)
35. de Miguel, M.Á., Moreno, F.M., Marín-Plaza, P., Al-Kaff, A., Palos, M., Martín, D., Encinar-Martín, R., García, F.: A research platform for autonomous vehicles technologies research in the insurance sector. Appl. Sci. **10**, 5655 (2020)
36. https://www.nxp.com/docs/en/release-note/ATZelektronik-Worldwide-062017_NXP-Domains-Article.pdf. 05 May 2021
37. Mariașiu, F., Iclodean, C.: Managementul Motoarelor cu Ardere Internă, Editura Risoprint, Cluj-Napoca (2013). ISBN 978-973-53-1004-2
38. http://www.tecchannel.de/a/flexray-flexibler-highspeed-datenbus-im-auto,1735234,7. 30 July 2020
39. Iclodean, C.: Rețele de comunicație pentru autovehicule, Risoprint Ed., Cluj-Napoca, Romania (2017). ISBN: 978-973-53-1992-2
40. https://www.csselectronics.com/screen/page/lin-bus-protocol-intro-basics. 30 July 2020
41. https://www.csselectronics.com/screen/page/simple-intro-to-can-bus. 30 July 2020
42. Bosch, R.: CAN Specification Version 2.0. Robert Bosch GmbH, Postfach 50, D-7000 Stuttgart 1 (1991)
43. Bosch, R.: CAN with Flexible Data-Rate Specification Version 1.0. Robert Bosch GmbH, Postfach 50, D-7000 Stuttgart 1 (2012)
44. FlexRay Consortium: FlexRay Communications System Protocol Specification Version 3.0.1 (2010)
45. Paret, D.: Multiplexed Networks for Embedded Systems CAN, LIN, FlexRay, Safe-by-Wire…. Wiley (2007). ISBN 978-0-470-03416-3
46. Paret, D.: FlexRay and its Applications Real Time Multiplexed Network. Wiley (2012). ISBN: 978-1-119-97956-2

47. https://www.snapon.com/EN/US/Diagnostics/News-Center/FlexRay-Networks. 23 Apr 2022
48. Grzemba, A.: MOST The Automotive Multimedia Network. Franzis Verlag, Poing (2008). 978-3-7723-5316-1
49. Robert Bosch GmbH: Bosch Automotive Electrics and Automotive Electronics. Springer, Heidelberg (2014) https://doi.org/10.1007/978-3-658-01784-2_7
50. MOST Media Oriented Systems Transport Multimedia and Control Networking Technology, MOST Specification, MOST Cooperation, Germany, Rev. 3.0 E2, 07/2010
51. https://support.ixiacom.com/sites/default/files/resources/whitepaper/ixia-automotive-ethernet-primer-whitepaper_1.pdf. 17 Sept 2020
52. https://www.ti.com/lit/wp/szzy009/szzy009.pdf?ts=1596122064880. 13 Aug 2020
53. He, Q., Meng, X., Qu, R., Xi, R.: Machine learning-based detection for cyber security attacks on connected and autonomous vehicles. Mathematics **8**, 1311 (2020)
54. https://www.bosch-mobility-solutions.com/en/mobility-topics/ee-architecture/. 23 Apr 2022
55. Lee, J., Lee, K., Yoo, A., Moon, C.: Design and implementation of edge-fog-cloud system through HD map generation from LiDAR data of autonomous vehicles. Electronics **9**, 2084 (2020)
56. https://searchdatamanagement.techtarget.com/definition/Hadoop-Distributed-File-System-HDFS. 5 Aug 2020
57. https://studio.apollo.auto/login?url=%2F%3Flocale%3Den-us. 5 May 2021
58. https://github.com/ApolloAuto/apollo/blob/master/docs/specs/Dreamland_introduction.md. 5 Aug 2020

Chapter 4
Safety and Cybersecurity

4.1 Safety

4.1.1 Functional Safety

Functional safety is an essential component of a vehicle's hardware and software development cycle, which is designed to identify system-wide malfunctions and to specify actions, techniques, and methods to reduce potential risks and damages caused by errors. ISO 26262 standard (Road Vehicles Functional Safety) [1] addresses the need for an international standard specific to the automotive field that focuses on critical safety components and uses a system of measures to manage functional safety and to regulate product development at the hardware and software level (Fig. 4.1) [2, 3]. The purpose of functional safety is to reduce the risk to an acceptable level and to reduce the negative impact of this risk.

The increasing complexity of technology in the automotive industry requires the existence of a unique safety standard for all manufacturers of systems integrated in a vehicle. ISO 26262 is a common standard that assesses the degree of safety of a system, in particular regarding its purpose in the composition of a vehicle. The main systems covered by ISO 26262 are: ADAS (Advanced Driver Assistance Systems), X-by-Wire systems, active systems, passive systems, electronic control system etc.

The classification of risks according to the ISO 26262 standard is based on four ASIL (Automotive Safety Integrity Levels) ranging from level A for the most permissive requirements for the integrity of car safety to level D for the strictest car safety integrity requirements. The ISO 26262 standard is structured in ten parts that manage functional safety and that aim to develop the product at system, hardware, and software level [1–4].

The safety protocols implemented in the command and control of the autonomous driving system comply with the following criteria [5]:

(1) The autonomous vehicle will stop if a fault is detected in any of the sensors of the autonomous driving system.

C. Iclodean et al., *Autonomous Vehicles for Public Transportation*, Green Energy and Technology, https://doi.org/10.1007/978-3-031-14678-7_4

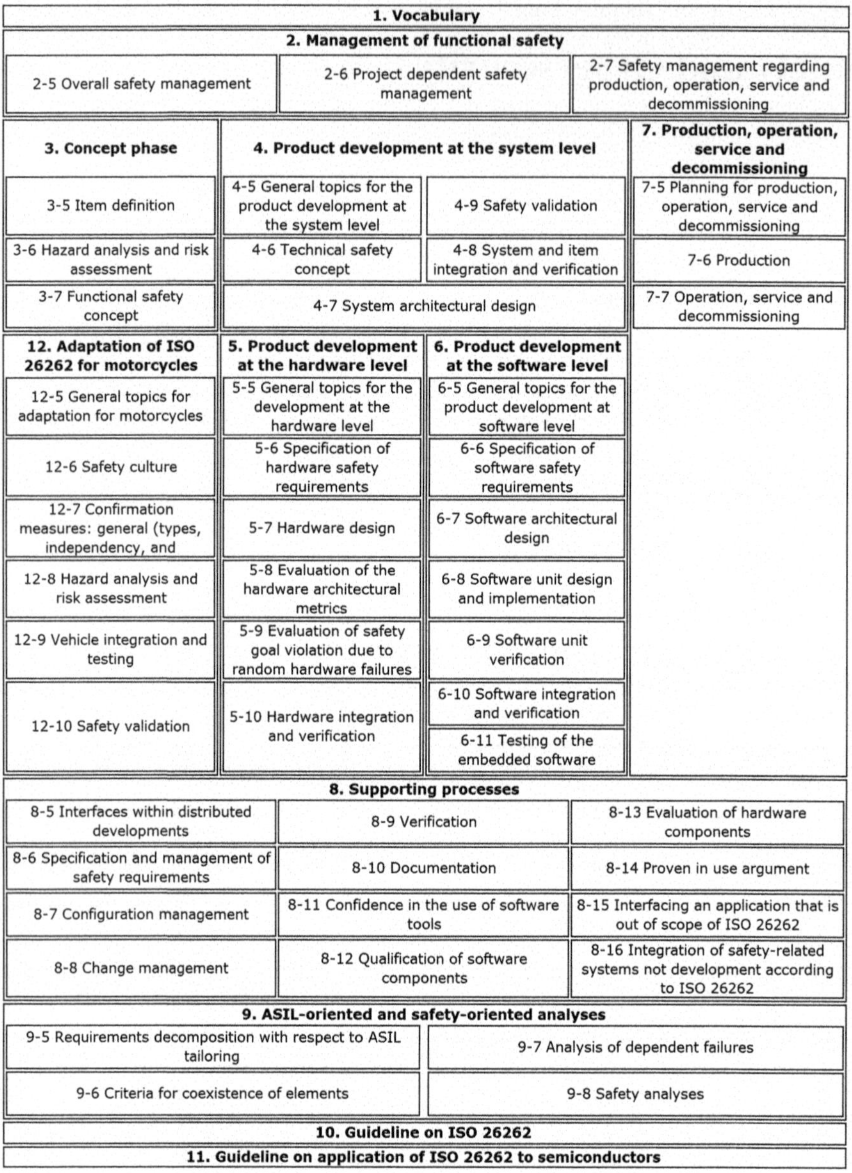

Fig. 4.1 Structure of standard ISO 26262

(2) The autonomous vehicle will stop if a communication error is detected between the components of the autonomous driving system.
(3) The autonomous vehicle will stop if the power supply to the autonomous driving system is interrupted.
(4) The autonomous vehicle will stop if the autonomous driving system detects an anomaly.

ISO 26262 is a risk-based safety standard in which the risk of some hazardous operational situations is qualitatively assessed, and safety measures are defined to prevent or control systematic failures and to detect or control random hardware failures or to reduce their effects. The objectives of the standard are [1]:

(1) Defining the life cycle from the point of view of car safety (management, development, production, operation, service, decommissioning) and supports the adaptation of the necessary activities during these phases of the life cycle.
(2) Covering the functional safety aspects of the entire development process (requirement specification, design, implementation, integration, verification, validation, and configuration).
(3) Defining a risk-based approach specific to the automotive field for determining the ASIL risk classes.
(4) Using the ASIL risk classes to specify safety requirements for achieving an acceptable residual risk.
(5) Defining the requirements for validation and confirmation measures to ensure a sufficient and acceptable level of safety.

ISO 26262 is intended for safety systems which include one or more E/E Architecture, and which are installed in standard road vehicles. The provisions of the standard address the potential hazards caused by malfunctions of the E/E behavior regarding the safety systems, including the interaction between these systems. The structure of the standard consists of the following parts [1]:

Part 1 (Vocabulary) defines the language of ISO 26262: terms, abbreviations, acronyms, vocabulary of terms used in the ISO 26262 series of standards.

Part 2 (Management of Functional Safety) is a guide that focuses on managing safety requirements, both from a project and organizational point of view.

Part 3 (Concept Phase) defines the conceptual phase. This step refers to the initial definition of the project, the establishment of safety requirements and criteria for the project and the initiation of the functional safety cycle.

Part 4 (Product Development at the System Level) detailed requirements analysis, system synthesis, functional and logical allocation, system evaluation, validation, and verification of functional safety integrity levels.

Part 5 (Product Development at the Hardware Level) hardware aspects of system design and implementation for functional safety.

Part 6 (Product Development at the Software Level) software aspects of the design and implementation of the system for functional safety.

Part 7 (Production, Operation, Service, and Decommissioning) detailed requirements for production, operation, installation, service, decommissioning etc.

Part 8 (Supporting Processes) defines the requirements for development processes, management process, documentation standards, instrument classification, verification, and validation of all support processes.

Part 9 (ASIL) oriented and safety-oriented analyses provides requirements and guidance on functional safety analysis, in particular, all aspects related to ASIL oriented functional safety requirements.

Part 10 (Guidelines on ISO 26262) is a guide to the application of the ISO 26262 standard.

Part 11 (Guidelines on Application of ISO 26262 to Semiconductors) is a guide to the application of ISO 26262.

ASIL standard defines functional safety as the absence of unreasonable risk due to hazards caused by malfunctioning behavior of electrical or electronic systems. ASILs establish safety requirements based on the probability and acceptability of harm for automotive components to be compliant with ISO 26262.

Systems like airbags, anti-lock brakes, and power steering require an ASIL-D grade the highest rigor applied to safety assurance because the risks associated with their failure are the highest. On the other end of the safety spectrum, components like rear lights require only an ASIL-A grade. Head lights and brake lights generally would be ASIL-B, while cruise control would generally be ASIL-C (Fig. 4.2) [6].

ASILs are established by performing hazard analysis and risk assessment. For each electronic component in a vehicle three specific variables are measured:

(1) Severity and the type of injuries to the autonomous driving system and the passengers.
(2) Exposure how often the vehicle is exposed to the hazard.

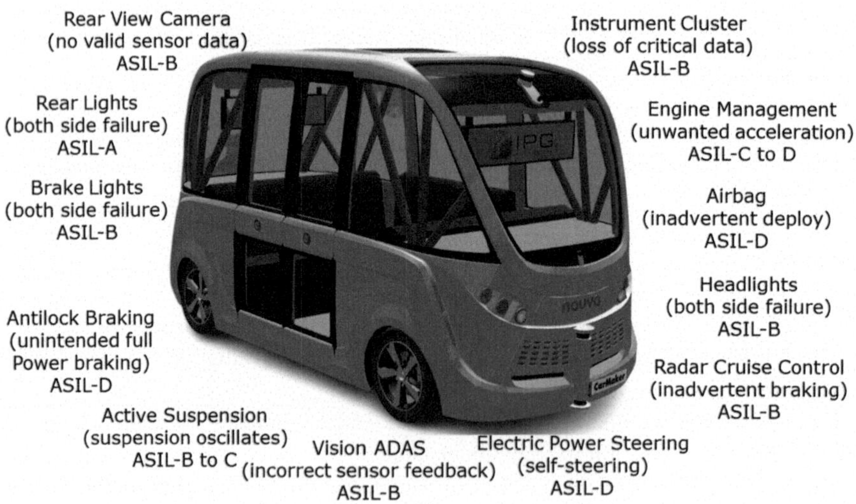

Fig. 4.2 ASIL risk classification systems

(3) Controllability how much the autonomous driving system can do to prevent the injury.

ASIL varies from the most permissive car safety integrity requirements to the strictest car safety integrity requirements. ASIL is determined by three factors, namely: severity S, exposure E and controllability C [1]:

Severity measures how severe is the damage caused by a system failure:

(S0) There are no injuries.
(S1) Mild to moderate lesions.
(S2) Serious life-threatening injuries (likely to survive).
(S3) From serious injuries with uncertainty of survival to fatal injuries.

Exposure is the probability of the conditions that a certain malfunction would result in a hazard:

(E0) It is incredibly unlikely.
(E1) Low probability (injury can only occur under operating conditions).
(E2) Low probability.
(E3) Average probability.
(E4) High probability (injury can occur in most operating conditions).

Controllability is a measure of the probability that when a dangerous condition occurs it can be avoided:

(C0) Generally controllable.
(C1) Simply controllable.
(C2) Normally controllable (most drivers could act to prevent injury).
(C3) Difficult to control or uncontrollable.

Functional safety according to ISO 26262 defines the following potential sources of errors in the operation of an ECU [4]:

(1) Requirement specification and communication:
(a) The software application runs correctly according to the specifications, but they were incorrectly defined in the scope.
(b) The specifications were misunderstood/interpreted by the automotive software developer.
(2) Software design and coding errors:
(a) Embedded software code based on a poorly developed structure.
(b) Time, syntax or algorithm error, error in displaying the errors of the results, lack of self-tests, manipulation failures.
(3) Errors due to software changes:
(a) Changes in the code of the developed software without being validated can generate errors.
(b) Configuration control process errors.
(4) Errors due to inadequate testing:
(a) The software application works correctly when testing the hardware unit, but the test fails due to the critical level of functional safety.

(5) Errors in timings:
(a) The software application works correctly when testing the hardware unit, but at the wrong time or in the wrong conditions.

Daimler, together with experts from 11 companies, leaders in the field of autonomous driving technology (Aptiv, Audi, Baidu, BMW, Continental, Fiat Chrysler, HERE, Infineon, Intel, and Volkswagen) have developed a study on the definition of basic principles of the functional safety of the functional generic architecture of autonomous vehicles entitled SaFAD (Safety First for Automated Driving) [7]. These principles are:

(A) Safe operation highlights the reaction of the autonomous driving system in case the critical components become unstable or cease to function.
(B) Safety layer highlights the mechanism by which the autonomous driving system recognizes its limits of operability and minimizes the risk by returning control to the human operator who accompanies the autonomous vehicle and who will be hereinafter referred to as the operator.
(C) ODD highlights the operating conditions under which the autonomous driving system is designed to operate.
(D) Behavior in traffic highlights the predictability and easy understanding of the behavior of the autonomous driving system for all traffic participants.
(E) User responsibility highlights the user's availability for the procedure of taking control of the autonomous driving system.
(F) Autonomous driving-initiated handover highlights the ability of the autonomous driving system to perform a maneuver that reduces any risk in the event that the operator does not respond to a request for takeover.
(G) User initiated handover highlights the activation and/or deactivation of the autonomous driving system following the request of the operator.
(H) Effect of automation highlights the effects of the degree of automation on the need for operator intervention on the functional safety of the autonomous driving system.
(I) Safety assessment highlights the verification and validation of safety objectives implemented in the autonomous driving system.
(J) Data recording highlights the recording of relevant data (in compliance with privacy legislation) in the event or an incident involving the autonomous vehicle.
(K) Security highlights the measures required to protect the autonomous driving system when there are threats to its security.
(L) Passive safety highlights the adaptation of the operation of the autonomous vehicle to the changes of the damage scenarios caused by the automation of the vehicle.

Elements necessary for the implementation of the basic principles of functional safety of the autonomous driving system consist of possible real-world situations that meet the safety capabilities of FS (Fail-Safe) in case of the appearance of degraded capabilities FD* (Fail-Degraded) (Fig. 4.3) [7]. These capabilities are:

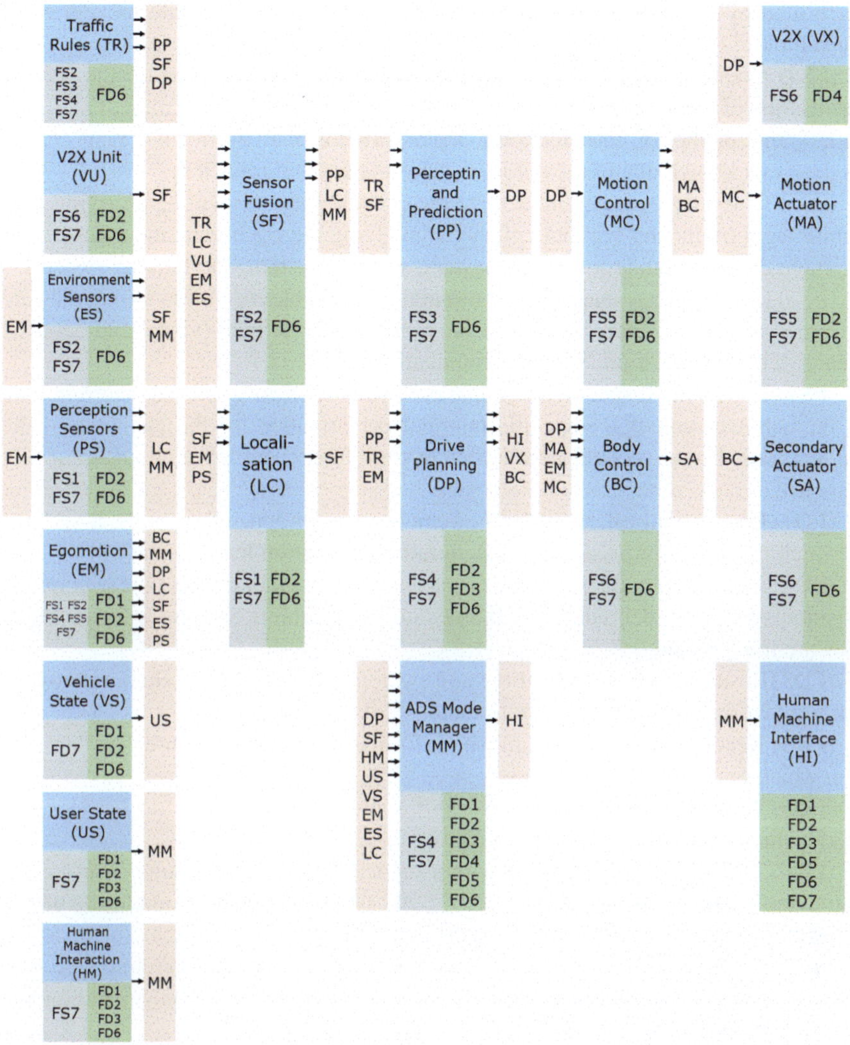

Fig. 4.3 Functional generic architecture of autonomous vehicles

(FS1) Determining the location of the autonomous vehicle based on information received from sensors on the current location, superimposed over the coordinates of the existing HD maps in the autonomous driving system.

(FS2) Perceive relevant static and dynamic objects in proximity to the automated vehicle with maximum priority over entities that present a high risk of collision based on information from sensors, V2X connections etc.

(FS3) Predict the future behavior of relevant objects in the vicinity of the autonomous vehicle, based on their dynamic evaluations. These evaluations are performed by the autonomous driving system to determine the safe movement of the vehicle.

(FS4) Create a collision-free and lawful driving plan is based on the precise localisation of the autonomous vehicle, respectively on the identification of all traffic participants, on the correct detection and interpretation of all traffic rules, respectively on the exact knowledge of the autonomous vehicle travel trajectory.

(FS5) Correctly execute and actuate the driving plan—executing a driving plan within legal limits and without collisions involving the command and control over the main drive systems (powertrain system, steering system, and braking system).

(FS6) Communicate and interact with other vulnerable road users so that they are informed about the future actions of the autonomous vehicle (visual and acoustic means of communication, V2X communications etc.).

(FS7) Determine if specified nominal performance is not achieved determination of the achievement of the specific nominal performance for the elements of the autonomous driving system in a sufficient time interval in order to ensure a safe response.

(FD1) Ensure controllability for the vehicle operator depending on the SAE level in which the autonomous driving system falls (automation levels 3 and 4).

(FD2) Detect when degraded performance is not available detect the occurrence of situations where degraded capabilities are not available and the ways in which they are implemented.

(FD3) Ensure safe mode transitions and awareness ensuring the switching between different working modes by the autonomous driving system.

(FD4) React to insufficient nominal performance and other failures via degradation which may occur.

(FD5) Reduce system performance in the presence of failure for the degraded mode which may occur.

(FD6) Perform degraded mode within reduced system constraints ensuring the automatic operation of the autonomous driving system in degraded mode at nominal capacities with new limits defined for the system.

Planned autonomous driving is coordinated by a modular architecture that includes functions of object detection, data processing, information generation and decision making (Fig. 4.4) [8].

4.1.2 Minimal Risk Operations

The safety protocols implemented in the command and control of the autonomous driving system meet the following criteria [5, 9]:

(1) The autonomous vehicle will stop if a fault is detected in any of the sensors of the autonomous driving system.
(2) The autonomous vehicle will stop if a communication error is detected between the components of the autonomous driving system.

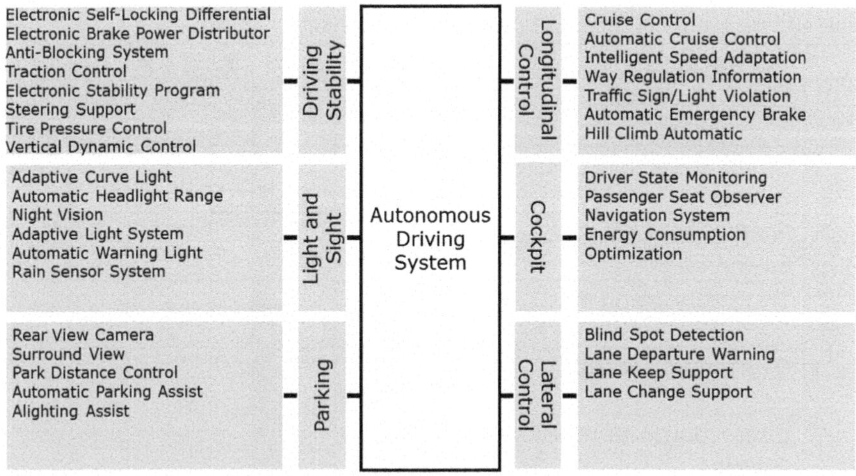

Fig. 4.4 The architecture of planned autonomous driving system

(3) The autonomous vehicle will stop if the power supply to the autonomous driving system is interrupted.
(4) The autonomous vehicle will stop if the autonomous driving system detects an anomaly.

The protocols implemented for the verification, evaluation, and validation of the functional safety of the autonomous driving system are based on a set of steps corresponding to the safety criteria and respectively to the performance criteria (Table 4.1) in which the blue color represents the active safety criteria and the yellow color represents the active performance criteria [9, 10].

These protocols aim to achieve a balance between functional safety criteria and performance criteria. Of the two types of criteria, functional safety criteria will always have priority. Thus, at any deviation from these criteria, according to the defined protocols, the autonomous driving system will command the prompt stop of the autonomous vehicle.

The steps corresponding to the safety criteria, respectively the performance criteria are the following [9–11]:

(1) The safety level of the autonomous driving system must recognize the limits for which it is permitted to switch from manual to automated driving and vice versa, under minimum risk conditions.
(2) Cyber security ensures the protection of electronic systems, communication networks, control algorithms and software products against cyber-attacks and unauthorized access.
(3) Safety assessments by verifying, evaluating, and validating the operation of all autonomous vehicle systems so as to ensure that all general safety objectives are met.

Table 4.1 The matrix of performance criteria versus safety criteria

Criteria/Steps		1	2	3	4	5	6	7	8	9	10	11
Safety criteria	Vehicle localisation											
	Object detection/classification											
	Prediction of object behavior											
	Trajectory planning											
	Overtaking control											
	Interaction with other vehicles											
Performance criteria	Maximum performance (fully autonomous)											
	Nominal performance (high automation)											
	Low performance (conditional automation)											
	Minimum performance (partial automation)											
	Nominal performance (manual)											
	Minimum performance (manual)											

(4) Safe operation by certifying that the operating condition of the autonomous driving system corresponds to a level of automation that allows the safe operation of the vehicle.

(5) Operational field by the operating conditions under which the driving system is designed to operate and which reflects the technological capacity of the autonomous driving system.

(6) The traffic behavior of the autonomous driving system must be predictable for all road traffic participants, and the traffic rules must be understood and respected by the system.

(7) Manually assisted when the autonomous driving system controls the longitudinal or lateral movement of the vehicle, a function automatically deactivated by the intervention of the operator.

(8) Partial automation when the autonomous driving system controls the longitudinal or lateral movement of the vehicle, a function automatically deactivated by the intervention of the operator.

(9) Conditional automation when the autonomous driving system is activated by the operator, and it independently controls the movement of the vehicle.
(10) High automation when autonomous driving system is activated by the operator, and it independently controls the movement of the vehicle.

Fully autonomous when the autonomous driving system is activated automatically according to the implemented autonomous driving algorithm, and it independently controls the movement of the vehicle.

4.2 Cybersecurity

4.2.1 Security Objectives

In an interconnected world such as the one we live in, ensuring the cybersecurity of the vehicle must be approached by carefully monitoring all activities on the communication channels of the internal and external networks of the vehicle.

Autonomous vehicles are equipped with hundreds of sensors and dozens of electronic command and control units, which are controlled by software code sequences and interconnected via LIN bus, CAN bus, FlexRay and MOST communication networks and Ethernet and they need to be protected against cyber-attacks through cybersecurity solutions. The autonomous vehicle is commanded and controlled by a command-and-control unit that respects the algorithm of the implemented autonomous driving system.

The future implementation of 5G technology in the management and control of autonomous vehicles will lead to an increase in the response speed of the autonomous driving system to both safety protocols and safety criteria. This requires a strict protocol on computer security, in particular as a measure to protect against unauthorized access to the communication network of the autonomous vehicle. As a protection measure, the communication with the autonomous vehicle will be possible only from the management platform through a secure connection using a VPN channel through mobile data transmission technology (4G/5G), and the information transmitted on this channel will be encrypted.

The cybersecurity control mechanism is built on the following security objectives [10, 12, 13]:

(1) Integrity:
(a) Cybersecurity at the online level ensured by managing the integrity of access rights.
(b) Cybersecurity at the vehicle level ensured by securing the communications and by separating the critical systems.
(c) Cybersecurity at the system component level ensured by controlling the integrity of the data flow.
(2) Authenticity:

(a) Cybersecurity at online level ensured by controlling access to the platform, respectively by securing communications.
(b) Cybersecurity at the vehicle level ensured by codes for authentication.
(c) Cybersecurity at the system component level provided by secured initialization sequences with public keys.
(3) Availability:
(a) Cybersecurity at online level ensured by detection and response mechanisms in case of attacks and intrusions.
(b) Cybersecurity at the vehicle level ensured by controlling the data flow in the gateway/router.
(c) Cybersecurity at component-level ensured by limiting network access points.
(4) Confidentiality:
(a) Cybersecurity at online level ensured by controlling the access to the databases.
(b) Cybersecurity at the vehicle level ensured by encrypting data flows, through SSL (Secure Sockets Layer), TSL (Transport Layer Security), IP security.
(c) Cybersecurity at component-level ensured by data encryption and storage unit security.

John D. Howard in cooperation with Longstaf in [14] classified Internet security incidents into seven categories (Fig. 4.5), which need to be reported and investigated by CERT/CC (Computer Emergency Response Team/Coordination Center).

In terms of physical access to the network and autonomous vehicle systems, the OBD port is a physical port that allows access from the outside and can become a means of initiating security incidents. Depending on the construction model (port for physical connection or port with wireless access), respectively depending on the degree of protection against unauthorized access, this port may represent a greater or lesser danger for unauthorized access. Other systems that allow physical access

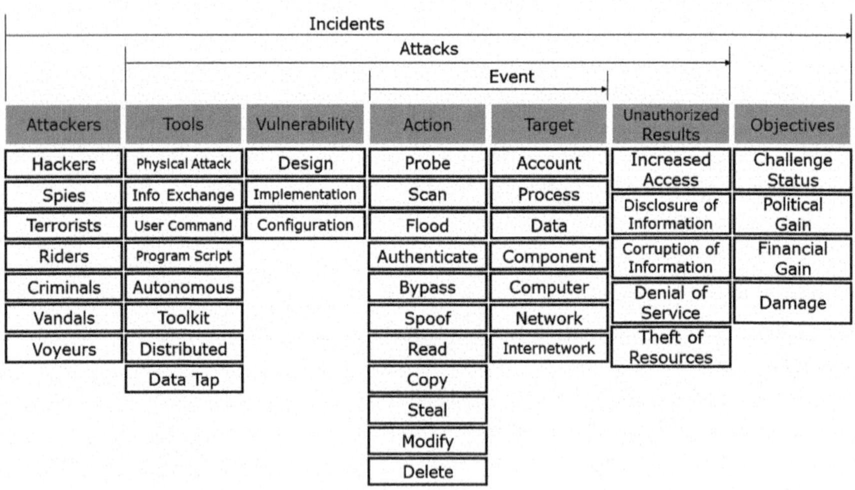

Fig. 4.5 Types of security incidents (Sommer et al.)

to the databases of an autonomous vehicle and that must be protected as such are: infotainment systems, telematics systems, TPMS (Tire Pressure Monitoring System) etc. [15].

Incidents:

(1) Attackers: hackers, spies, terrorists, corporate raiders, professional criminals, vandals, voyeurs.
(2) Objectives: challenge status, thrill, political gain, damage.
(3) Attacks:
(a) Tool: physical attack, information exchange, user command, script of program, autonomous agent, toolkit, distributed tool, data tap.
(b) Vulnerability: design, implementation, configuration.
(4) Event:
(a) Action: probe, scan, flood, authenticate, bypass, spoof, read, copy, steal, modify, delete.
(b) Target: account, process, data, component, computer, network, internetwork.
(5) Unauthorized results: increased access, disclosure of information, corruption of information, denial of service, theft of resources.

4.2.2 *Online Security*

The main access ways in the information system of an autonomous vehicle, which are vulnerable in terms of unauthorized access are presented below (Fig. 4.6) [16]:

(1) The command-and-control unit of the autonomous control system, connected through the communication networks (LIN, CAN, FlexRay, MOST, Ethernet) with the other command and control systems (vehicle access system, steering

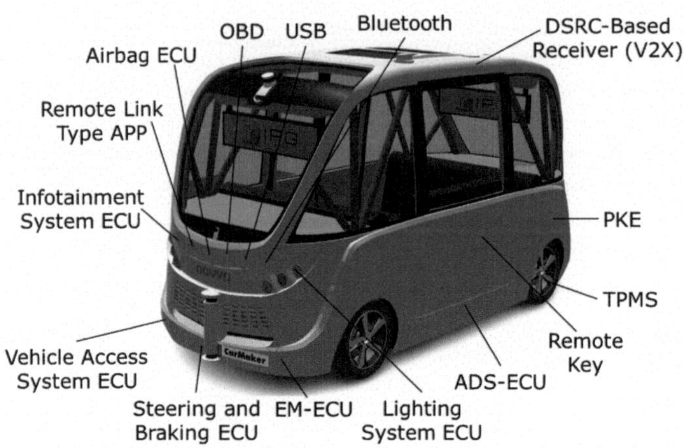

Fig. 4.6 Access ways in the information system

and braking, motor and transmission, lighting system, TPMS, keyless system, airbag etc.). The risk of unauthorized access may occur to any of these systems that can be accessed online from the outside.

(2) The OBD connector that facilitates the access to the command-and-control systems that manage information, and through the communication networks (LIN, CAN, FlexRay, MOST, Ethernet) can facilitate access to any subsystem of the autonomous driving system.
(3) The DSRC receivers that facilitate the V2X connections.

Cybersecurity at the component level of software applications compared to lifecycle management is dependent on the following criteria:

(1) Periodic update of the operating system.
(2) Installing patches for the software applications.
(3) Installing new versions for the software applications.
(4) Periodic replacement of technologically deprecated components.
(5) Periodic data backup.
(6) Periodically update encryption algorithms and keys.

An autonomous vehicle connected to various intelligent architectures (V2X: V2V, V2I, V2P) is vulnerable to various cyberattacks that can take place through online intrusions, through the following access routes [17]:

(1) Wireless communications include radio signals (AM, FM etc.), IEEE 802.11 data transmissions, V2X communications via IEEE 802.11p, and GSM communications (LTE, 4G/5G), IEEE 802.15.4 communications, or ZigBee IoT (Internet of Things). The systems of an autonomous vehicle that interact through these communication systems are vulnerable to multiple online cyber-attacks DoS (Denial of Service) attack V2X, eavesdrop, replay, MITM (Man-In-The-Middle) interception, information leak (location, identity etc.), Sybil attacks, Wormhole attacks.
(2) Physical inputs and outputs existing in the constructive architecture of an autonomous vehicle (USB, OBD, audio etc.)—by connecting to these ports one can achieve physical access to the electronic systems of the vehicle through remote connections.
(3) Internal communications on communication networks (CAN, FlexRay, MOST, Ethernet), through which electronic systems that are interconnected to these buses can be accessed and compromise their security.
(4) Internal/external sensors are very sensitive elements that have an important role in controlling the operation and movement of the autonomous vehicle, and if a cyberattack takes control of the sensors, the vehicle becomes exposed to major risks regarding the safety of its movement.
(5) Data storage—an autonomous vehicle generates a large volume of data that must be stored on internal or external media. This data contains information about firmware and software for command and control of the autonomous driving system, maps and information for navigation, multimedia files for the entertainment system etc. This data will be stored in different internal

locations of the vehicle or even transmitted to a few platforms (cloud, edge) for data storage, where they will be selectively grouped according to the source and purpose of this data. The cyberattacks that may target this data are violate integrity manipulate data, violate confidentiality extract data, violate availability delete data, remote software and/or firmware updates.

(6) Data analysis—the data provided by the sensors will be analyzed by the autonomous driving system in order to decide on the basis of their content the commands sent to the execution elements. The cyberattacks that can target the analysis results for these data are inducing a faulty analysis, copying the analysis results, DoS type attacks etc.

(7) Localisation—a very important factor in controlling the movement of an autonomous vehicle is its precise location. The localisation is made with information from the GNSS/GPS sensors and taking control of these sensors following a cyberattack can influence the correct positioning of the autonomous vehicle on the map.

(8) Object identification, static and dynamic objects detected in particular by the video camera sensor that equips the autonomous vehicle must be identified and classified. The identification and classification process are based on machine learning algorithms that are used to perform visual identification. These algorithms make the connections between a set of inputs and a set of outputs, based on a set of predefined instructions. Machine learning algorithms are subject to cyberattacks that by changing the data from the video camera sensor can change the identification and classification mechanism leading to erroneous results.

(9) Sensor's fusion to synchronize information from all data received from sensors, this data merges through a Kalman-type filter providing the autonomous control system a volume and maximum quality of recorded data. If cyberattacks occur on the data resulting from the sensors fusion, they can produce inaccurate results, followed by incorrect decisions from the autonomous driving system.

(10) Energy system supplies power to the propulsion systems of an autonomous vehicle and by connecting to the charging stations it also supplies power to the electric batteries. If the power supply system is compromised as a result of a cyberattack, then it may result in improper use (by overcharging the batteries or generating power losses during charging) which could lead to damage to the vehicle's charging system.

(11) Powertrain, steering, and braking are drive systems that can perform actions with an impact on the movement of the autonomous vehicle. If a cyberattack can take control of one of these systems, it could jeopardize the functional safety of the autonomous vehicle.

(12) Monitoring and logging are important aspects from the point of view of the scenarios of the correct operation of the autonomous vehicle, the analysis of the decisions taken in the history of operation and the management of the maintenance programs.

4.2.3 Security at the Vehicle Level

Vehicles in general and autonomous vehicles in particular are increasingly using computer systems to control and command the powertrain, steering and braking systems, respectively for the navigation, assistance, and Infotainment systems.

Thus, all new vehicles manufactured after March 31, 2018, are connected due to the obligation to equip them with the eCall system (for emergency call 112), according to Regulation (EU) 2015/758 of the European Parliament and of the Council of April 29, 2015 on approval requirements for the installation of the in-vehicle 112-based eCall system and amending Directive 2007/46/EC [18].

Cybersecurity in the case of autonomous vehicles requires compliance with the following rules [19]:

(1) Risk management based on periodically updated risk analyzes.
(2) Carrying out a compliance audit and intrusion tests before putting into service new models of autonomous vehicles.
(3) Regular upgrades to integrate cybersecurity throughout the development cycle of new autonomous vehicle models.
(4) The adoption of cybersecurity principles starting with the design stage for both software applications and equipment (components, devices, systems, vehicle).
(5) To establish operational security objectives to detect and prevent cyberattacks.
(6) Cybersecurity monitoring throughout the life cycle of the autonomous vehicle through security updates.

There are many ways in which autonomous vehicles are exposed to cyberattacks. Depending on the access point in the vehicle's computer system and the systems that can be accessed following a potential cyberattack, the vehicle vulnerability degree can be classified as low, moderate, and high (Fig. 4.7).

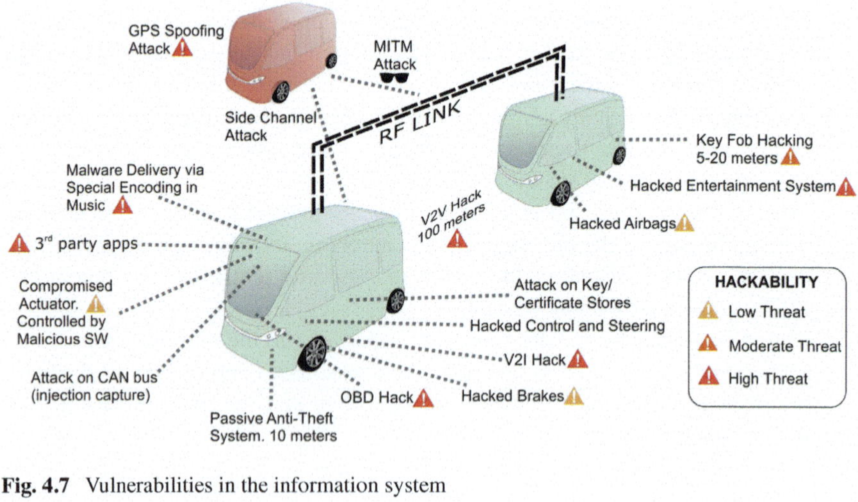

Fig. 4.7 Vulnerabilities in the information system

In order to prevent these attacks, cybersecurity at the vehicle level must meet the following common requirements [20]:

(1) Confidentiality is the provision of access to an entity's proprietary information only on the basis of a security certificate, that no unauthorized user can access the vehicle's control systems, or the data stored by it.
(2) Integrity consists in ensuring the transmission of a message to a receiving node so that it cannot be modified to guarantee a secure transmission of proprietary information. The receiving node must be able to verify and confirm whether the data received is corrupt or legitimate.
(3) Availability consists in ensuring the access and authorized use of proprietary information, in strict compliance with the conditions of confidentiality and integrity of this information.
4) Authenticity consists in ensuring the possibility of verifying and confirming the identity that a user claims to have.
(5) Non-repudiation consists in ensuring the necessary measures so that after sending/receiving information through a secure communication system, the sender/recipient cannot deny (falsely) that he/she sent/received that information.
(6) Real-time constraints consist in ensuring the transmission of information in real time in order to comply with some deadlines.
(7) Flexibility consists in ensuring the characteristics of the secure communication system to accept structural and modular changes according to the changes of the properties of the environment in which the autonomous vehicle evolves.

Communication networks that are part of the constructive architecture of an autonomous vehicle are used to transmit large volumes of data and they allow access to and control over electronic systems that play a key role in the operation of this vehicle. The most important communication networks that are found in the structure of an autonomous vehicle are the following: LIN, CAN, FlexRay, MOST, Ethernet. The main vulnerabilities (exploitation) that can be used to take control over the command-and-control systems of an autonomous vehicle through communication networks are presented below [20].

4.2.3.1 LIN Bus System Exploitation

LIN is a communication network that has been introduced as a low-cost alternative to CAN, used to command and control systems for which packet data transmission time is not critical. The network uses a single wire on which data packets are transmitted at a maximum speed of 20 kbps. The main vulnerabilities (exploitation) that can be used to take control of a command-and-control systems of an autonomous vehicle through the LIN communication network are [15]:

(1) Message spoofing attacks the situation in which an identification is made using a false identity, for the transmission of a message (data packet), to a targeted system within the autonomous vehicle, and this system accepts the reception and access of the transmitted message.

(2) Response collision attacks.
(3) Header collision attacks.
(4) Countermeasures in order to prevent the attacks described above.

4.2.3.2 CAN Bus System Exploitation

CAN is a network that uses a communication protocol with low cost and reliability, based on a network of master/slave nodes that triggers the transmission of data packets on the bus based on some events with communication speeds up to 1 Mbps (up to 12 Mbps for CAN FD) and generally connects the electronic units that are responsible for the command and control of the powertrain, steering and braking systems (but not limited to these systems).

The CAN protocol uses a packet data broadcast feature, which means that when a node transmits a packet, it will reach all the other nodes in the network. The transmitted packets do not contain identification data of the sending nodes or of the destination nodes, each node will decide for itself according to the data content how to proceed with the received data. Each data packet transmitted in the CAN bus has an ID, which is a unique number that indicates the priority of data packet delivery. The lower the ID values, the higher the transmission priority. This feature of the CAN protocol can be exploited by generating DoS attacks on the communication network [15].

The main vulnerabilities (exploitation) that can be used to take control of the command-and-control systems of an autonomous vehicle through the CAN communication network are [15]:

(1) Masquerading attacks that occur by using a false identity to gain unauthorized access to the command-and-control systems of a vehicle by identification based on legitimate access. Masked attacks are performed either by using valid, stolen authentication data, or by locating vulnerabilities in the authentication process.
(II) Eavesdropping attack which is information theft and occurs when data is transmitted over an unsecured network. Eavesdropping attacks are passive, a situation in which data transfer is only monitored through malware introduced in a device connected to the network, which has the role of recording data traffic on the network and directing it to the attacker's servers, or active, situation in which data packets can be modified before being delivered to the recipient, an attack also called MITM.
(3) Injection attacks that occur by providing a trusted entry to a program, which is interpreted as part of a command or query that modifies the execution code of that program. Injection attack is considered to be a major problem especially for web security.
(4) Replay attacks that occur by intercepting the transmission of data packets, stopping them, or redirecting them to the wrong address. Replay attack is based on stealing legitimate information about the user's login credentials, then logging in instead of the user, making the system believe that the user is authentic.

(5) The DoS attacks that have the role of stopping the activity in a network, making it inaccessible to its users by targeted flooding of the network with a very large amount of information that triggers the blocking of traffic within the network.
(6) Bus-off attack are DoS type attacks that use the error handling mechanism of packet data on CAN bus to send cyclical erroneous messages that increment the transmission error counter for each transmitted message, thus blocking the bus.

4.2.3.3 FlexRay System Exploitation

FlexRay is a communication network designed for the transmission of data packets on two channels with a communication speed of 10 Mbps redundantly, respectively 20 Mbps non-redundantly.

The main vulnerabilities (exploitation) that can be used to take control of the command-and-control systems of an autonomous vehicle through the FlexRay communication network are [15]:

(1) Static segment attacks of the transmitted message.
(2) Eavesdropping attacks (see 4.2.3.2).

4.2.3.4 MOST System Exploitation

MOST is a high-speed network developed based on a communication protocol for the transmission of large data packets via fiber optics between nodes in the Infotainment system. A peculiarity of the MOST protocol is the fact that each data packet transmitted in the MOST network contains both the address of the sending node and the address of the destination node, a feature that is a disadvantage in case of a cyberattack.

The main vulnerabilities (exploitation) that can be used to take control of the command-and-control systems of an autonomous vehicle through the MOST communication network are [15]:

(1) Synchronization disruption attack are packet-based time synchronization attacks.
(2) Jamming attacks are attacks that block network communication.

4.2.3.5 Ethernet System Exploitation

The main vulnerabilities (exploitation) that can be used to take control of the command-and-control systems of an autonomous vehicle via the Ethernet communication network are [15]:

(1) Network security attacks on the communication network.
(2) Traffic confidentiality attacks.
(3) Traffic integrity attacks.
(4) DoS attack (see 4.2.3.2).

4.2.4 Security at the Component Level

Computer security at the component level of software applications related to lifecycle management is dependent on the following criteria:

(1) Periodic update of the operating system.
(2) Installing patches for software applications.
(3) Installing new versions for the software applications.
(4) Periodic replacement of technologically deprecated components.
(5) Periodic data backup.
(6) Periodically update encryption algorithms and keys.

Because all software products are vulnerable to cyberattacks during updates, when certain ports are open outside the system, and some software modules are not working, it is recommended that updates be performed using an external drive, without any active internet connection.

Petit et al. in [21] investigated the potential cyberattacks specific to the main components of autonomous vehicles, highlighting the main vulnerabilities due to the possibilities of access to vehicle systems through communication networks and potential connection points to these networks (OBD, USB (Universal Serial Bus), DSRC, wireless connections etc.).

4.2.4.1 Lidar Attack

The Lidar sensor can be attacked by blocking its operation (jamming), an attack that has very high chances of success and it does not require physical access to the sensor; the possibilities of detection by the operator are low, but the possibilities of detection by the autonomous control system are high [21].

The success probability of the attack can be estimated as an average, and the direct consequences resulting from the attack consist in stopping the operation of the Lidar sensor (turn off Lidar, degraded mode), which creates situations of loss of orientation for the autonomous vehicle.

The technique that can be used to counteract the attack and to reduce losses is to filter the signals, and if the situation requires, the use of other source of data to replace or deactivate the compromised sensor.

Another way of cyberattacking the Lidar sensor can be the false induction of static or dynamic objects in traffic by simulating smart materials (absorbent or reflective of Lidar signals), an attack that has very high chances of success and does not require physical access to the sensor; the possibilities of detection by the operator are average and the possibilities of detection by the autonomous driving system are average [21].

The success probability of the attack can be estimated as an average, and the direct consequences resulting from the attack consist in the false detection of static and dynamic objects, which creates situations of traffic disturbance.

The technique that can be used to counteract the attack and to reduce losses is to filter the signals, and if the situation requires, the use of other sources of data to replace or deactivate the compromised sensor.

4.2.4.2 Radar Attack

The radar sensor can be attacked by misleading it, an attack that has average chances of success and it does not require physical access to the sensor; the possibilities of detection by the operator are medium, but the possibilities of detection by the autonomous driving system are high [21].

The success probability of the attack can be estimated as an average, and the direct consequences resulting from the attack are the degradation of the functioning of the radar sensor (degraded mode), which creates situations of traffic disturbance.

The technique that can be used to counteract the attack and to reduce losses consists in filtering the signals, and if the situation requires, the use of other sources of data to replace or deactivate the compromised sensor.

Another way of cyberattacking the radar sensor can be the false induction of static or dynamic objects in traffic by simulating smart materials (non-reflective surfaces for radar signals, which hide objects in traffic (invisible object)), an attack that has low chances of success and it does not require physical access to the sensor; the chances of detection by the operator are average and the chances of detection by the autonomous driving system are low [21].

The success probability of the attack can be estimated as an average, and the direct consequences resulting from the attack consist in not detecting the reflection of radar waves of static and dynamic objects in traffic (no detection of surroundings), which can lead to traffic collisions.

The technique that can be used to counteract the attack and to reduce losses is to use other sources of data to replace or deactivate the compromised sensor.

The third way of cyberattacking the radar sensor can be the false induction of a blockage of static or dynamic objects in traffic (jamming (saturation with noise)), attack that has high chances of success and it does not require physical access to the sensor; the possibilities of detection by the operator is low, but the possibilities of detection by the autonomous driving system are very high [21].

The success probability of the attack can be estimated as an average, and the direct consequences resulting from the attack consist in stopping the operation of the radar sensor (turn off radar, degraded mode) which creates situations of traffic disturbance.

The technique that can be used to counteract the attack and to reduce losses is to filter the signals, and if the situation requires, the use of other sources of data to replace or deactivate the compromised sensor.

The fourth way of cyberattacking the radar sensor can be the false induction of a non-existent vehicle in traffic (ghost vehicle (signal repeater)), attack that has high chances of success and it does not require physical access to the sensor; the possibilities of detection by the operator are average and the possibilities of detection by the autonomous driving system are average [21].

The success probability of the attack can be estimated as an average, and the direct consequences resulting from the attack consist in the false detection of static and dynamic objects, which creates situations of traffic disturbance. The technique that can be used to counteract the attack and to reduce losses consists in filtering the signals, and if the situation requires, the use of other sources of data to replace or deactivate the compromised sensor.

4.2.4.3 Vision Attack

The video camera sensor (vision) can be attacked by reducing the visibility to zero (blind), by stopping the flow of information from the sensor—attack that has high chances of success, and it does not require physical access to the sensor; the possibilities of detection by the operator are average, but the possibilities of detection by the autonomous driving system are high [21].

The success probability of the attack can be estimated as high, and the direct consequences of the attack are the degradation of the camera sensor (degraded mode), which creates disturbances to the proper functioning of the autonomous driving system (autonomous driving system disturbance). The technique that can be used to counteract the attack and to reduce losses consists in equipping the autonomous vehicle with a larger number of video cameras connected independently to the autonomous driving system and located at different angles on the vehicle (multiple cameras with different angles) and if necessary, isolate and deactivate the compromised video camera.

Another way to attack the video camera sensor (vision) can be by reducing the visibility to zero (blind), using streams of virtual information, false others source of information—an attack that has high chances of success and it does not require physical access on the sensor; the detection possibilities by the operator are average, but the detection possibilities by the autonomous control system are high [21].

The success probability of the attack can be estimated as high, and the direct consequences of the attack are to turn off the camera, which creates disruptions to the proper functioning of the autonomous driving system (autonomous driving system disturbance).

The technique that can be used to counteract the attack and to reduce losses consists in equipping the autonomous vehicle with a larger number of video cameras connected independently to the autonomous driving system and located at different angles on the vehicle (multiple cameras with different angles), isolate and deactivate the compromised video camera.

The third way to attack the video camera sensor (vision) can be by inducing false images of fictitious situations in the road traffic (fake pictures with emergency brake light etc.), using virtual information flows, false other sources of information—the attack that has a low chance of success and it does not require physical access to the sensor; the possibilities of detection by the operator are average, but the possibilities of detection by the autonomous control system are low [21].

The success probability of the attack can be estimated as an average, and the direct consequences resulting from the attack consist of false reactions from the autonomous driving system (fake reaction), which creates situations of disruption of the proper functioning of the autonomous vehicle (autonomous vehicle disturbance).

The technique that can be used to counteract the attack and to reduce losses is to use other sources of data to replace the compromised sensor and, if necessary, isolate and deactivate the compromised video camera.

Another way of attacking the video camera sensor (vision) that recognizes the traffic signs and consists in changing their interpretation (change sign (fake, irrelevant))—attack that has low chances of success, and it does not require physical access to the sensor; the possibilities of detection by the operator are high, but the possibilities of detection by the autonomous driving system are low [21].

The success probability of the attack can be estimated as low to medium, and the direct consequences resulting from the attack consist of false reactions from the autonomous driving system (fake reaction), which creates situations of traffic disturbance.

The technique that can be used to counteract the attack and to reduce losses is to create a digital map of traffic signs on the infrastructure to be updated in case of any changes, respectively reporting to the operator accompanying the vehicle in case of observation of any anomalies in terms of traffic signs.

Also, in the category of recognizing traffic signs, another way to attack the video camera sensor (vision) is to delete information about certain traffic signs that have a key impact on the safe conduct of road traffic, such as the sign for STOP (remove sign)—attack that has a high chance of success and it does not require physical access to the sensor; the possibilities of detection by the operator are high, but the possibilities of detection by the autonomous driving system are low [21].

The successful probability of the attack can be estimated as low to medium, and the direct consequences resulting from the attack are the lack of reactions from the autonomous driving system (fake reaction), which creates situations of traffic disturbance, with the possibility of crash hazards. The technique that can be used to counteract the attack and reduce losses is to create a digital map of traffic signs on the infrastructure to be updated in case of any changes, respectively reporting to the operator accompanying the vehicle in case of any anomalies regarding the signs traffic.

4.2.4.4 GPS/GNSS Attack

The GPS/GNSS sensor can be attacked by blocking its operation (jamming), an attack that has very high success chances and it does not require physical access to the sensor; the possibilities of detection by the operator are low, but the possibilities of detection by the system autonomous leadership are from medium to high [21].

The success probability of the attack can be estimated as high, and the direct consequences resulting from the attack are the inaccurate location of the autonomous vehicle (no accurate positioning information available), which requires the need to stop the autonomous vehicle if there is no possibility for the localisation of its position from other secure sources (need to stop autonomous vehicle unless other location source available).

The technique that can be used to counteract the attack and to reduce losses is to use GPS anti-jam techniques, respectively a precision IMU system as a source of information on the localisation of the autonomous vehicle (high quality IMU).

Another way of cyberattacking the GPS/GNSS sensor can be the false induction of data on the localisation of the autonomous vehicle (spoofing)—attack that has very high chances of success and it does not require physical access to the sensor; the possibilities of detection by the operator are low and the chances of detection by the autonomous driving system are medium [21].

The success probability of the attack can be estimated as high, and the direct consequences resulting from the attack consist in the wrong positioning of the autonomous vehicle, which creates situations of traffic disturbance or even the possibility of crash hazards.

The technique that can be used to counteract the attack and to reduce losses consists in authentication to the source that transmits the localisation data (satellites, zonal relay).

Another way to attack the GPS/GNSS sensor that transmits information on the location of the autonomous vehicle in the digital HD map is to modify the delimitation of the route—attack that has a low chance of success, and it does not require physical access to the sensor; the possibilities of detection by the operator are average, but the possibilities of detection by the autonomous system are low [21].

The success probability of the attack can be estimated as low, and the direct consequences resulting from the attack consist in the autonomous vehicle travelling on the wrong direction of travel, which creates situations of traffic disturbance (traffic disturbance).

The technique that can be used to counteract the attack and to reduce losses is to report to the operator accompanying the vehicle in case of any anomalies regarding the traffic signs (user reporting).

4.2.4.5 IMU Attack

The IMU sensor can be attacked by blocking its operation by inducing a magnetic flux (magnetic attack)—an attack that has very high chances of success, but it requires physical access to the sensor; the possibilities of detection by the operator are low and also the possibilities of detection by the autonomous driving system are low [21].

The success probability of the attack can be estimated as an average, and the direct consequences resulting from the attack consist in the inaccurate localisation of the autonomous vehicle (wrong position navigation), which creates situations of traffic disturbance.

The technique that can be used to counteract the attack and reduce losses is to use other sources of data to replace the compromised sensor.

The second way of cyberattacking on the IMU sensor can be launched by a local thermal overheating effect on the gyroscope (thermal attack of gyroscope), an attack that has average chances of success, but it requires physical access to the sensor; the possibilities of detection by the operator are low and the possibilities of detection by the autonomous driving system are also low [21].

The success probability of the attack can be estimated as low, and the direct consequences of the attack are the inaccurate localisation of the autonomous vehicle (wrong position navigation), which creates situations of traffic disturbance.

The technique that can be used to counteract the attack and to reduce losses is to use a casing that provides additional thermal protection, or to use other sources of data to replace the compromised sensor.

Another way of attacking the IMU sensor that records the velocity of the autonomous vehicle, is to change the velocity value, making it impossible to set this velocity (alter (change speed) make it unreadable)—attack that has a high chance of success, and it does not require physical access to the sensor; the possibilities of detection by the operator are high, but the possibilities of detection by the autonomous driving system are low [21].

The success probability of the attack can be estimated as low to medium, and the direct consequences resulting from the attack consist of false reactions from the autonomous driving system (fake reaction), which creates situations of traffic disturbance.

The technique that can be used to counteract the attack and to reduce losses consists in creating a digital map of the traffic rules, respectively the reporting to the operator accompanying the vehicle in case of observing any anomalies regarding the compliance with the traffic rules.

4.2.4.6 In-Vehicle Device Attack

Any of the electronic command and control units that are part of the E/E architecture of the autonomous vehicle can be attacked by a cyberattack, which can lead to its malfunction. One way to attack an electronic unit is by injecting a malware code (inject malware), an attack that has average chances of success, but it requires physical access to the communication bus to which the drive is connected (USB, OBD, wireless etc.); the possibilities of detection by the operator are low and the possibilities of detection by the autonomous driving system are average [21].

The success probability of the attack can be estimated as an average, and the direct consequences of the attack depend on the destructive capabilities of the malware, the resulting risks also depend on the destructive capabilities of the malware code.

The technique that can be used to counteract the attack and to reduce losses consists in separating the powertrain, steering, and braking systems from the infotainment systems by separating the communication buses (separation infotainment/safety buses), respectively detecting intrusions in the system by using antivirus-type solutions, respectively a firewall (intrusion detection system, antivirus, firewall).

Another type of attack is by direct attack—autonomous driving system head unit attack, an attack that has average chances of success, but it requires physical access to the communication bus to which the unit is connected (USB, OBD, wireless etc.); the possibilities of detection by the operator are high and the possibilities of detection by the autonomous driving system are average [21].

The success probability of the attack can be estimated as an average, and the direct consequences resulting from the attack may consist of the display of unexpected information, and the direct consequences resulting from the attack consist of situations of traffic disturbance.

The technique that can be used to counteract the attack and to reduce losses consists in separating the powertrain, steering, and braking systems from the infotainment systems by separating the communication buses (separation infotainment/safety buses), respectively detecting intrusions in the system by using antivirus-type solutions, respectively a firewall (intrusion detection system, antivirus, firewall).

References

1. https://www.iso.org/obp/ui/#iso:std:iso:26262:-1:ed-1:v1:en. 20 Aug 2020
2. https://content.intland.com/blog/navya-case-study-overcoming-challenges-of-autonomous-driving-innovation. 20 Aug 2020
3. https://www.ni.com/ro-ro/innovations/white-papers/11/what-is-the-iso-26262-functional-safety-standard-.html. 20 Aug 2020
4. http://www.sesamo-project.eu/content/iso-26262-functional-safety-draft-international-standard-road-vehicles-background-status-and. 20 Aug 2020
5. https://navya.tech/shuttle/. 11 Apr 2020
6. https://www.synopsys.com/automotive/what-is-asil.html. 20 Aug 2020
7. https://www.daimler.com/documents/innovation/other/safety-first-for-automated-driving.pdf. 20 Aug 2020
8. https://www.semagarage.com/assets/pdf/advanced-driver-assistant-system-paper.pdf. 07 Sept 2020
9. Iclodean, C., Cordos, N., Varga, B.O.: Autonomous shuttle bus for public transportation: a review. Energies **13**, 2917 (2020)
10. https://newsroom.intel.com/wp-content/uploads/sites/11/2019/07/Intel-Safety-First-for-Automated-Driving.pdf. 20 Aug 2020
11. https://www.nhtsa.gov/sites/nhtsa.dot.gov/files/documents/13882-automateddrivingsystems_092618_v1a_tag.pdf. 13 Apr 2020
12. https://www.iiconsortium.org/news/joi-march-2019.htm. 20 Apr 2020
13. https://www.automotive-iq.com/content-auto-download/5bcecf6ed1d92e3b5f3ec797. 13 Apr 2020
14. Sommer, F., Dürrwang, J., Kriesten, R.: Survey and classification of automotive security attacks. Information **10**, 148 (2019)
15. Sun, J., Iqbal, S., Arabi, N.S., Zulkernine, M.: A classification of attacks to In-Vehicle Components (IVCs). Veh. Commun. **25**, 100253 (2020)
16. https://www.mcca.com/wp-content/uploads/2018/04/Autonomous-Vehicles.pdf. 06 July 2020
17. Maple, C., Bradbury, M., Le, A.T., Ghirardello, K.: A connected and autonomous vehicle reference architecture for attack surface analysis. Appl. Sci. **9**, 5101 (2019)
18. https://eur-lex.europa.eu/legal-content/EN/TXT/?uri=CELEX:02015R0758-20180331. 03 Oct 2020

19. https://www.vie-publique.fr/sites/default/files/rapport/pdf/184000287.pdf. 02 Oct 2020
20. El-Rewini, Z., Sadatsharan, K., Selvaraj, D.F., Plathottam, S.J., Ranganathan, P.: Cybersecurity challenges in vehicular communications. Veh. Commun **23**, 100214 (2020). ISSN 2214-2096. https://doi.org/10.1016/j.vehcom.2019.100214. 19 May 2022
21. Petit, J., Shladover, S.E.: Potential cyberattacks on automated vehicles. IEEE Trans Intelli Transp Syst **16**(2), 546–556 (2015)

Chapter 5
Route Specification

5.1 Local Map Generator

5.1.1 Plane Map: Construction and Optimization

Autonomous vehicles (especially autonomous shuttle buses) operate like a vehicle on rails that travels on a well-established route and stops at stations on the route. The main difference is the road, which in the case of autonomous vehicles is a virtual route, a line defined by geographical coordinates (latitude, longitude, altitude).

The virtual route is implemented in a HD digital map which is implemented in the memory of the autonomous driving system. This map is constantly updated online, and the position of the autonomous vehicle is determined by its coordinate system. The movement of the autonomous vehicle depends on the planning of the driving behavior (determined by the implemented route and the information perceived by the sensors), respectively on the generation of the optimal trajectory [1].

The digital map that defines the road network and the optimal trajectory planning can be developed with the help of an ArcGIS type application [2]. Optimal planning of the virtual route for an autonomous vehicle guarantees the safe operation of the autonomous driving system without human intervention (Fig. 5.1).

The GPS/IMU system transmits localisation information (global planning), which determines the position of the autonomous vehicle by providing data on velocity and direction of travel, respectively the global position (global path planning), relative to geographical coordinates (global reference points).

Perception manages the local position of the autonomous vehicle in relation to the position of static and dynamic objects detected and tracked in the environment (local planning). The driving behavior planning of the vehicle determines its trajectory depending on any static or dynamic objects existing on the travel route, a route defined on a GIS (Geographic Information System) map. The high-resolution GIS map is built according to the GPS coordinates of the route, and the ArcGIS analysis tool is used to determine the optimal route (optimal trajectory) relative to the current

C. Iclodean et al., *Autonomous Vehicles for Public Transportation*, Green Energy and Technology, https://doi.org/10.1007/978-3-031-14678-7_5

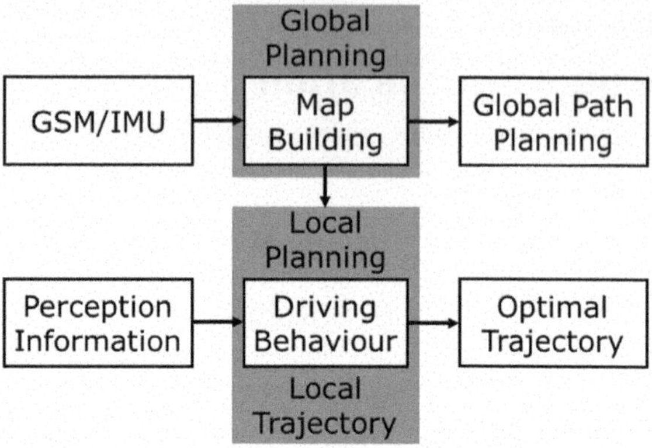

Fig. 5.1 Virtual route planning

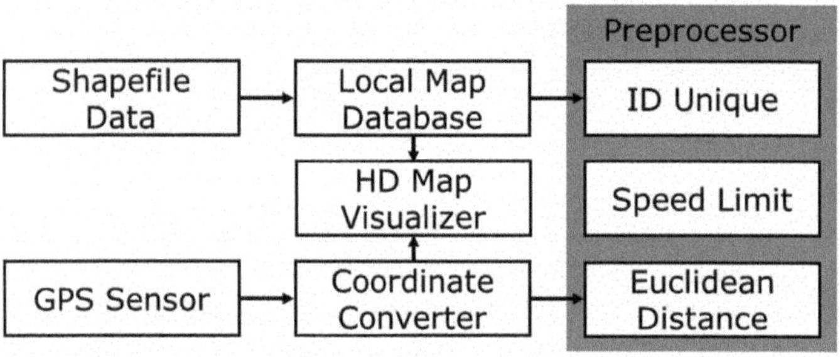

Fig. 5.2 HD map development methodology

location of the autonomous vehicle (local trajectory), obtained based on the data provided by GPS/IMU systems.

The methodology for developing the digital map in HD format is shown in Fig. 5.2 [3] and it contains the following elements:

(1) Shapefile data generator takes the structure of the sector considered from Google Maps (local map database) and converts it into UTM coordinates and based on these coordinates it generates the geometry (shapefile) of the road.
(2) Preprocessor takes the map road geometry (shapefile) and associates it with the data characteristic to the digital map:
    (a) Unique ID assigned to each item.
    (b) Speed limits characteristic for each road sector according to the legal regulations in the respective areas.
    (c) Euclidean distance used to calculate the optimal route between two travel points on the map.
    (d) GPS sensor that generates the geographical coordinates of the considered points (latitude, longitude).
    (e) Coordinate converter that converts position coordinates to UTM (grid zone) coordinates.

(f) HD map visualizer that reproduces the HD map in points in the Cloud Computer platform [3].

The general path consists of road segments defined by GPS coordinates, respectively nodes connected through edges (Fig. 5.3).

To reduce the number of points defined by the GPS coordinates, the initial data is filtered by setting a value (d) of the distance between nodes [1]. Turns, intersections or ramps are defined by GPS coordinates between two nodes: entry point and exit point, over which the path generated in ArcGIS is superimposed and stored in the GIS database (Fig. 5.4) [1].

The road network map consists of two basic elements: the edges and the nodes, which form a topographic network structure. The direction of travel on the lanes, the markings and traffic rules are established to guide the movement of the autonomous vehicle from the initial position to the destination based on the network topology. According to the methodology presented by Yu and Kong in [1] the road network map is based on the WGS-84 coordinate system, which will be converted into a geographical coordinate system: longitude (Log), latitude (Lat) and altitude (Alt),

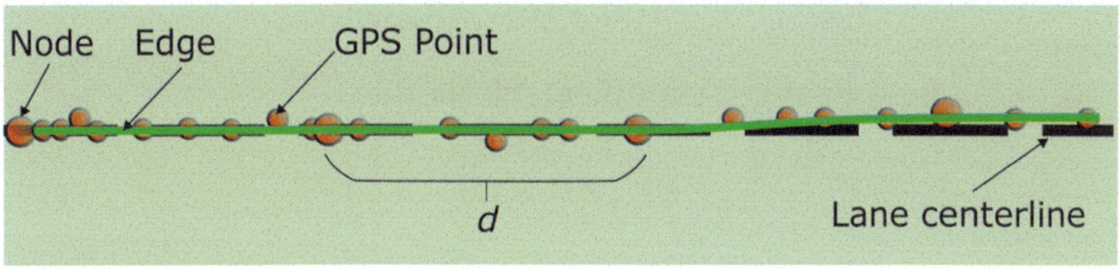

Fig. 5.3 General path

Fig. 5.4 The turns, intersections, or ramps

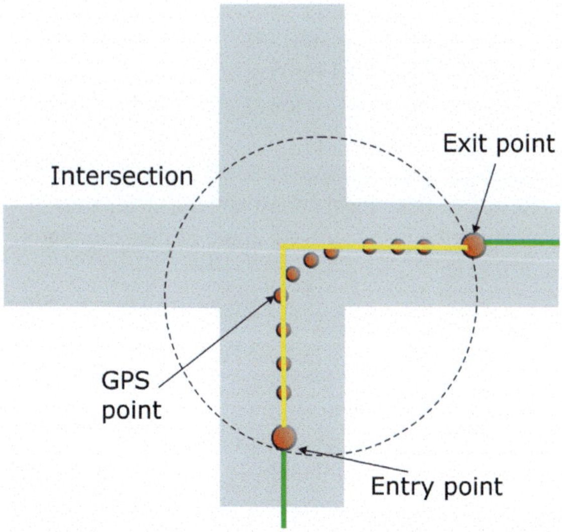

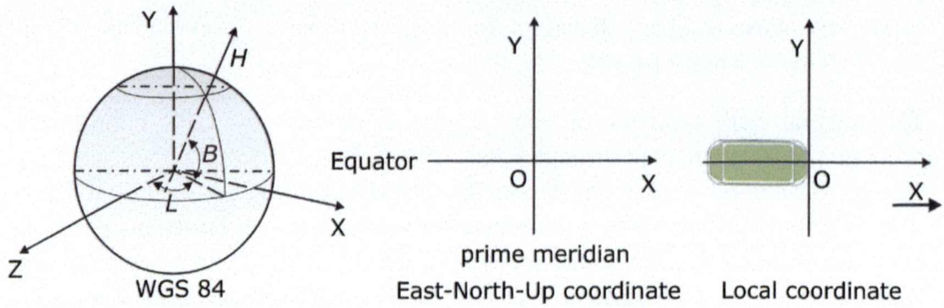

Fig. 5.5 Coordinate system conversion

which represent the GPS coordinates received from the GNSS/GPS sensors, respectively from IMU (Fig. 5.5) and which are converted into the $P(X_P, Y_P, Z_P)$ coordinate system based on the equations:

$$\begin{bmatrix} X_P \\ Y_P \\ Z_P \end{bmatrix} = \begin{bmatrix} (R_n + Alt)\cos(Log)\cos(Lat) \\ (R_n + Alt)\cos(Log)\sin(Lat) \\ \left[(R_n + H) - e^2 R_n\right]\sin B \end{bmatrix} \tag{5.1}$$

$$R_n = a^2/\left(a^2\cos^2(Log) + b^2\sin^2(Log)\right)^{1/2} \tag{5.2}$$

$$e = \left(a^2 - b^2\right)/a \tag{5.3}$$

where $R_n(m)$ is the radius of the vertical curve (the eccentricity of Earth's ellipsoid), a and b are the axes of the ellipsoid and $P(X_Q, Y_Q, Z_Q)$ are points in rectangular coordinates. The East-North-Up coordinates are given by the equation [1]:

$$\begin{bmatrix} X_{QP} \\ Y_{QP} \\ Z_{QP} \end{bmatrix} = \begin{bmatrix} -\sin(Lat) & \cos(Lat) & 0 \\ -\sin(Log)\cos(Lat) & -\sin(Log)\sin(Lat) & \cos(Log) \\ \cos(Log)\cos(Lat) & \cos(Log)\sin(Lat) & \sin(Lat) \end{bmatrix} \begin{bmatrix} X_Q - X_P \\ Y_Q - Y_P \\ Z_Q - Z_P \end{bmatrix} \tag{5.4}$$

Resulting from local coordinate transformations (x', y') that can be used for route planning (yaw is the steering angle of the autonomous vehicle):

$$\begin{aligned} x' &= x_{op}\cos yaw - y_{op}\sin yaw \\ y' &= y_{op}\cos yaw + x_{op}\sin yaw \end{aligned} \tag{5.5}$$

The SLAM technique allows the simultaneous mapping and localization within this map of the environment based on virtual reality. SLAM is a recursive estimation process, based on a probabilistic approach in the construction of a navigation map for a vehicle. In every traveled moment the probabilistic form of SLAM is calculated according to the probability distribution for the considered motion model (see Fig. 5.3) [4, 5]:

$$P\left(x_{k|k-1}, M_{k|k-1} | Z_{0:k-1}, U_{0:k}\right) = \int P\left(x_{k|k-1}, x_{k-1|k-1}, u_k\right) \cdot P\left(x_{k-1|k-1}, M_{k-1|k-1} | Z_{0:k-1}, U_{0:k-1}\right) \cdot dx_{k-1|k-1} \quad (5.6)$$

where $x_{k|k-1}$ is the state vector describing the position of the autonomous vehicle at the time (k) given the knowledge of the previous state at the time (k − 1), $M_{k|k-1}$ is the map estimated at the time (k) given the knowledge of the map characteristics at the previous time (k − 1), $Z_{0:k-1}$ is the set of observations from the time (0) to the time (k − 1), $U_{0:k}$ is the set of control inputs from the time (0) to the time (k), $x_{k-1|k-1}$ is the state vector describing the position of the autonomous vehicle at the time (k − 1) given the knowledge of the previous state at the time (k − 1), u_k is the control vector applied at time (k − 1) to move the vehicle to the state x_k, $M_{k-1|k-1}$ is the map estimated at time (k − 1) given the knowledge of the map characteristics at the previous time (k − 1), $x_{k|k}$ is the state vector describing the position of the autonomous vehicle at the time (k), $M_{k|k}$ is the estimated map at the time (k), $Z_{0:k}$ is the set of observations from the time (0) to the time (k), $Z_{i,k}$ is the set of observations for the landmark (ith) considered at the time (k), $U_{0:k-1}$ is the set of control inputs from the time (0) to the time (k − 1).

Similar to Eq. (5.1) is defined the perception model based on the data provided by the sensor regarding the detection and at the moment [5]:

$$P\left(x_{k|k}, M_{k|k} | Z_{0:k}, U_{0:k}\right) \alpha P\left(z_{i,k} | x_{k|k-1}, M_{k|k-1}\right) \cdot P\left(x_{k|k-1}, M_{k|k-1} | Z_{0:k-1}, U_{0:k}\right) \quad (5.7)$$

The autonomous vehicle movement prediction is calculated as an iterative estimate, based on a recursive Bayesian method for the implementation of SLAM by combining the relations (5.1) and (5.2) [5]:

$$P\left(x_{k|k}, M_{k|k} | Z_{0:k}, U_{0:k}\right) \alpha P\left(z_{k,i} | x_{k|k-1}, M_{k|k-1}\right) \cdot \int P\left(x_{k|k-1}, x_{k-1|k-1}, u_k\right) \cdot P\left(x_{k-1|k-1}, M_{k-1|k-1} | Z_{0:k-1}, U_{0:k-1}\right) \cdot dx_{k-1|k-1} \quad (5.8)$$

The autonomous vehicle steering control is based on the methodology applied to the MPC, which involves building a method of linearizing the vehicle model using prediction information to minimize deviation of the sidewalk tracking. The MPC is linearized by a succession of steering angles, which are obtained assuming that the vehicle will move on the set route, reaching the destination and maintaining solid state [6].

5.1.2 Grid Map: Multi-layer Generation

Generating the digital map at the grid level involves drawing a series of vertical and horizontal lines that are used to divide a map vertically and horizontally into lines, columns, edges, and so on. In the case of the autonomous vehicle, the Lidar sensor draws lines along the travel route using the following detection algorithm [7].

The data contained in the point cloud generated by the Lidar sensor marks the surface of the road to be traveled: $P = \{x_p, y_p, z_p\}$, resulting in a set of parameters $\{u_x, u_y, u_z, x_0, y_0, z_0\}$. The algorithm considered for detecting the line on which the autonomous vehicle is to move (road line detection) can be written in code as follows [7]:

```
/*  Initialization                                                        */
for k ← 1 to K ≥ 2L − 1 do
    Initialize parallel line parameters as
                    u_xk ← 0, u_yk ← 1.0, u_zk ← 0,
                    x_0k ← x_leftmost + 0.5wk, y_0k ← 0, z_0k ← 0,
                    where w is the width of the lane
end
/*  EM - like algorithm to estimate line parameters                       */
for k ← 1 to 100 do
    Expectation: Assign each points to the nearest line by calculating point
    Maximization: Update x_0k for each line as Σ_Nk x_p /N_k
    where N_k is the number of poits assigned to line k and
    Σ_Nk x_p /N_k is the sum of the x - coordinate of the assigned N_k points
end
/*  Line selection                                                        */
for k ← 1 to K do
    Calculate the final cluster size N_k for each line
        Calculate the standard deviation σ_k of the y
                    - coordinate of the assigned N_k points as
    σ_k = sqrt( Σ_Nk (y_p − ȳ_p)^2 / (N_k − 1) )
end
Set i = 0
for k ← in the order of having largest N_k do
    if σ_k > 2.5 and Δx_0 < 0.4 then
        Select k as a reliable set of line parameters
        i ← i + 1
    end
    if i == L then
        break;
    end
end
```

5.1.3 Local Map: Static and Dynamic Objects

The DTM (Digital Terrain Model) or DEM (Digital Elevation Model) is a 3D digital representation of the terrain surface, in the form of a grid, in which for each pixel that forms the map is assigned a unique elevation value. DTM presents the altitude profile of the terrain without any other objects on its surface (static or dynamic objects) [8].

The DSM (Digital Surface Model) is a 3D digital representation of the terrain surface, in the form of a grid, with all static and dynamic objects on this surface [9]. The detection of objects detected by the sensors of the autonomous vehicle involves a 3D projection of them on the terrain surface. The video camera will detect and project large objects (cars, cyclists, people etc.) on the surface, and the Lidar sensors will detect and project the edges of the road on the surface.

For the evaluation of local maps, the values for the following deviations are used, which highlight the differences between the estimated values and the determined values: RMSE, MAE, iRMSE, and iMAE, which are expressed by the following relations (where: d_{pred} are the estimated values of the deviations, and d_{gt} are the determined values of the evaluated deviations) [10]:

$$\mathrm{RMSE} = \sqrt{\frac{1}{n}\sum \left(d_{pred} - d_{gt}\right)^2} \tag{5.9}$$

$$\mathrm{MAE} = \frac{1}{n}\sum \left|d_{pred} - d_{gt}\right| \tag{5.10}$$

$$\mathrm{IRMSE} = \sqrt{\frac{1}{n}\sum \left(\frac{1}{d_{pred}} - \frac{1}{d_{gt}}\right)^2} \tag{5.11}$$

$$\mathrm{IMAE} = \frac{1}{n}\sum \left|\frac{1}{d_{pred}} - \frac{1}{d_{gt}}\right| \tag{5.12}$$

5.2 Virtual Route Digitization

5.2.1 Real Map: Coordinates of the Selected Route

The digitization process consists in transforming a real road sector into a digital map format by assigning coordinates for each registered point. The resolution of a map in vector system, represents the smallest distance between two points that is perceived by the coordinate system, the two points being different. In the raster system, the resolution of the map represents the maximum size in the terrain that corresponds to a pixel.

In the process of digitizing a map, the general path is represented by a straight line that has no sharp turns. The GPS coordinates of the actual route are imported into ArcGIS to build the road network of the selected route. The resolution of the map is given by the number of imported GPS coordinates and the distance between the nodes [1].

The virtual route is a digitized representation of a real route that is generated for the movement of the autonomous vehicle. The digital route can be generated in two ways:

(1) Combining individual road segments (straight and curved segments) to form a road sector. For each defined road sector, the following characteristics can be specified: the length, width, road angle, slope, friction coefficient etc. At the same time, friction zones, road bumps, road signs, obstacles etc. can be configured.
(2) Using digitized data based on the geographical coordinates of an existing road sector that can be exported from a web application such as Google Earth, Bing Maps, OpenStreetMap, HERE WeGo, TomTom etc.

In the first stage of the digitization of the route to be implemented for the movement of the autonomous vehicle, the global routes are determined by a high-resolution digital map based on a localisation system. Global planning methods include an APF (Artificial Potential Field), the cells that make up the considered area, the visibility graph, and the probabilistic road maps. All these methods establish the model for the environmental maps through the network of nodes and on their basis the optimal route that the autonomous vehicle has to travel is established.

Online route planning is done through the PRM (Probabilistic Roadmap Method) combined with global route planning through MOMA (Multi-Objective Memetic Algorithms). The local planning of the route contains two parts: the planning of the driving behavior and the trajectory generation. The driving behavior that guides the autonomous vehicle is determined by the global planning and perception coordinates [1].

To estimate the localization error derived from the characteristics of the environment, respectively from the elements of the road map, Wong et al. in [11] proposed the following factors for evaluating these maps for a more accurate localisation of the autonomous vehicle:

(1) Auxiliary layers that consist of view line and view polygon starts in order to model the behavior of the Lidar sensor in the localization process. For each sampling point, 360 view lines are generated, each with a length of 20 m, radiating outwards from the center at an interval of 1°. The view lines are cut where they intersect with the surface layer of static objects (buildings, walls etc.), according to the 30° (± 15°) vertical field of view of a Lidar sensor.
(2) The number of features that are positioned near the autonomous vehicle, where the main geometric features used for 2D vector mapping are points, lines, and polygons. To consider the differences in geometric representations, the number of features is calculated based on the points of intersection between layers and view lines or view polygons.

(3) The ADF (Angular Dispersion of Features) is a measure of the positioning of the characteristics of the objects around a test point according to the concept that if the characteristics of the nearby objects are scattered and unordered, they must provide better positioning accuracy in the map development. This factor describes the dispersion uniformity and is:

$$\mathrm{ADF} = \sqrt{\left(\frac{\sum \sin \alpha}{n}\right)^2 + \left(\frac{\sum \cos \alpha}{n}\right)^2} \tag{5.13}$$

where α (rad) is the azimuth of the characteristics of the objects and has a value between 0 and 1, where 0 indicates a uniform dispersion and 1 indicates a complete concentration in a certain direction.

(4) View line mean length for each sampling point—the average length of each view line that intersected with the surface of the objects near the autonomous vehicle was calculated.

(5) Area of view polygon represents the perception surface of the Lidar.

(6) Perimeter of view polygon represents the perception perimeter of the Lidar sensor. The dimensions of this perimeter depend on the number of characteristics of nearby objects that are perceived by the Lidar sensor.

(7) Compactness of view polygon is the ratio between the area of the view polygon and the area of a circle with the same radius. Low compactness is the result of a large number of characteristics of objects near the autonomous vehicle at a higher resolution, and high compactness is the result of a small number of characteristics of objects near the autonomous vehicle at a lower resolution.

5.2.1.1 OpenStreetMap

OpenStreetMap is an open-source collaborative project providing user-generated street maps of the world, using GPS traces, local knowledge, and aerial imagery (Fig. 5.6).

Features added to OpenStreetMap are modeled and stored as tagged geometric primitives (points, lines, and polygons). In certain areas, OpenStreetMap is more complete and more accurate than corresponding proprietary datasets. Because mapping data are abstract representations of the world, it can be argued that a map can never truly be complete. Nevertheless, the mapping data should be collected, specific for the required application.

To model the surrounding environment, OpenStreetMap data layers were used: (1) building polygon, (2) barrier polygon, (3) barrier line, (4) barrier point, and (5) natural point. The buildings layer was selected as they represent the most stable feature within the urban landscape. Buildings are least affected by seasonality or time of day. For vehicle localization, the use of buildings alone for map matching may be insufficient. In these scenarios, roadside features become increasingly important for localization. Within OpenStreetMap, these are classified under the barriers tag [11].

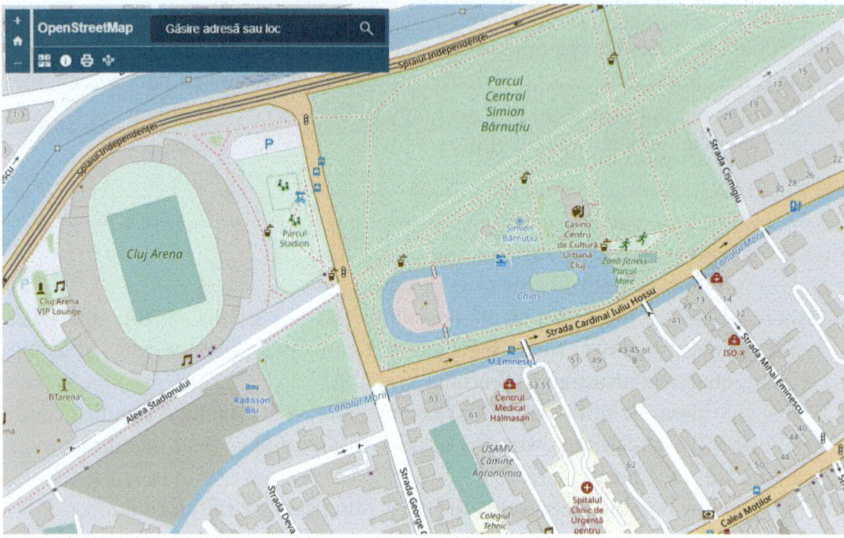

Fig. 5.6 OpenStreetMap

5.2.1.2 Google Earth

Google Earth is a computer program, formerly known as keyhole Earth viewer, that renders a 3D representation of Earth based primarily on satellite imagery (Fig. 5.7). The program maps the Earth by superimposing satellite images, aerial photography, and GIS data onto a 3D globe, allowing users to see cities and landscapes from various angles. Users can explore the globe by entering addresses and coordinates, or by using a keyboard or mouse. Users may use the program to add their own data using KML (Keyhole Markup Language) files and uploading them through various sources. Google has revealed that Google Earth now covers more than 98% of the world and has captured 10 million miles of street view imagery, a distance that could circle the globe more than 400 times. In addition to Earth navigation, Google Earth provides a series of other tools through the desktop application, including a measure distance tool.

Google Earth engine is a Cloud Computing platform used for processing satellite imagery and other geospatial and observation data. It provides access to a large database of satellite imagery and the computational power needed to analyze those images. Google Earth engine allows observation of dynamic changes in agriculture, natural resources, and climate using geospatial data from the Landsat satellite program, which passes over the same places on the Earth every sixteen days. Google Earth engine has become a platform that makes Landsat and Sentinel-2 data easily accessible to researchers in collaboration with the Google Cloud Computer storage. Google Earth engine provides a data catalog along with computers for analysis; this

Fig. 5.7 Google Earth

allows scientists to collaborate using data, algorithms, and visualizations. Google Earth engine has been used multiple times as a tool for tracking deforestation [12].

5.2.1.3 Bing Maps

Bing Maps (previously Live Search Maps, Windows Live Maps, Windows Live Local, and MSN Virtual Earth) is a web mapping service provided as a part of Microsoft's Bing suite of search engines and powered by the Bing Maps for Enterprise framework. Bing Maps was a continuation of previous Microsoft technologies such as Microsoft MapPoint and TerraServer. Its original stand out feature was the aerial imagery. The original version lacked many of its distinguishing features, including birds' eye view and 3D maps (Fig. 5.8).

Users can browse and search topographically shaded street maps for many cities worldwide. Maps include certain points of interest built in, such as metro stations, stadiums, hospitals, and other facilities. It is also possible to browse public user-created points of interest. Searches can cover public collections, businesses or a certain type of business, locations, or people. Five street map views are available: road view, aerial view, bird's eye view, street side view, and 3D view.

Road view is the default map view, and it displays vector imagery of roads, buildings, and geography. The data from which the default road map is rendered is licensed from Navteq. In certain parts of the world, road view maps from alternative data providers are also available. A Bing Maps app is available that will display road data from OpenStreetMap [13].

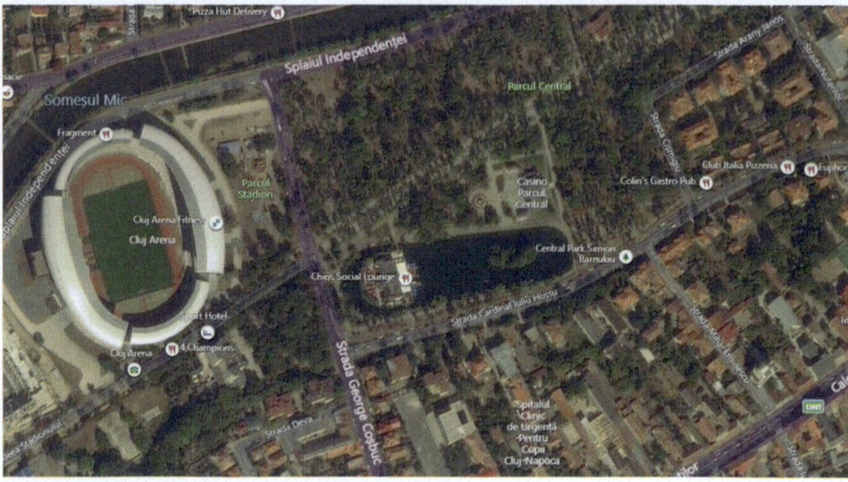

Fig. 5.8 Bing Maps

5.2.1.4 HERE Maps

HERE WeGo is a web mapping and navigation service, operated by Here Technologies (Fig. 5.9). The data is provided in-house by HERE, which includes satellite views, traffic data, and other location services. HERE WeGo provides turn-by-turn navigation in both offline and online modes. Users can enter a destination address, landmark, or business name, and then the app automatically calculates directions and distance to the destination. Real-time traffic data is also factored in, and a prediction is made on the arrival time to the destination.

Public transportation data for transport options such as buses, trams, subways, and trains are displayed by the Here app when planning a route and the user is online.

HERE WeGo had real-time information on traffic conditions and incidents available for many countries, and real-time traffic for reversible express lanes across the United States and Europe.

HERE WeGo can download maps of countries to internal storage so that a data connection is not needed for navigation. Online usage provides live traffic data and additional venue information. The maps include floor-by-floor details of some large indoor venues such as shopping malls and airports. If any vehicle is equipped with a CarPlay connectivity component, HERE WeGo application can be projected to the vehicle in-dash navigation screen and controlled as any other built-in application [14].

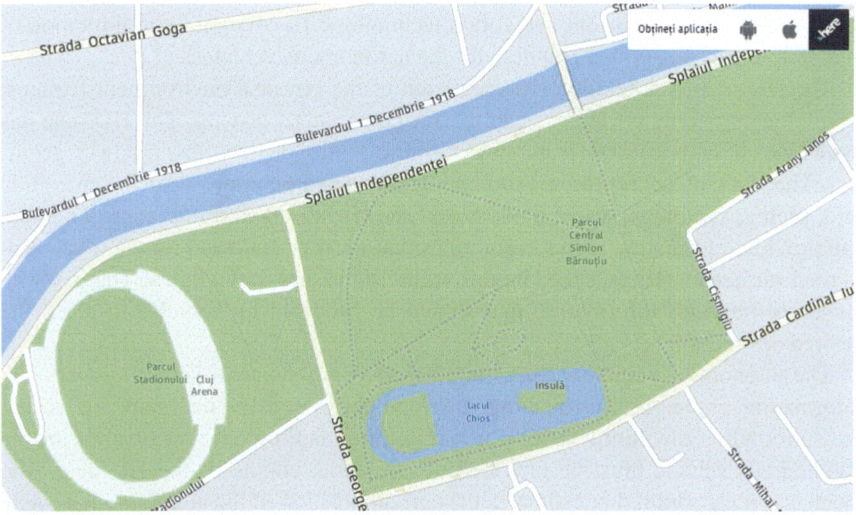

Fig. 5.9 HERE WeGo maps

5.2.2 Full 3D Map: Map of the Environment

HD map is a detailed representation of the characteristics of the physical environment, which autonomous vehicles can use for autonomous driving. HD map contains information on the following content details: static objects, road traffic control devices and road geometry.

Static objects consist of entities that have a fixed position within the map and that must be avoided by the autonomous vehicle in its movement (buildings, trees, poles, barriers etc.).

Traffic control devices provide information on traffic rules that must be followed by all traffic participants (markings, speed limits, traffic signs, traffic lights etc.).

Autonomous vehicles can use the HD map for autonomous driving by accessing information about the environment contained in the HD map. To access the information regarding the environment on the map, the autonomous vehicle must provide the map search engine with its current position and direction of travel. This requirement is met by the autonomous driving system localization module based on information provided by GPS/GNSS sensors and Lidar sensors, respectively [15].

The autonomous vehicle requires the surrounding environment information, (objects, traffic control devices, and roadway geometry) to perform the planning and control.

Perception sensors, such as cameras, radars, and lidars, can provide the information regarding the surrounding environments in real time, but the sensors have limitations for the perception range and recognition performance. The perception sensors cannot detect the physical environments that are located far from the ego-vehicle or are blocked by obstacles. Also, not all physical environments on the road can be recognized using the current perception technology. To overcome the limitation of perception, we can apply a HD map for the autonomous vehicles.

The HD map is a detailed representation of the physical environment features, which the autonomous vehicles can use for autonomous driving. The HD map has three types of map features: static object, traffic control devices, and roadway geometry. The static object represents something that the autonomous vehicle can collide into, such as buildings, walls, trees, poles, and barriers. The traffic control devices can provide information about traffic rules that must be followed on the road, such as road surface markings, speed bumps, traffic signs, and traffic lights. The roadway geometry provides information that should be followed by vehicles to reach the desired destination.

The autonomous vehicle can use the HD map for autonomous driving by accessing information about the surrounding environment stored on the map. To access the surrounding environment information on the map, the current position of the autonomous vehicle must be provided to the search engine of the HD map. The localization algorithm that estimates the current position of the autonomous vehicle is an essential component for the intelligent driving system using the HD map.

The localization can estimate the position using a DR* (Dead Reckoning) of vehicle motion sensors or GNSS. The HD map can also be used for accurate localization by aligning landmark perception with the HD map landmarks. For instance, the autonomous vehicle can estimate the position of ego-vehicle by matching the traffic sign perception from the camera and the traffic sign position data in the HD map [15].

The most significant problem regarding the accuracy of the information provided by an HD map is the change of the physical characteristics of the terrain. It can provide incorrect information about the environment, and the autonomous driving system can make decisions with a negative impact on autonomous driving in complete safety. Localization based on alignment of the landmark perception within the HD map may be affected by the degradation of the accuracy and the reliability resulted from changing the HD map details. All HD map changes must be detected and managed by the autonomous driving system in real-time and with maximum accuracy to prevent performance degradation of the autonomous driving.

For the safe operation of autonomous vehicles, the command-and-control unit of the autonomous driving system analyzes and processes the data from the sensors. Then, based on decision-making algorithms, it creates a virtual model of the environment through which the autonomous vehicle travels. This virtual travel environment reproduces the following features of the real environment [4]:

(1) Road condition and type, including road signs and signals, speed limits, road lane width, traffic conditions etc.
(2) Geographical area, longitudinal and transverse inclination of the road, altitude profile of the route etc.
(3) Climate, atmospheric conditions, brightness (day, night), air temperature, precipitation (fog, rain, snow etc.).

The reference data on the geographical coordinates (latitude, longitude, and altitude) of the selected and implemented route (local map) are stored in a memory block of the command-and-control unit and are constantly updated with data from sensors or through periodic updates. The programming of the selected and implemented route allows the autonomous vehicle to move only on this route. The autonomous vehicle is programmed by the autonomous driving system algorithm to detect the passing outside the selected route and to treat this situation as a condition of minimum risk, realizing a correction of the movement or a stop in safe conditions [4].

The orientation of the autonomous vehicle on the selected route is achieved by the following navigation techniques [4, 16, 17]:

(1) SLAM that monitors the environment using sensors and it provides the perception and localisation in real time.
(2) HD map, which overlaps the existing objects over a detailed representation of the environment.

The methodology for generating an extended map of the environment (full 3D map) is presented in Fig. 5.10. The data from the Lidar sensors generate a virtual image of the environment in which the autonomous vehicle moves, an image made up of static and dynamic objects.

An important aspect in the digitization and implementation of the selected route is the quality of the GPS/GNSS signal that ensures a safe movement of the autonomous vehicle [4, 17–19]. This aspect can be improved by transferring the GNSS/RTK differential corrections from the base stations to the autonomous vehicles. The two overlapping maps, 2D and 3D, respectively, will generate a local map that will contain all existing objects (pedestrians, vehicles etc.). This map of the environment superimposed with the map of the selected route that is implemented in the internal memory

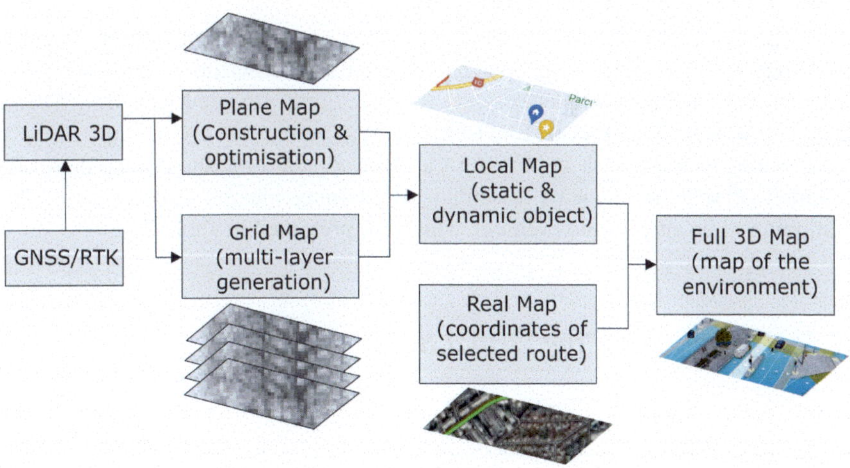

Fig. 5.10 Generation of the extended map (full 3D map)

Table 5.1 Specific elements for the development of a full 3D map

No	Name	Explanations	Type
1	Lane marking	Left–right lanes, lane color, centerline, bus lane	Line
2	Road facilities	Vehicle stop line, pedestrian stop line	Line
3	Centerline	Highway, national road, county road, local road, communal road	Line
4	Signs	Attention, regulation, indication, subsidiary sign	Point
5	Sign	Sign	Point
6	Direction of the road	Road signs, turn signals	Point
7	Guide	Left or right turn guideline	Line
8	Crosswalk	No stopping areas, guidance area, sloping road, crosswalks	Area
9	Autonomous driving	Protected areas (children, people with disabilities, elderly people etc.)	Area
10	Roadside	Roadside facilities, traffic lights and road signs	Point
11	Nodes	Nodes	Point

of the command-and-control unit of the autonomous driving system generates a full 3D map [4, 20].

GPS-based vehicle monitoring systems have a low accuracy (between 10 and 30 m), insufficient to locate an autonomous vehicle at the level of a full 3D map. The sensor systems that equip an autonomous vehicle have the role of determining its location in real time, while stationary or while moving, with maximum accuracy at the level of a full 3D map. This maximum accuracy depends on the accuracy of the RTK of the autonomous vehicle, correlated with the DOP (Dilution Of Precision), which is a calculated correction of the estimated position considering the positioning coordinates of the satellite constellation [9].

Full 3D maps for autonomous vehicles are developed based on the base maps of companies that offer public data (standard or premium subscription-based services) for localisation and other related services (Google Earth, Google Maps, Bing Maps, OpenStreetMap, HERE WeGo, TomTom).

According to Ham et al. in [16], for the development of a digital map there are a number of specific elements, essential in obtaining maximum accuracy of the positioning of an autonomous vehicle, which are presented in Table 5.1.

5.2.3 Digitization of the Virtual Road

IPGRoad is a software library used by the scenario editor to define and develop a road network. IPGRoad road networks are built according to a hierarchical structure, consisting of multiple lanes connected by junctions. The entire road network is divided into several interconnected sectors (links). These sectors are restricted by nodes arranged at each end. Interconnected sectors can be broken down into lanes separated by lane sections (Fig. 5.11) [21–23]:

(1) Road:
    (a) Junction (0,1) (road element used to join different links).
    (b) Link (0,1) (section of the road network, usually one road).
        (i) Node (0,1).
        (ii) Lanesection (0,1) (sections of a link with constant number of lanes).
        (iii) Laneleft (0,1).
        (iv) Laneright (0,1).
    (c) Route (trajectory defining the entire desired course of the vehicle travel on the road, consists of one or more paths).

All roads can be modeled to have lateral and longitudinal elevation and camber profiles. The height, slope and curvature of different, individual roads that meet at a junction are considered during the creation of the crossroads surface. Furthermore, there are various possibilities of altering the roads local surface available, like adding bumps or beams. All these components will be adequately described in the following chapters, including an expanded tree structure that will show all other components and attributes as well.

For better visualization, Fig. 5.12 shows the road network itself displayed as a graph. The nodal points represent junctions, while the directional edges display the links [23, 24].

Road network consists of all interconnected road sectors (see Fig. 5.9).

Route defines the entire route of the vehicle being made up of one or more paths.

Path is the predefined trajectory on which the vehicle moves, on the driving side established by the lane. Three types of paths can be defined while modeling a route:

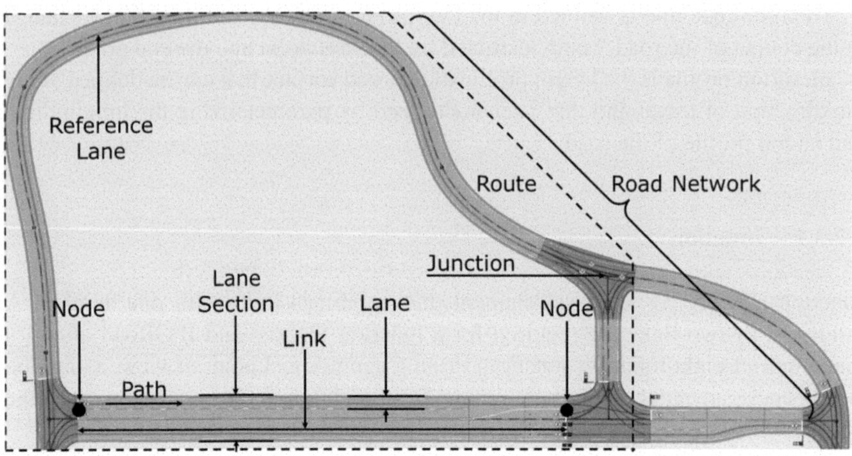

Fig. 5.11 IPGRoad elements

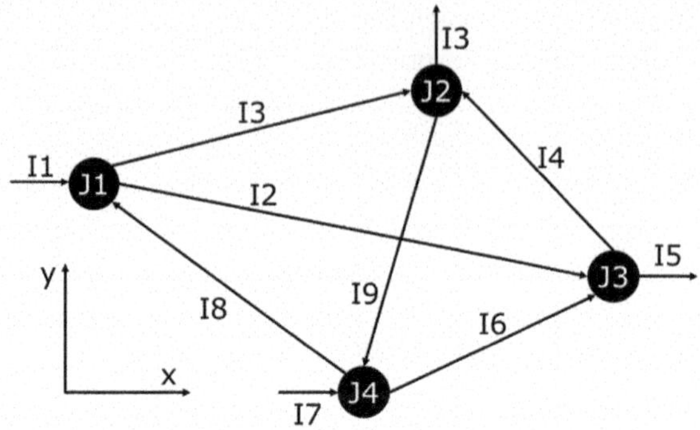

Fig. 5.12 Road network (graph)

(1) The lane path allows the user to determine on which lane the vehicle is to travel.
(2) The connector paths each consist of a start point and an end point and are used to connect other existing paths.
(3) The user path provides maximum flexibility by using randomly placed nodes to define an individual path along the direction of travel.

Lane is the lateral extension of a link between two nodes. A maximum of 10 lanes can be defined on a road section to form a lane section.

Node is a start point and/or an end point for a link.

Lane section is the portion of a link with a constant number of lanes. A new lane section will only be defined when the number of lanes changes.

Link is a section of a road network that connects two nodes and is specified using interconnected lane segments.

Junction is a road element used to unite different links. From a geometrical point of view, the junction facilitates the smoothing of the surface of the transitions between the links.

The reference line is defined in the (x, y, z)-coordinates and it models a change in the course of the road, being restricted by the start point and the end point.

Elevation profile is the height profile of the road surface that can be defined along the direction of travel and that is characterized by parameterizing the longitudinal and lateral profile of the road.

5.2.3.1 Junction

Junction (Fig. 5.13) is a road element that combines links with one another. A minimum of two links are required for a junction to exist and IPGRoad allows a maximum of eight links per junction. From a geometrical point of view, a junction is especially advantageous when aiming for maximum surface smoothness of the transitions between links [23, 24].

There are three types of junctions, and these can be distinguished depending on the angle between their arms and the approach in which the road geometries are connected:

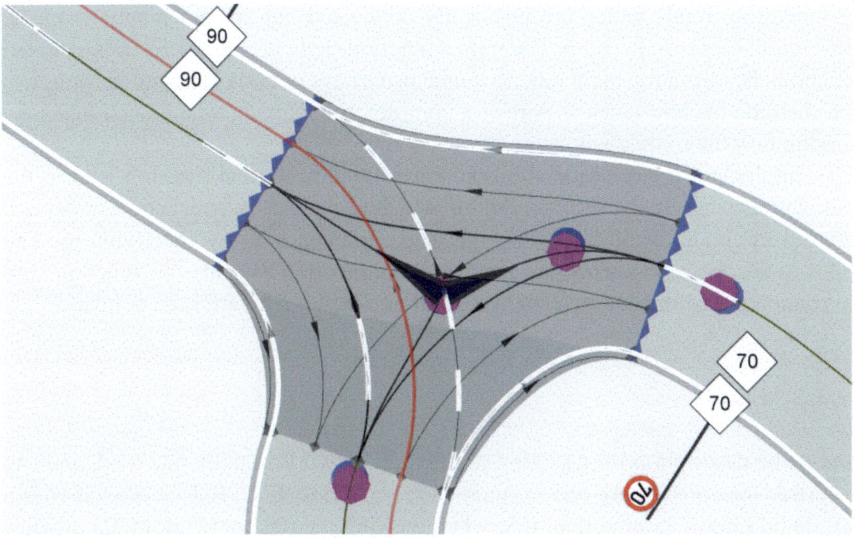

Fig. 5.13 Junction

(1) Type 1 normal junctions that can have up to eight arms.
(2) Type 2 highway entrances and exits where the angle between the arms is either zero or 180 degrees. This junction type is direct junction.
(3) Type 3 virtual junctions to connect secondary roads that branch out from the primary, without altering the continuity of the primary element.

Within the normal junctions, there are two different possibilities to model the junction's surface area, namely simple junctions, and extended junctions. Simple junctions use Bezier-patch-mapping functions to describe the junction geometry, which includes the junction's extent/boundaries as well as the 3D surface of the junction geometry. Extended junctions decouple the junction area's 3D surface model from the calculation of the junction boundaries.

For direct junctions, the 3D profile of the junction area can be additionally defined using the elevation, slope and camber and corresponding gradient parameters [23].

The direct junction element enables the user to build highway entries and exits. This feature defines a junction entry that allows multiple road segments to be connected with one another, without having an actual, visible junction region. The direct junction also acts as a border between links, as a result of which roads going outward from a direct junction are defined as separate links, just like when working with regular junctions.

The direct junction can be included at the end node of an existing link, by selecting the corresponding direct junction icon in the road segment tab of the scenario editor.

Virtual junction allows defining a special type of connection between links, in which an additional link can be connected with the main link without altering or interrupting the main link's geometry. The virtual junction approach is intended for modelling roads which are physically connected but are not junctions in the conventional sense of the term. The virtual junction element enables the user to model road branches from the main link, to define driveways or entry/exit into parking lots or to include cycling lanes as part of a complex roundabout, where such lanes are merging into the roundabout geometry.

To implement a virtual junction, the corresponding virtual junction icon in the road segment tab of the scenario editor will be selected. A connection is defined between the node of the branching link and the main link, by specifying the end points of the arm's boundaries. Since the virtual junction has only one junction arm, the connection is finished as soon as the left and right boundaries are defined.

5.2.3.2 Link

Link is the description for a section of the road network, usually one road, and can be further specified using one or many lane segments. Each link is labeled with a link ID that plays an important role when viewing the road network in 3D preview technical drawing mode.

The course of a link is defined by its reference line (Fig. 5.14). The reference line is displayed in global (x, y, z)-coordinates and lies in the (x, y)-level. It can be subdivided into multiple segments that form between points, which are set when modeling a change in the road course. The link is restricted by the start point (node 0) and the end point (node 1). The s- and t-axes that travel on the reference line, along the link [23, 24] are of great importance for the entire simulation.

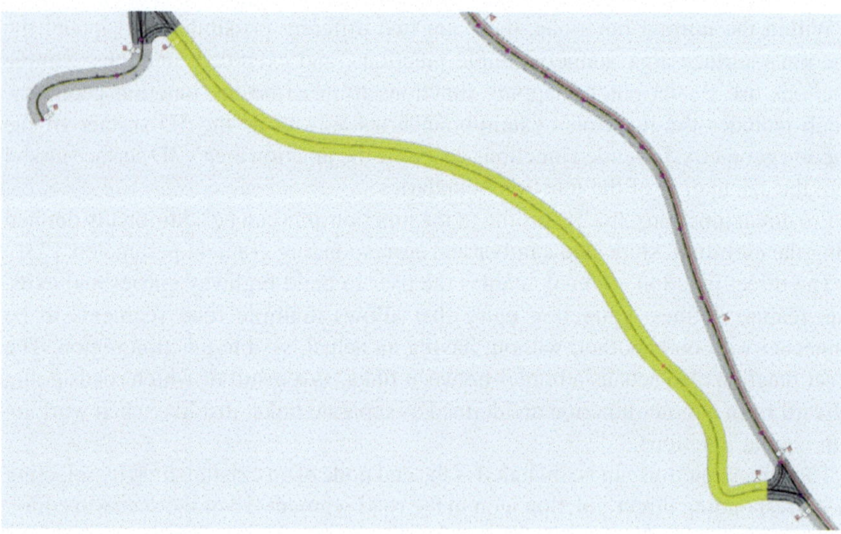

Fig. 5.14 Link

The origin of the (s, t)-coordinate system lies in node 0, the s-axis moves along the reference line and the t-axis points left of the s-axis at a right angle. The s-axis is important when it comes to adding road obstacles. Obstacle positions are specified depending on their distance from either the beginning or the end of a link or junction entrance or exit along the s-axis. The advantage of implementing road obstacles this way is, that when links are altered prior to having added obstacles, their positions are adapted accordingly [23].

5.2.3.3 Lane

Lanes are lateral extensions of a link in the s, t-level (Fig. 5.15). A maximum of ten lanes can be defined to form a lane section and the width of each lane can be individually adjusted by the user.

This means that the supporting points for the lane widths are independent of each other and are defined relative to the lane section. Values can be parameterized for the width, width difference and width interpolation. There are three methods that can be used to conduct the interpolation: cubic, linear and step.

Setting the lane width to zero is explicitly desired since it depicts a simple and valid method of modeling lanes that are either merging or diverging.

Furthermore, lanes can be assigned a specific type. Some of these types are road, boarder, roadside, and traffic island. Thus, not every lane is accessible for the test car [23, 24].

Straight segments are defined by specifying their length. The segment can be either horizontal or at an angle to the horizontal level, and it is created by clicking on the center window and selecting the start and end point. The start point of a segment can be the ending of a previous segment, a junction arm, or a new point altogether.

Fig. 5.15 Lane

The segment type turn is parameterized using the radius and angle of the turn. Whether the turn is a left or right turn it depends on the definition of the start direction and inclination of the curve that are determined when setting the second point.

Clothoids are similar to turn segments but are characterized by a different start and end radius, as well as the turn angle. Similar to the turn segment, whether the clothoid is of the clothoid-right or clothoid-left type is dependent on the selection of the start direction and the inclination of the curve when setting the second point.

5.2.3.4 Lane Section

Lane sections are sectors of a link with a constant number of lanes (Fig. 5.16). A link can consist of any number of lane sections and the start point of each lane section is defined relatively to the appending links (node 0). The start of a new lane section terminates the previous one.

A new lane section should only be defined when the number of lanes changes. If it is just the lane width that alters, it is sufficient to define the width as function of the link direction [23, 24].

To split any given link into a lane section, the corresponding feature is selected from the road segment panel. On the link that is to be split, the start and end point of the lane section are selected by mouse click. This defines the offset of the lane section start. The start offset of the specific lane section can be modified in the right tab after selecting the lane section in the center window.

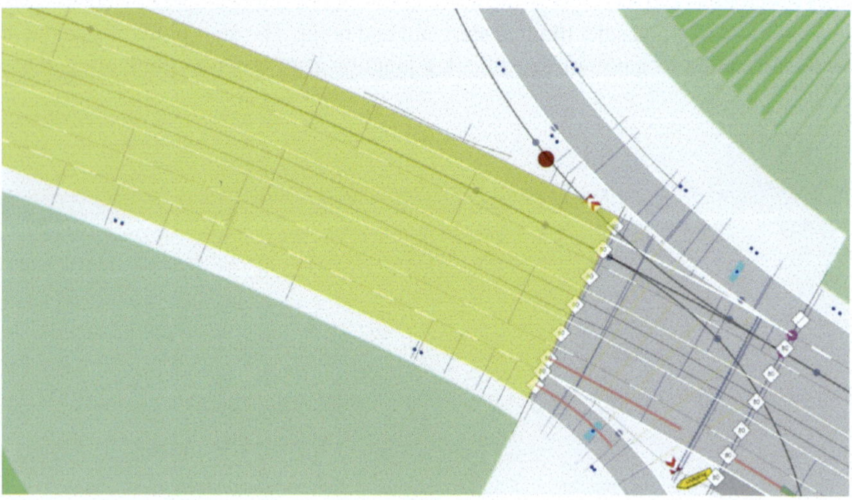

Fig. 5.16 Lane sections

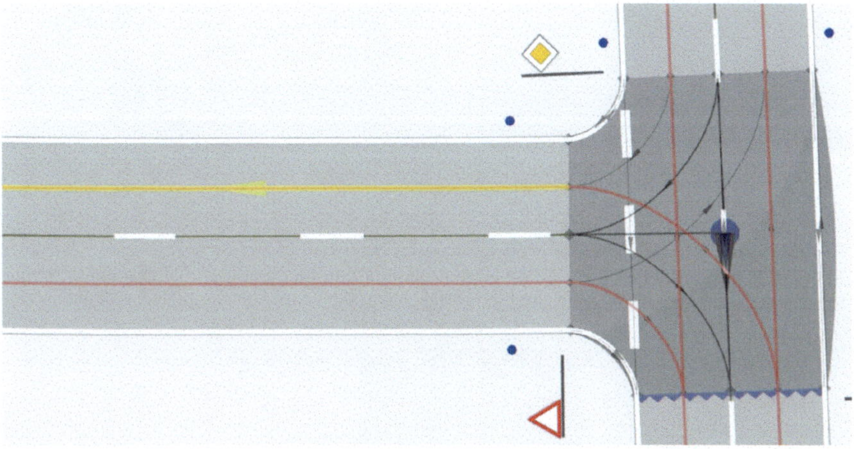

Fig. 5.17 Path

5.2.3.5 Path

A path is a trajectory section on a link that is used for defining routes. There are three types of paths to choose from for the route definition: lane path, connector path, and user path (Fig. 5.17).

Lane path follows the course of a lane and there is a separate preset for every lane within every lane section. This allows the user to easily decide on which lane the vehicle is to drive on. Lane paths run directly along the middle of the lane, which means that they may change course depending on the lane's width to the left and to the right [23, 24].

Connector paths consist of a start and an end point and are used to connect other existing paths between them using a spline. It is also possible to define connector paths on a junction. This type of path is mainly relevant for lane changes.

User path gives the user maximum flexibility whereas supporting node points can be placed randomly to define an individual path that goes in one direction.

5.2.3.6 Route

Route is a sequence of consecutive paths with its own (s, t)-coordinate system. The paths can be of any type and can be connected in any order (Fig. 5.18). The route shows not only a possible road map for the vehicle, but also where exactly on the road the driver and vehicle will be located. This takes corner cutting and driving on a specific lane into account [23, 24].

Fig. 5.18 Route

5.2.3.7 Elevation Profile

The height of the road's surface can be defined along the s-direction of a link and is characterized by parameterizing the road's longitudinal and lateral z-profile (Fig. 5.19). This is achieved by setting values for elevation, slope, and camber.

The elevation is used to adjust the road's longitudinal elevation profile by defining height and gradient values in relative distance to (node 0). Slope and camber define the road's lateral height profile by specification of absolute slope and slope differences, as well as absolute curvature and curvature differences. These values are also stated

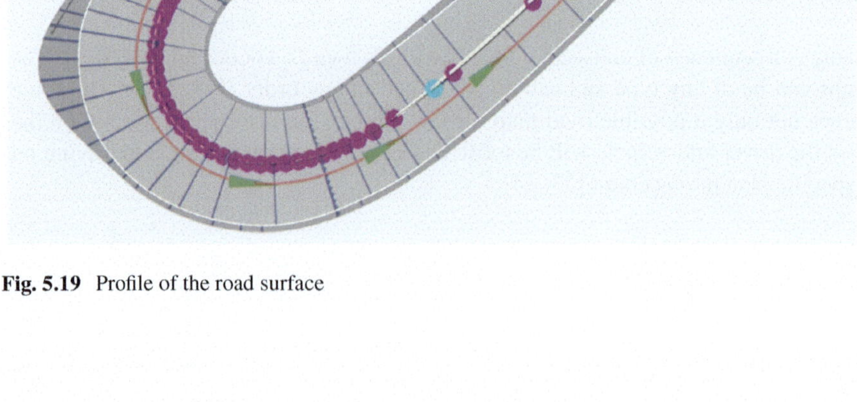

Fig. 5.19 Profile of the road surface

in relative distance to (node 0). The number of supporting points of the appending z-profile, as well as their s-coordinates, are independent of each other [23, 24].

The slope profile provides lateral slope to the road in relation to its reference line. To add this feature, the 3D surface is selected from the tools panel and then the slope profile. The different nodes for the slope profile are selected in reference to the links' start point on the appendant link.

5.2.4 *Scenario for Virtual Road*

The scenario editor is the graphical user interface that assists the user in the following tasks [23, 24]:

(1) Building road networks.
(2) Modifying road surfaces by adding different kinds of road bumps.
(3) Adding traffic signs and traffic lights.
(4) Defining different routes that can each be assigned to either the ego vehicle or other traffic objects.
(5) Adding other miscellaneous objects to beautify surroundings, as well as generate terrain.

The digitization of a virtual path uses the scenario editor, which performs the following tasks through the graphical interface (Fig. 5.20):

(6) Building of road networks in different configurations.

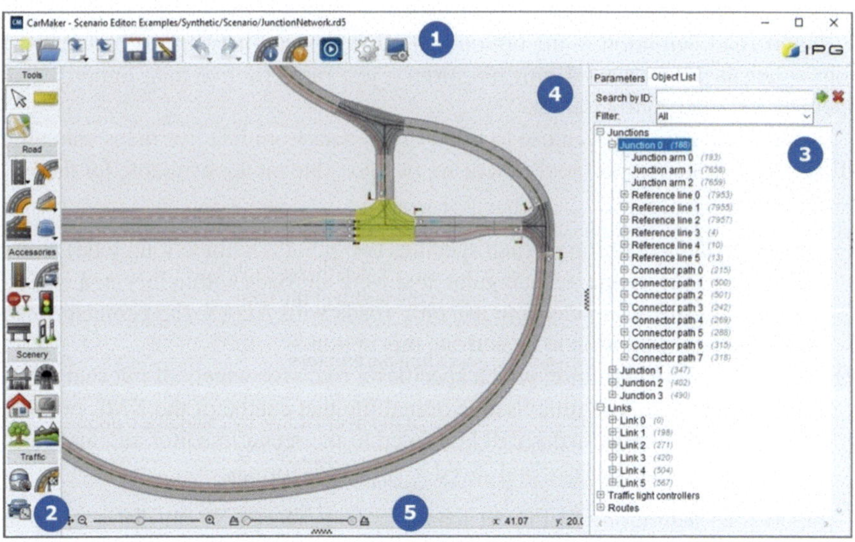

Fig. 5.20 Graphical interface of the scenario editor

(7) Implementation of traffic signs and traffic lights.
(8) Defining different routes for moving the virtual vehicle or for other traffic objects.

The main elements of the scenario editor graphical interface are:

(1) Menu tab—allows one to upload files with road routes, to modify and save general settings of the travel scenarios.
(2) Left sidebar—which allows access to the tools needed to build road networks and to add or modify specific road attributes.
(3) Right sidebar—which allows access to the parameters menu and is used to configure the details of the selected road elements, respectively access to the object list menu which provides a list of all the road elements in the current scenario in hierarchical order.
(4) Center window presents a 2D perspective of the road network, offering the possibility to modify the view parameters.
(5) Bottom window section allows access to the height profile menu which provides a side view of the road network, respectively to the messages menu that shows warnings and errors generated in case of improper parameterization of the road network [22, 23].

New road definition option clears an existing road network in the scenario editor dialog, so that a new road network can be built from scratch. A pop-up appears in case the road network that is going to be cleared has not been saved.

Load road file option can be loaded from previously created road files in format (.rd5) or (.road) in the scenario editor. Road configurations generated in OpenDRIVE can be imported (.xodr). User defined road files need to be saved in the folder data/road in the CarMaker project folder.

Import road definition is the option that allows a road network to be imported from either an external road info file, from a test run, HD live map import or an OpenDRIVE road configuration.

Import road definition can use to import road data from HD live maps using the HERE HD live map import tool. There are two possible modes available for the HD live map import:

(1) Import by area definition which specifies two points (latitude/longitude) in the HD live map to define a rectangular area were all roads within this area will be imported. If the user selects to use only roads with ADAS, the geometry data must be high resolution to be activate this option.
(2) Import by route definition which specifies a route for which all relevant roads will be exported. The route is an external file that can be of the KML or GPX file type. The road network will be loaded to the scenario editor automatically after it has been generated and saved in the (.rd5) format.

Import road definition can import road data from Google Earth maps, then the altitude profile is added with GPSPrune. From the utility (parameters/scenario/road) select the option (road-segment alt-r) and (file). In the work page (segment from file:

start direction xxx …) is selected and (choose digitized road file) opens automatically from where the file (name.KML) can be chosen. Follow (road/lane alt-l) and in the area on the right side of the road section enter the driving side (yellow).

To run the simulation, select (traffic)—(route alt-u) and double click on the route to be formed.

References

1. Yu, L., Kong, D.: A driving behavior planning and trajectory generation method for autonomous electric bus. Future Internet **10**, 51 (2018)
2. https://www.arcgis.com/index.html. 06 Aug 2020
3. Arshad, S., Sualeh, M., Kim, D., Nam, D.V., Kim, G.-W.: Clothoid: an integrated hierarchical framework for autonomous driving in a dynamic urban environment. Sensors **20**, 5053 (2020)
4. Iclodean, C., Cordos, N., Varga, B.O.: Autonomous shuttle bus for public transportation: a review. Energies **13**, 2917 (2020)
5. Debeunne, C., Vivet, D.: A review of visual-LiDAR fusion based simultaneous localization and mapping. Sensors **20**, 2068 (2020)
6. Sun, C., Zhang, X., Xi, L., Tian, Y.: Design of a path-tracking steering controller for autonomous vehicles. Energies **11**, 1451 (2018)
7. Jung, J., Bae, S.-H.: Real-time road lane detection in urban areas using LiDAR data. Electronics **7**, 276 (2018)
8. https://www.sciencedirect.com/topics/earth-and-planetary-sciences/digital-terrain-model. 26 Aug 2020
9. Haris, M., Hou, J.: Obstacle detection and safely navigate the autonomous vehicle from unexpected obstacles on the driving lane. Sensors **20**, 4719 (2020)
10. Liu, H., Liao, K., Lin, C., Zhao, Y., Liu, M.: PLIN: a network for pseudo-LiDAR point cloud interpolation. Sensors **20**, 1573 (2020)
11. Wong, K., Javanmardi, E., Javanmardi, M., Kamijo, S.: Estimating autonomous vehicle localization error using 2D geographic information. ISPRS Int. J. Geo-Inf. **8**, 288 (2019)
12. https://en.wikipedia.org/wiki/Google_Earth. 23 Apr 2021
13. https://en.wikipedia.org/wiki/Bing_Maps. 23 Apr 2021
14. https://en.wikipedia.org/wiki/Here_WeGo. 23 Apr 2021
15. Jo, K., Kim, C., Sunwoo, M.: Simultaneous localization and map change update for the high definition map-based autonomous driving car. Sensors **18**, 3145 (2018). https://doi.org/10.3390/s18093145
16. Ham, S., Im, J., Kim, M., Cho, K.: Construction and verification of a high-precision base map for an autonomous vehicle monitoring system. ISPRS Int. J. Geo-Inf. **8**, 501 (2019)
17. Ilci, V., Toth, C.: High-definition 3D map creation using GNSS/IMU/LiDAR sensor integration to support autonomous vehicle navigation. Sensors **20**, 899 (2020)
18. Im, J.-H., Im, S.-H., Jee, G.-I.: Extended line map-based precise vehicle localization using 3D LIDAR. Sensors **18**, 3179 (2018)
19. https://patents.google.com/patent/US9612123/en. 13 Apr 2020
20. Dimitrievski, M., Veelaert, P., Philips, W.: Behavioral pedestrian tracking using a camera and LiDAR sensors on a moving vehicle. Sensors **19**, 391 (2019)
21. IPG Automotive: Users Guide Version 10.1 (2021)
22. IPG Automotive: Quick Start Guide Version 10.1 (2021)
23. IPG Automotive: User Guide Version 10.1 (2021)
24. IPG Automotive: ROAD5 File Format Description Version 10.1 (2021)

Chapter 6
Virtual Model

6.1 Virtual Model Development Methodology

Note: Validation via Simulation in a Virtual Environment
Before reaching the asphalt, an autonomous vehicle must struggle to validate its drivability and decision-making strategy in a closed loop simulation environment, millions of kilometers of digitalized road populated with real life scenarios are the environment where the control algorithms are evaluated, maneuvers are performed, and safety scenarios are implemented.

It would seem a somehow atypical from a classical vehicle perspective, but the **legal framework of various countries is making the virtual validation process mandatory before releasing the acceptance of road-testing process to be stared for the homologation procedures to take place**.

Before presenting the concept of the authors on the matter of the simulation procedures, we must answer to a typical problem of why and how.

The **why** matter is a straightforward approach, autonomous driving means form the beginning that most of the control is taken from the human factor and placed in the responsibility of the digital decision-making algorithms. These Artificial Intelligence decisions represent a replacement of human selective criteria and are based on the perception coming from various of sensors that are the information feeding elements of the decision-making algorithms. Validation in the virtual environment is not necessary, the process represents the mandatory birthplace of these automation processes feed with scenarios of the programmer itself. AI and forced learning software structures are becoming nowadays the key element in generic autonomous driving learning strategies. Running the autonomous vehicle in the virtual environment represents the first step in the validation process, running millions of kilometers represents a milestone and provides annalists major feedback for improvement before running live the real world. Authors call this process the learning playground and it represents the base foundation of the pre-validation process. Overall, you cannot present a vehicle on the market with autonomous driving characteristics without a proper validation of the control algorithms.

C. Iclodean et al., *Autonomous Vehicles for Public Transportation*, Green Energy and Technology, https://doi.org/10.1007/978-3-031-14678-7_6

The **how** approach is a little bit more complex because it represents a matrix of simulation environments able to provide solid feedback. There are various validation methodologies, on element level and on the complete vehicle level. A validated methodology is presented in the current material using simulation based on physical validated modes of sensors and dynamical validated model of the vehicle body. All items being put together in a simulation environment capable of validation of a complete autonomous vehicle in a closed-loop environment populated with real life traffic scenarios and scalable from several km routes to thousands of km of routes of real road coming for digitized roads imported from Google Maps environment.

The presented methodology comes from years of simulation experience in automotive environment and its aim is to provide scientist and researchers a clear path and methodology for an autonomous vehicle validation process and procedures.

What is also important from authors perspective is that the presented methodology is a unique selection of layers of validation arranged in an original manner with the aim of underlining the true nature of the simulation processes that becomes valid only if it is combined with the mathematical models of the validated sensors that operate in a dynamically balanced body.

Compared to real car models, virtual models used in computer simulations have a number of advantages, such as the ability to modify constructive and functional parameters, respectively the constructive architecture of the virtual model at any step of the development process. Computer simulation is performed using software applications that aim to translate a real car model into a virtual application to estimate its behavior and performance and to set limits for improving performance while maintaining solid state.

Computer simulation is the study of reality in a virtual environment, with the help of a theoretical model, which is a digital image of a real existing model which is to be improved. The virtual model is developed according to the aspects tracked by the user, and the complexity of this model must correspond to the reality of the evaluated system, as complex as necessary and as simple as possible, so that the results obtained from computer simulations would be validated by the experimental results [1].

The virtual model concept includes all the constructive, functional and performance characteristics of the real vehicle to be cloned into an identical, virtual vehicle. The software application in which the virtual model is to be developed is CarMaker, an application developed by the German company IPG Automotive [2]. CarMaker is an application used to simulate the RT operation of the main systems in a car at the system level and is dedicated to researching the transient conditions of offline operation in desktop applications, or online in in-the-Loop simulation environments. The application covers three principal areas:

(1) The concept development phase in which the CarMaker application allows the rapid configuration and simulation of any constructive variants of the virtual vehicle model by evaluating its main performances.
(2) The design phase of the powertrain system in which the electric motor is connected to the driveline.

(3) Virtual vehicle testing phase in which the model developed in CarMaker operates as a unitary system allowing the development, calibration, and testing of the control functions of a virtual vehicle.

A virtual vehicle is a computer modeled representation of a real vehicle with a behavior that matches that of its real-world counterpart. With CarMaker, the virtual vehicle is made up of mathematical models that contain the equations of motion, kinematics along with other mathematical formulas that define the multibody system.

The model is then parameterized with real data that relates directly to the vehicle to be studied. By this approach it is possible to use CarMaker to test any vehicle with a validated parameter set. The virtual vehicle contains all parts of a real vehicle, including powertrain, chassis, other main systems, and auxiliary systems. The virtual vehicle has real automotive controllers (ABS, ESP, ACC) or software modeled controllers integrated using hardware or software in-the-Loop methodology.

A virtual road is a digitized or computer modeled representation of a real road. In CarMaker, the road can be generated in either of two ways [3]:

(1) Combining individual road segments, like as straights and curves, to form a larger road. For each road section defined there is a specific length, width, angle, slope, pitch, friction coefficient, which can be specified.
(2) Using digitized data of an existing road that has been collected. With this approach, the measured data that is taken by survey or some other method is contained in a data file that is read by CarMaker and used as the road or test track during simulation.

A virtual driver is a computer driver, which simulates the actions of a real driver. Everything that would normally be controlled by a real driver is controlled by the virtual driver. There are two approaches that can be taken to perform the driver actions [3]:

(1) Simple control, the actions are performed by specifying in advance what will happen at a certain time or distance.
(2) IPGDriver, the course is controlled by a smart computer driver, which tries to maintain a course on a virtual road, and which is also capable of planning. The driver can be modified to operate within specified limits.

The following parameters are used only if the option autonomous is selected for the longitudinal motion of a traffic object. When a traffic object drives autonomously, it considers speed limit markers, traffic lights, road curvature and other traffic objects, but also the ego-vehicle.

The HDM (Human Driver Model) is a model which uses the IIDM (Improved Intelligent Driver Model) as the basis model. The target acceleration for the following object is calculated as follows [4].

The desired gap distance g^{des} (m) to the object in front depends on the velocities of the following object v (km/h) and the leader object v_l (km/h):

$$g^{des} = s_O + vT - \frac{v(v_l - v)}{2\sqrt{a^{max} d^{comf}}} \tag{6.1}$$

With the current gap distance, the ratio is determined:

$$z = \frac{g^{des}}{g} \tag{6.2}$$

For the case if the current velocity is lower than the desired velocity $v \leq v^{des}$:

$$a^{free} = a^{max}\left(1 - \left(\frac{v}{v^{des}}\right)^{\delta}\right) \tag{6.3}$$

$$a = \begin{pmatrix} a^{max}\left(1 - z^2\right) z \geq 1 \\ a^{free}\left(1 - z^{(2a^{max})/a^{free}}\right) \end{pmatrix} \tag{6.4}$$

For the case if the current velocity is higher than the desired velocity $v > v^{des}$:

$$a^{free} = -d^{conf}\left(1 - \left(\frac{v^{des}}{v}\right)^{(a^{max}\delta)/d^{conf}}\right) \tag{6.5}$$

$$a = \begin{pmatrix} a^{free} + a^{max}\left(1 - z^2\right) z \geq 1 \\ a^{free} \end{pmatrix} \tag{6.6}$$

IPGTraffic allows additional road users and ambience to be added to the TestRun which replicates a typical traffic scenario as seen in reality. This can be primarily classified into two different modes, as follows [4]:

(1) Movable when the traffic object is a dynamic part of the TestRun, which can be parameterized to have an influence on the TestRun. The physics behind the dynamic motion of the traffic object is performed with the use of five different motion models which can be selected. Also, a specific set of Minimaneuvers can also be defined for every movable traffic object. The maneuver steps are limited by a duration and/or a dynamic end condition where each step can have its own longitudinal and lateral motion. Alternatively, a traffic object can be also defined to have a free motion maneuver which does not follow a motion model as such but is rather dependent on the information as specified using input from file for its longitudinal and lateral motion. It should be noted that, the main difference between the motion of the traffic vehicle is that for the motion model the steering dynamics are not calculated. It is rather based on a simple approach for motion models:

(a) Parking vehicles.
(b) Slow driving cars.
(c) Vehicles that change the driving lane abruptly.
(d) Vehicles that collide at a crossing.
(e) Bicycle riders.
(f) Pedestrians crossing the road.

(2) Stationary for traffic objects which are used to create an ambience of a driving scenario, which aid in creating realistic visual representation. Common use cases that can be included in the TestRun are the following:

    (a) Residential buildings.
    (b) Office buildings.
    (c) Parking garages.
    (d) Gas stations.

6.2 Virtual Model for Navya Shuttle Bus

6.2.1 Simulation Tools CarMaker

The main menu of the CarMaker simulation application or GUI (Graphical User Interface) (Fig. 6.1) is structured on modules that allow the definition of the parameters necessary for the development of the vehicle model.

(1) Car allows the editing of the vehicle constructive parameters (vehicle data set) in the configuration established by the user, using the initial data collected according to the considered methodology.
(2) Tires allows the loading of predefined models for the tires of the virtual vehicle (data/tire/project/examples) or allows the editing and modification of the constructive dimensions (model parameters) or general parameters [3, 5].

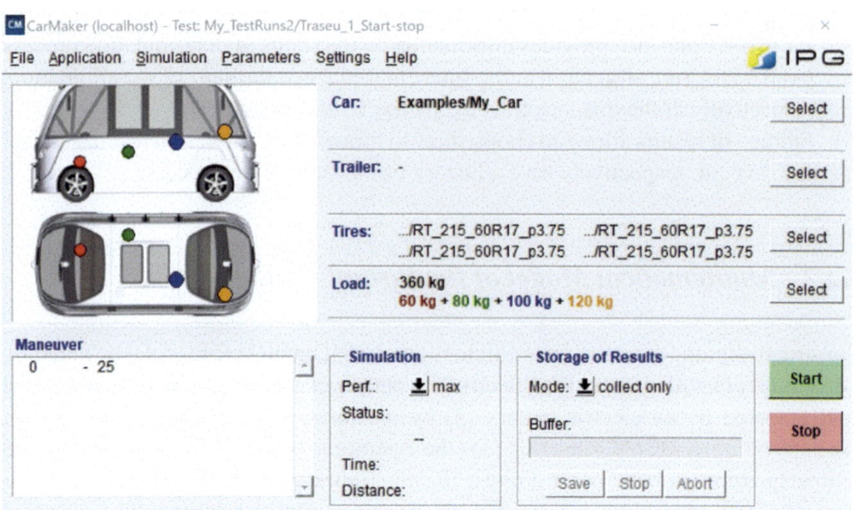

Fig. 6.1 Graphical User Interface

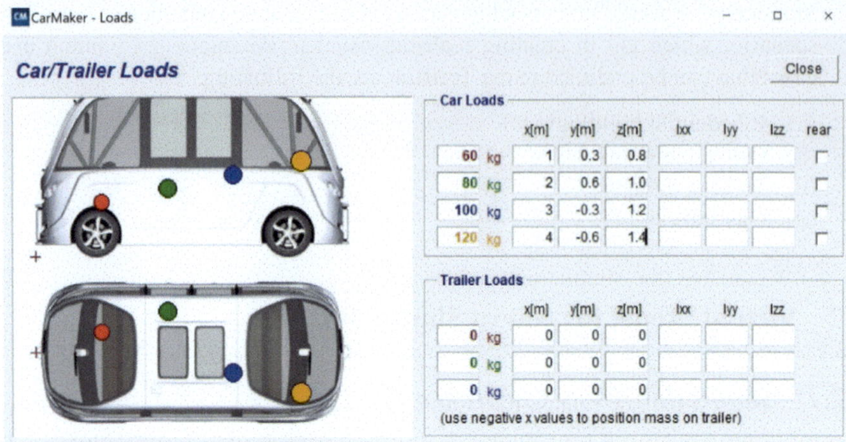

Fig. 6.2 CarMaker loads module

(3) Loads allows the loading of the virtual vehicle with four load elements that can be distributed 3D (x/y/z) over its entire surface and that have a direct influence on driving performance (Fig. 6.2).

(4) Maneuver allows to define a set of instructions that describe the movement scenario of the virtual vehicle highlighting the specification of maneuver step, the longitudinal dynamics characteristics, respectively the lateral dynamics characteristics. Maneuver allows the selection of predefined models or customizable driver models (Fig. 6.3).

(5) Simulation allows to define the running time of the simulation process from 0.2 RT to 5 RT and also provides information on the status of the simulation process (wait, start, run, stop etc.), at the time elapsed since the start of the simulation, respectively at the distance traveled by the virtual vehicle.

(6) Storage of results allows to choose how to process the simulation results: collect all, save all, respectively save a history of the simulation process.

6.2.2 Mathematical Model of the Virtual Configuration

Constructively, the autonomous vehicle is an electric vehicle (Fig. 6.4), powered by a drive unit (1) with one or more electric motors (2), arranged axially or in the wheel hub, powered by an electric battery (3) by means of an electronic power system (4) controlling a DC/AC inverter (5), the operation of which is managed by the following command and control units: BCU* (Battery Control Unit), MCU (Motor Control Unit), respectively TCU. Electric batteries can be charged via a standard power connector (6). The autonomous vehicle consists of systems characteristic of an electric vehicle (powertrain system, steering system, braking system, auxiliary system) [4].

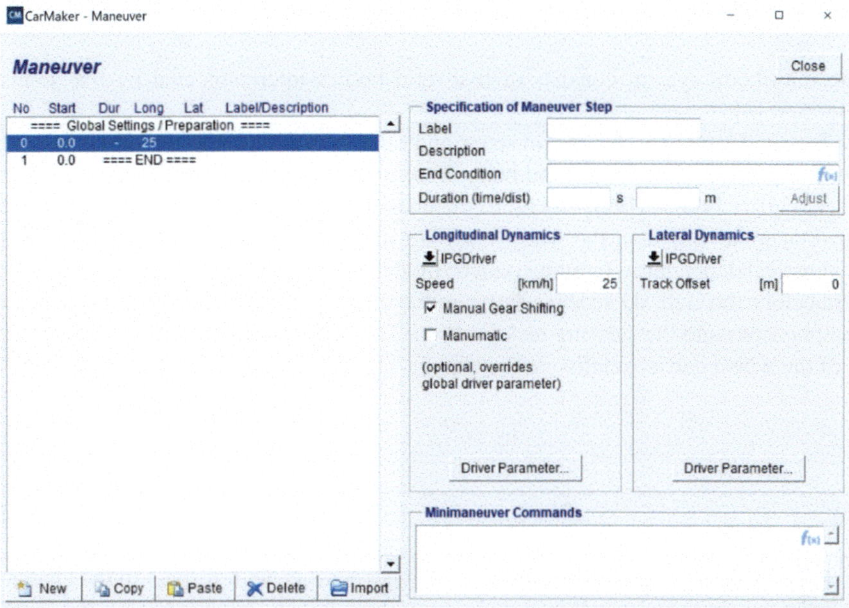

Fig. 6.3 CarMaker maneuver module

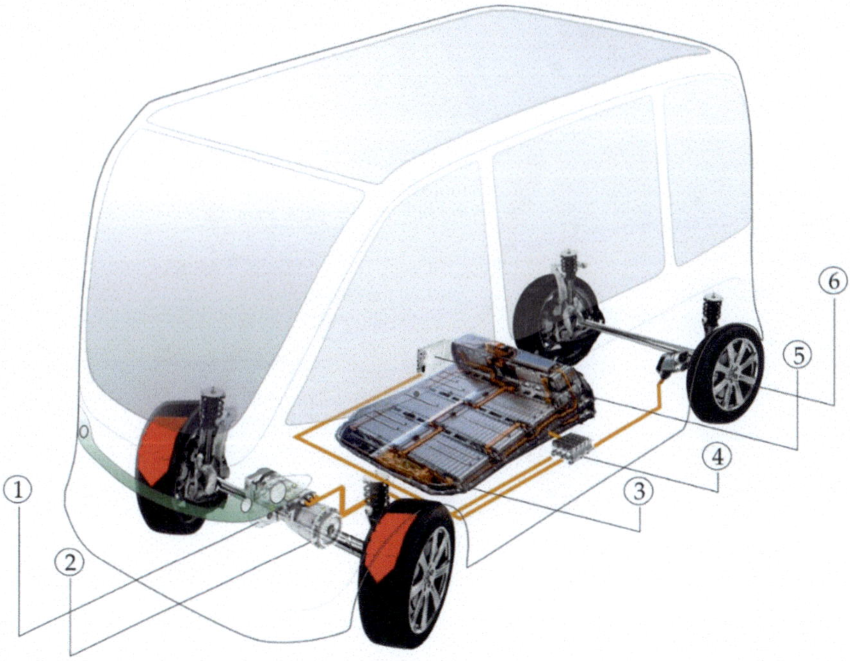

Fig. 6.4 Constructive architecture of the autonomous vehicle

6.2.2.1 Vehicle Multibody System

The multibody system consists of five rigid bodies interconnected by five joints (Fig. 6.5) [4].

The rigid bodies may move in space and relative to each body. The chassis body (Fr1) is connected to the ground body (Fr0) with a 6 degree of freedom joint. The wheel carrier bodies (Fr2ij) relate to the chassis body with a complex 1 degree of freedom joint. Between the front (rear) left and front (rear) right wheel carrier a kinematical dependance (coupler) can be modelled. The wheel carrier center follows a trajectory through 3D space, relative to the chassis, as function of the suspension compressions and the steering rack position. The relative position t_{ij} and orientation r_{ij} of the wheel carrier relative to the chassis has a kinematical part [4]:

$$\bar{t}_{ij} = \bar{t}_{ij}^{kin} + \bar{t}_{ij}^{com} \tag{6.7}$$

$$\bar{r}_{ij} = \bar{r}_{ij}^{kin} + \bar{r}_{ij}^{com} \tag{6.8}$$

The kinematical relative position of the wheel carrier with respect to the chassis is dependent on the suspension compressions, both left and right side q_{OiL}, q_{OiR}, and for the front axle it is dependent on the steering rack position q_{Steer} [4]:

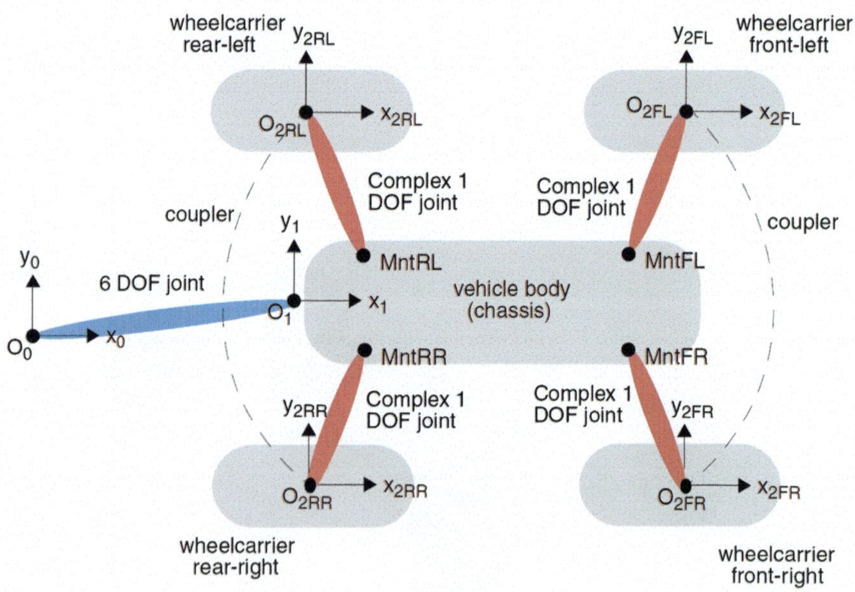

Fig. 6.5 Vehicle multibody system

$$\overline{t_{ij}^{kin}} = \overline{Mnt_{\iota J} O_{2\iota J}} = \begin{bmatrix} t_{ij,x}^{kin} \\ t_{ij,y}^{kin} \\ t_{ij,z}^{kin} \end{bmatrix}_1 = \begin{bmatrix} t_{ij,,x}^{kin}(q_{oiL}, q_{oiR}, q_{\text{steer}}) \\ t_{ij,y}^{kin}(q_{oiL}, q_{oiR}, q_{\text{steer}}) \\ t_{ij,z}^{kin}(q_{oiL}, q_{\text{oiR}}, q_{\text{steer}}) \end{bmatrix}_1 \tag{6.9}$$

The relative orientation of the wheel carrier with respect to the chassis contains three scalar Cardan angles to describe the orientation using the Z-X-Y rotating sequence. The kinematical relative orientation is dependent on the suspension compressions and for the front axle on the steering rack position [4]:

$$\overline{r_{ij}^{kin}} = \begin{bmatrix} r_{ij,x}^{kin} \\ r_{ij,y}^{kin} \\ r_{ij,z}^{kin} \end{bmatrix} = \begin{bmatrix} r_{ij,x}^{kin}(q_{OiL}, q_{OiR}, q_{Steer}) \\ r_{ij,y}^{kin}(q_{OiL}, q_{OiR}, q_{Steer}) \\ r_{ij,z}^{kin}(q_{OiL}, q_{OiR}, q_{Steer}) \end{bmatrix} \tag{6.10}$$

Additionally, to the relative position and orientation, the length of the suspension force elements also depends on the suspension compressions and the steering rack position [4]:

$$\begin{bmatrix} l_{ij}^{Spring} \\ l_{ij}^{Damper} \\ l_{ij}^{Buffer} \\ l_{ij}^{Stabi} \end{bmatrix} = \begin{bmatrix} l_{ij}^{Spring}(q_{OiL}, q_{OiR}, q_{Steer}) \\ l_{ij}^{Damper}(q_{OiL}, q_{OiR}, q_{Steer}) \\ l_{ij}^{Buffer}(q_{OiL}, q_{OiR}, q_{Steer}) \\ l_{ij}^{Stabi}(q_{OiL}, q_{OiR}, q_{Steer}) \end{bmatrix} \tag{6.11}$$

The dependence on the suspension compressions and the steering rack position is subjected to linear or nonlinear constraints. In the case of linear constraints, the relative position, orientation, and suspension force element lengths are calculated with constant partial derivatives [4]:

$$t_{ij,y}^{kin} = \frac{\partial t_{ij,y}^{kin}}{\partial q_{OiL}} q_{OiL} + \frac{\partial t_{ij,y}^{kin}}{\partial q_{OiR}} q_{OiR} + \frac{\partial t_{ij,y}^{kin}}{\partial q_{Steer}} q_{Steer} = c_0 q_{OiL} + c_1 q_{OiR} + c_2 q_{Steer} \tag{6.12}$$

$$r_{ij,y}^{kin} = \frac{\partial r_{ij,y}^{kin}}{\partial q_{OiL}} q_{OiL} + \frac{\partial r_{ij,y}^{kin}}{\partial q_{OiR}} q_{OiR} + \frac{\partial r_{ij,y}^{kin}}{\partial q_{Steer}} q_{Steer} = c_0 q_{OiL} + c_1 q_{OiR} + c_2 q_{Steer} \tag{6.13}$$

$$\begin{aligned} l_{ij,y}^{Spring} &= \frac{\partial l_{ij}^{Spring}}{\partial q_{OiL}} q_{OiL} + \frac{\partial l_{ij}^{Spring}}{\partial q_{OiR}} q_{OiR} + \frac{\partial l_{ij}^{Spring}}{\partial q_{Steer}} q_{Steer} \\ &= c_0 q_{OiL} + c_1 q_{OiR} + c_2 q_{Steer} \end{aligned} \tag{6.14}$$

In the case of nonlinear constraints, the relative position, orientation, and suspension force element lengths are deposited in nonlinear mappings:

$$t_{ij,y}^{kin}, r_{ij,y}^{kin}, l_{ij}^{Spring} = f(q_{OiL}, q_{OiR}, q_{Steer}) \tag{6.15}$$

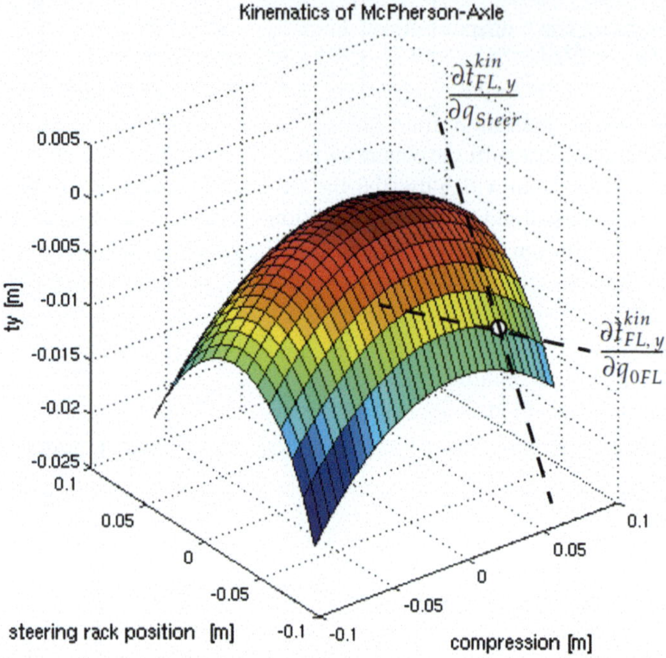

Fig. 6.6 Case of nonlinear mappings

Because the direction of the wheel trajectory determines how the tire forces are transmitted to the chassis body, the partial derivatives are also established in case of nonlinear mappings for further calculations in the working point (Fig. 6.6).

The other relative positions $t^{kin}_{ij,x}$, $t^{kin}_{ij,z}$, relative orientations $r^{kin}_{ij,x}$, $r^{kin}_{ij,z}$ and other suspension force element lengths l^{Damper}_{ij}, l^{Damper}_{ij}, l^{Buffer}_{ij}, l^{Stabi}_{ij} are calculated.

Suspension system joints are typically mounted with elastic bushings. These bushings allow small motions against the kinematically desired motion of each joint. These small motions express the deformation due to exciting forces and torques in the wheel carrier [4]:

$$t^{com}_{ij,x} = \left[\frac{\partial t^{com}_{ij,x}}{\partial F_{iL,k}}\right]\overline{F}^{WC}_{iL} + \left[\frac{\partial t^{com}_{ij,x}}{\partial F_{iR,k}}\right]\overline{F}^{WC}_{iR} + \left[\frac{\partial t^{com}_{ij,x}}{\partial T_{iL,k}}\right]\overline{T}(O_{2iL})^{WC} + \left[\frac{\partial t^{com}_{ij,x}}{\partial T_{iR,k}}\right]\overline{T}(O_{2iR})^{WC} \tag{6.16}$$

$$r^{com}_{ij,x} = \left[\frac{\partial r^{com}_{ij,x}}{\partial F_{iL,k}}\right]\overline{F}^{WC}_{iL} + \left[\frac{\partial r^{com}_{ij,x}}{\partial F_{iR,k}}\right]\overline{F}^{WC}_{iR} + \left[\frac{\partial r^{com}_{ij,x}}{\partial T_{iL,k}}\right]\overline{T}(O_{2iL})^{WC} + \left[\frac{\partial r^{com}_{ij,x}}{\partial T_{iR,k}}\right]\overline{T}(O_{2iR})^{WC} \tag{6.17}$$

where $\left[\overline{F}_{ij}^{WC}\right]_1$ total applied forces of the wheel carrier ij (i = F, R; j = L, R) onto the chassis, expressed in chassis frame Fr_1, $\left[\overline{T}(O_{2iL})^{WC}\right]_1$ total applied torques of the wheel carrier ij (i = F, R; j = L, R) onto the chassis in the wheel carrier center O_{2ij}, expressed in chassis frame Fr_1, $\left[\frac{\partial t_{ij,x}^{com}}{\partial F_{iL,k}}\right]_1$ partial derivation of the elastokinematical relative x-position between the wheel carrier and the chassis with respect to exciting force $F_{k=x,y,z}$ in the wheel carrier iL and iR, $\left[\frac{\partial r_{ij,x}^{com}}{\partial F_{iL,k}}\right]_1$ partial derivation of the elastokinematical relative x-rotation between the wheel carrier and the chassis with respect to exciting force $F_{k=x,y,z}$ in the wheel carrier iL and iR. The partial derivations $\frac{\partial t_{ij,x}^{com}}{\partial F_{ij,k}}$, $\frac{\partial t_{ij,x}^{com}}{\partial T_{ij,k}}$, $\frac{\partial r_{ij,x}^{com}}{\partial F_{ij,k}}$ and $\frac{\partial r_{ij,x}^{com}}{\partial T_{ij,k}}$ are defined constant parameters in the vehicle data set. The other directions $t_{ij,y}^{com}$, $t_{ij,z}^{com}$, such as $r_{ij,y}^{com}$, $r_{ij,z}^{com}$ are calculated in the same manner.

6.2.2.2 Electric Machine

The main element of the powertrain is the electric machine operating as engine during acceleration and as generator during braking (Fig. 6.7) [6].

The electric machine provides torque based on the load characteristic and the power losses. The operating characteristic of the powertrain is described by the diagram of torque versus speed, for all traction regimes (Fig. 6.8) [7]:

(1) Drive Forward, Traction or Motoring Mode, the engine speed has a positive value, and the torque is positive.
(2) Brake Backward, Generating or Braking Mode, the engine speed is negative, and the torque is positive.

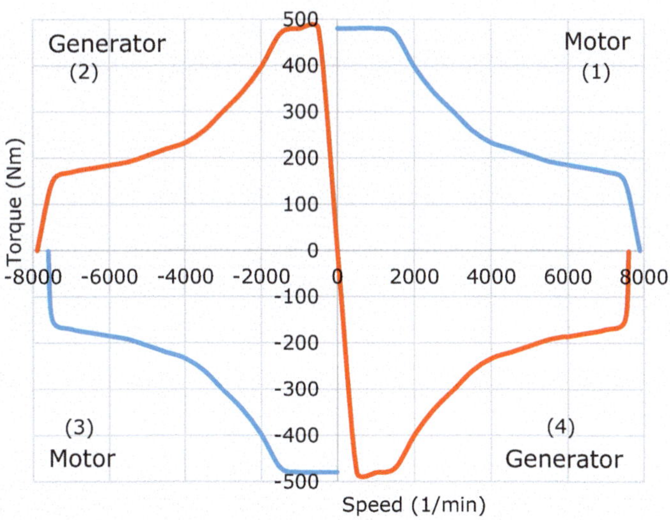

Fig. 6.7 Electric machine characteristics

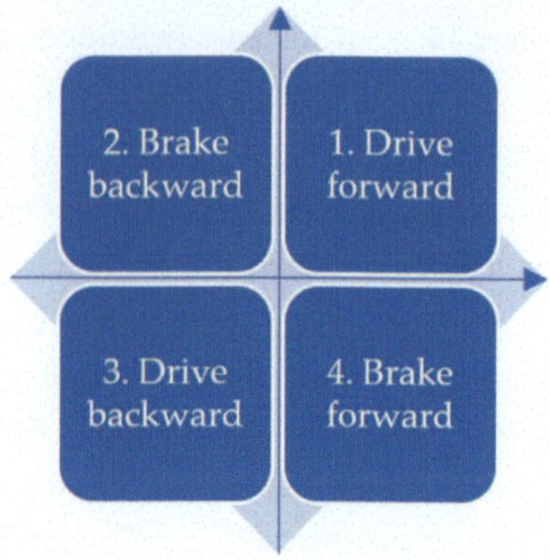

Fig. 6.8 Torque versus speed diagram

(3) Drive Backward Traction or Motoring Mode, the engine speed is negative, and the torque is negative.
(4) Brake Forward Generating or Braking Mode, the engine speed is positive, and the torque is negative.

The torque developed by the electric machine is maximum from the start and it remains uniform until a constant speed corresponding to the cruising speed is reached. The power of the electric motor increases linearly at the same time as the torque decreases (Fig. 6.9) [7].

In CarMaker, the model selected for the electric motor is the "1D Look-Up table", a model that contains the operating characteristics as motor and as generator of the electric machine (Fig. 6.10) [8].

The inertia of the electric motor sums up the inertia of all the rotating components of the motor. I_{xx} inertia is the corresponding inertia on the x-axis along the transmission components. The ratio parameter describes a constant transmission ratio between the output shaft of the electric motor and the shaft to which it is connected.

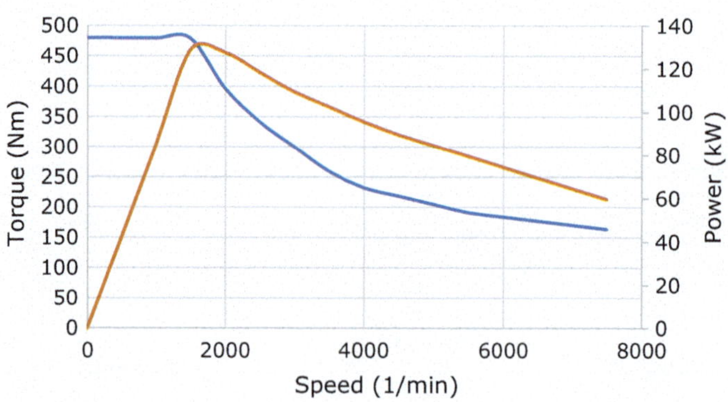

Fig. 6.9 Torque versus power

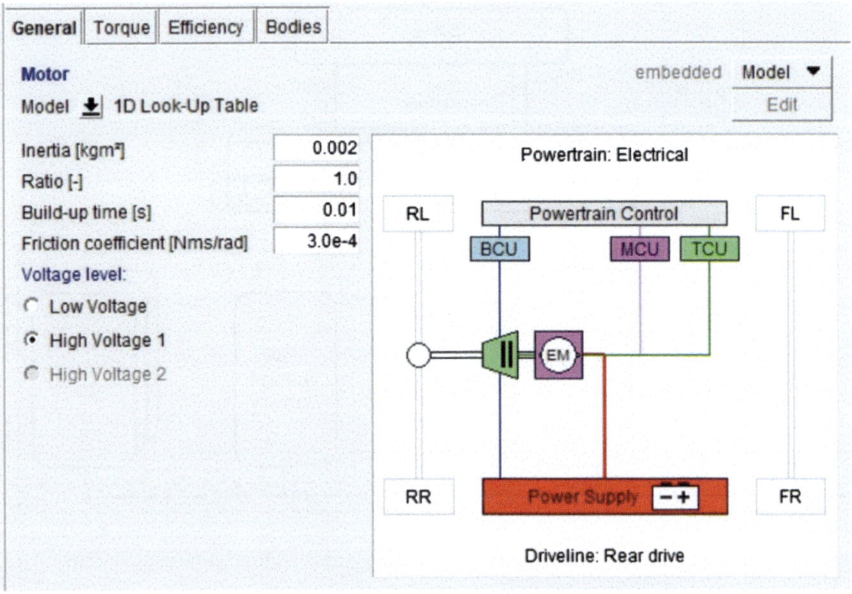

Fig. 6.10 Model 1D look-up table parametrization

The friction coefficient defines the friction of the engine elements which acts as a traction motor moment when the engine load is zero.

The efficiency of the DC/DC converter depends on the transmitted power divided by the maximum converter power.

6.2.2.3 Power Supply

The PS (Power Supply) represents the electric system of the vehicle and is part of the powertrain model. It includes all components of the vehicle's LV (Low Voltage) and HV (High Voltage) electric circuits and the batteries.

The PS interface is designed for models with the consumer and generator loads, delivered as electric powers. The interface distinguishes up to two different electric circuits (LV + HV1) (Fig. 6.11) [4].

The primary task of the PS subsystem is the accounting of the electric power flow to and between the electric circuits of the vehicle. It includes the battery models. The states of each electric circuit and battery are evaluated based on power consumption and generation of the electric motors and auxiliaries.

It evaluates the electric power taken from the LV circuit P_{LV} by summing the electric power of the auxiliary consumers/generators P_{AuxLV} and the consumer/generator power of the electric motor in this circuit $P_{MotorLVi}$ [4]:

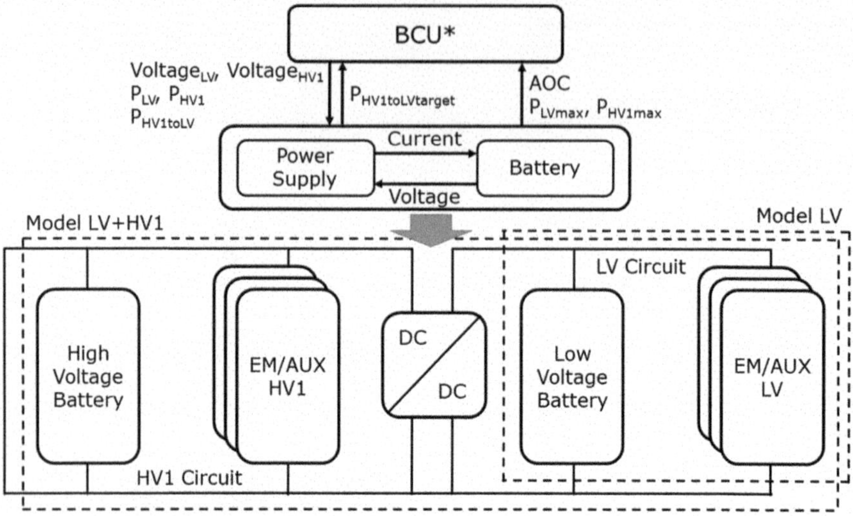

Fig. 6.11 PS interface

$$P_{LV} = P_{AuxLV} + \sum_i P_{MotorLVi} \tag{6.18}$$

The electric current of the battery model I_{LV} is calculated under the assumption that the electric circuit voltage U_{LV} is equal to its battery voltage:

$$I_{LV} = \frac{P_{LV}}{U_{LV}} \tag{6.19}$$

The PS model contains one LV electric circuit and one HV electric circuit. These circuits are interconnected by a DC/DC converter and each one contains a battery model.

The electric power taken from each circuit P_{XV} depends on the electric power of the auxiliary consumers P_{AuxXV}, the consumer/generator power of the electric motor in this circuit $P_{MotorXVi}$ and the power exchanged via the DC/DC converter $P_{HV1toLV}$. The power transferred between the circuits corresponds to the target power transferred between the circuits limited by the maximum DC/DC power. The model differentiates the cases [4]:

$$P_{LV} = P_{AuxLV} + \sum_i P_{MotorLVi} - P_{HV1toLV} \tag{6.20}$$

The HV circuit charges the LV circuit:

$$P_{HV1toLV} \geq 0\,W => P_{HV1} = P_{AUxHV1} + \sum_i P_{MotorHV1i} + \frac{P_{HV1toLV}}{\eta_{DCDC}} \tag{6.21}$$

The LV circuit charges the HV circuit:

$$P_{HV1toLV} < 0\,W => P_{HV1} = P_{AuxHV1} + \sum_i P_{MotorHV1i} + P_{HV1toLV}\eta_{DCDC} \tag{6.22}$$

The electric current of the battery model I_{XV} is calculated under the assumption that the electric circuit voltage U_{XV} is equal to its battery voltage.

$$I_{XV} = \frac{P_{XV}}{U_{XV}} \tag{6.23}$$

Voltage level specifies the electrical circuit of the PS to which the electric motor is connected. The electrical power consumed or generated by the motor is automatically considered in the power balance of the selected circuit.

Regenerative braking involves the partial recovery of kinetic energy and storage of this energy in the battery to increase the autonomy of electric vehicles. During regenerative braking, the electric car transforms into a generator and charges the batteries. The kinetic energy depends on the mass and speed of electric vehicles and only part of this energy can be recovered, depending on the capacity of the generator to produce energy and the charging capacity of the battery [7].

Battery acts as a storage unit for the electricity needed to operate propulsion engines. At the same time, the energy in the batteries must supply the auxiliary systems (cooling, heating, ventilation, lighting, multimedia, telematics etc.) in conditions of extreme values of ambient temperature, between − 30 and + 40 °C.

The base unit of the battery pack is the cell. A number of "n" cells arranged in a series–parallel group form a module [7]. The inverter converts the DC voltage supplied by the batteries into three-phase AC voltage (3 × 380 Vac), which powers the electric propulsion motors.

Most autonomous vehicle models are equipped with $Li_4Ti_5O_{12}$ (LTO) batteries, which are batteries with superior thermal stability, high cell energy storage capacity and long life (expressed by charge–discharge cycles) [7]. The electrical equations describe the characteristics of the cells, modules, and battery packs and were defined based on the virtual model for the base cell (Fig. 6.12) [7].

The electrical voltage of a cell (U_{cell}) is calculated using [7]:

$$U_{cell} = U_{ocv} - I_{cell} \cdot R_{ohmic} - \sum_{i=1}^{n} \frac{Q_i}{C_i} \tag{6.24}$$

where U_{ocv} (V) is the OCV (Open Circuit Voltage) of the cell, I_{cell} (A) is the current at the cell terminals, R_{ohmic} (Ω) is the internal resistance of the cell, Q_i (W) is the charging capacity of the cell, C_i (F) is the capacitance of the capacitor in the cell.

The charging capacity of the Q_i cell is expressed as follows [7]:

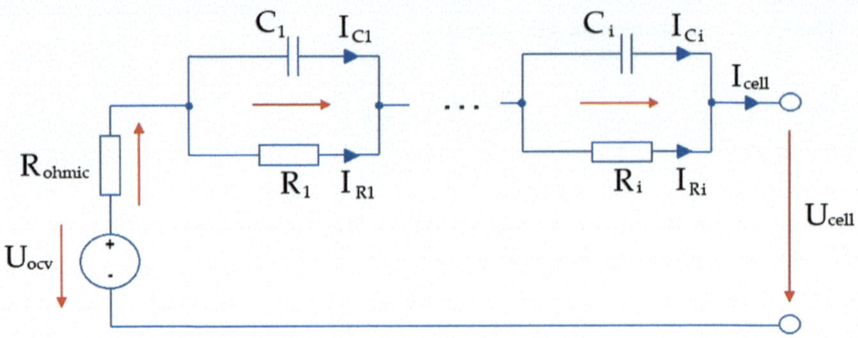

Fig. 6.12 Base cell of the battery pack

$$Q_i = \int I_{C_i} \cdot dt = \int \left(I_{cell} - \frac{Q_i}{R_i \cdot C_i} \right) \cdot dt \tag{6.25}$$

where I_{Ci} (A) is the current through the capacitor in the cell, R_i (Ω) is the resistance of each RC element of the cell.

The electrical voltage of a battery module (U_{module}) is calculated based on the number of cells connected in series:

$$U_{module} = U_{cell} \cdot n_{series} \tag{6.26}$$

The electric current of a battery module (I_{module}) is calculated based on the number of parallel cells:

$$I_{module} = I_{cell} \cdot n_{parallel} \tag{6.27}$$

The SOC of a battery module Q_{module} (%) is calculated based on the number of cells and the SOC of each cell Q_{cell} (%) [7]:

$$Q_{module} = Q_{cell} \cdot n_{parallel} = Q_{max} \cdot SoC \cdot n_{parallel} \tag{6.28}$$

where Q_{max} (%) is the SOC of the RC element of the cell.

The power of a battery module P_{module} (W) is calculated based on the number of cells and the power of each cell P_{cell} (W), respectively based on the number of cells connected in series:

$$P_{module} = P_{cell} \cdot n_{series} = U_{cell} \cdot I_{cell} \cdot n_{series} \tag{6.29}$$

The power loss of a battery module $P_{loss,module}$ (W) is:

$$P_{loss,module} = \left(I_{cell}^2 \cdot R_{ohmic} + \sum_{i=1}^{n} I_{R_i}^2 \cdot R_i \right) \cdot n_{series} \tag{6.30}$$

Instant battery charge is calculated based on the Coulomb-Counting algorithm, where Q_0 is the initial SOC of the battery:

$$Q(t) = Q_0 - \int_0^t I_{batt}(t)dt \tag{6.31}$$

The SOC of the battery is expressed as a percentage of the maximum charging capacity Q_{max} [7]:

$$SOC(t) = 100\% \cdot \frac{Q(t)}{Q_{max}} \tag{6.32}$$

Similarly, the DOD (Depth Of Discharge) of the battery is a percentage of the total consumed capacity [7]:

$$DOD(t) = 100\% \cdot \frac{Q_{max} - Q(t)}{Q_{max}} \tag{6.33}$$

6.2.2.4 Battery Management

Battery management controls the SOC of the batteries by assigning generator torques to the EM (Electric Motor) and managing the energy transfer between the electric circuits of the PS model. The functionality is closely linked to the architecture of PS, but its general intention is to keep each battery SOC at its target value. For this, battery management contains control loops over the total power in the electric circuits and the SOC of the batteries.

The virtual model contains a battery management for PS models with only one electric circuit. This battery management consists of one control loop. Calculation of the target generator torque [4]:

$$P_{target} = I_{SOC} \int (SOC - SOC_{target}) - I_P \int P_{tot_{LV}} \tag{6.34}$$

$$T_{EMtarget} = \frac{P_{target}}{\omega_{EM}} \tag{6.35}$$

The target generator torque is limited by the available surplus torque. The powertrain model contains one EM model and one MCU model. Its main tasks are:

(1) Determination of the maximum motor torque and the maximum generator torque at the current motor rotation speed.

(2) Control of the motor load to establish the target torque, target rotation speed or target load that is provided from powertrain control.
(3) Limit the rotation speed of the motors.

The MCU calculates torques and rotation speeds at the reference point that is located on the drive shaft.

The powertrain model may contain a LV and a HV battery and one BCU* model. This BCU* model controls the PS and all battery models. The main tasks are (Fig. 6.13) [4]:

(1) Calculation of the SOC and SOH of the batteries.
(2) Transfer of signals between powertrain control model and PS model.

The driveline model transfers the torque from each drive source output to the driven wheels. The calculation of wheel speeds and other rotation speeds is done in the computation part of this master module. The standard front and rear driveline configurations use one differential gear to distribute the torque from the drive shaft to the wheels (Fig. 6.14) [4].

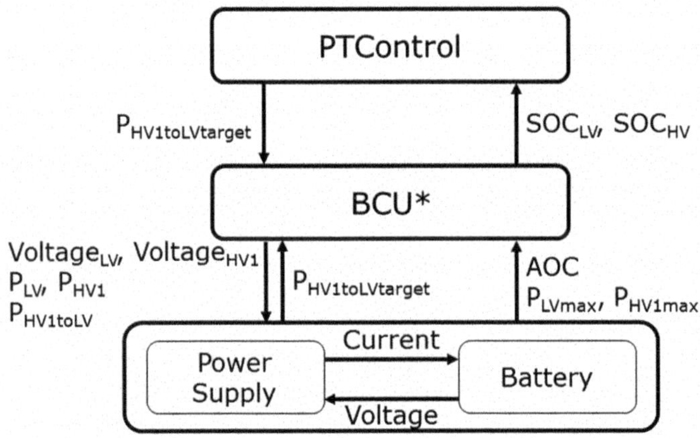

Fig. 6.13 BCU* model

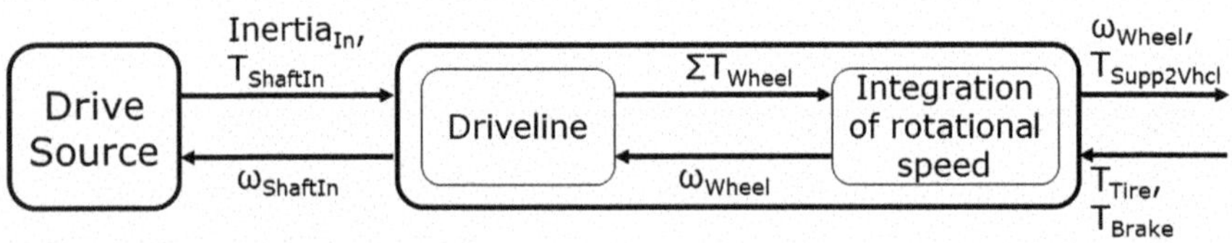

Fig. 6.14 Driveline model

6.2.2.5 Powertrain Control

The start button turns on the power, and the determination of operation state represents what is what happens on the side of the vehicle when the operator does so. The parameters for this model are set in the general tab (Fig. 6.15) [4].

The generic operation state machine differentiates six vehicle states:

(1) Absent driver is outside the vehicle, power off.
(2) Power off driver is inside the vehicle, power off.
(3) Power accessory electric power for accessories on.
(4) Power on electric power on (vehicle ignition).
(5) Start engine start.
(6) Driving ready for driving or driving.

The electric energy management in IPG Automotive powertrain control strategy controls the SOC of the batteries by assigning additional generator torques to the integrated starter motor and manages the energy transfer between different electric circuits. Its parameters can be set in the battery tab.

(1) Battery management deactivates the battery management completely. If it is activated, the parameters are available for LV/HV circuit.
(2) Maximum power of generator corresponds to the maximum admissible generator power for recharging the battery in LV/HV circuit.

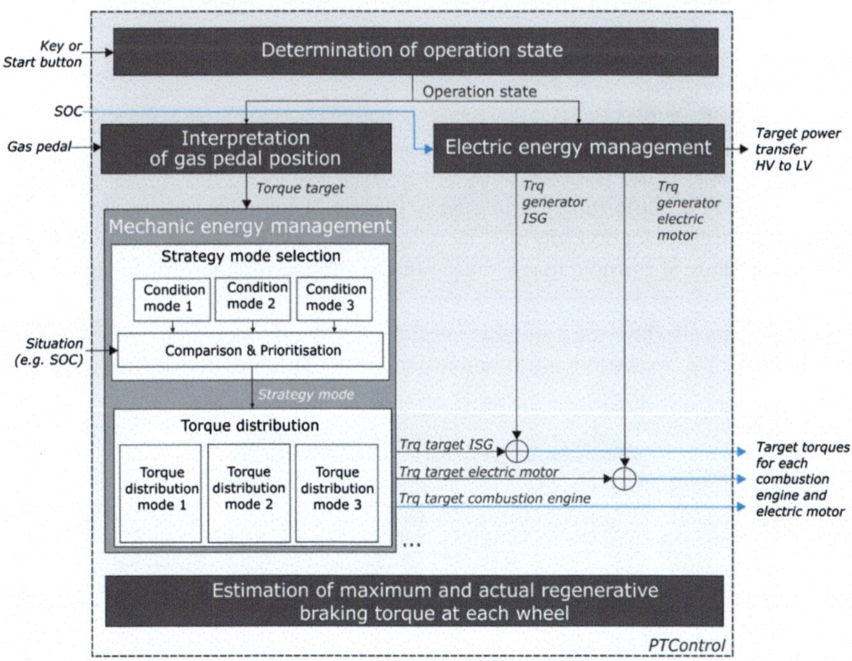

Fig. 6.15 Powertrain control

(3) Target SOC of LV/HV battery is used as target value for the controller algorithm of the battery management.
(4) SOC control I-value scales the influence of the difference between target and current battery SOC (integral element) on the additional generator torque demand.
(5) Consumer control I-value scales the influence of the auxiliary power consumption on the additional generator torque demand.

Regeneration tab contains the following elements (regeneration brake):

(1) SOC limit is the SOC of the main battery (HV battery) that needs to be exceeded to switch into this strategy mode. Depending on the strategy mode it refers to, this can be an upper (mode electric drive) or lower (mode regenerative braking/regeneration drag) limit.
(2) SOC limit range defines a zone around the SOC limit.
(3) Speed limit is the vehicle speed that needs to be exceeded to switch into this strategy mode. This is always an upper limit.
(4) Speed limit range defines a zone around the speed limit for mode electric driving:
    (a) Condition to enter mode electric drive (vehicle speed) < (speed limit) − 0.5 * (speed limit range).
    (b) Condition to stay mode electric drive (vehicle speed) < (speed limit) + 0.5 * (speed limit range).
(5) Torque limit (regeneration mode) sets the absolute value that the torque demand must exceed to activate regenerative braking or traction.

6.2.2.6 Sensor Models

The sensors extract the relevant information from the environment and deliver it in a form suitable to the function interface. In case of imaging sensors tracking of objects and target selection are also included. Since ideal sensors are technology independent, they are simple to use for parametrization with a low effort (Fig. 6.16) [4].

Ideal sensors also provide a suitable interface for user defined models for function development. The extracted information can be easily enhanced by user-specific

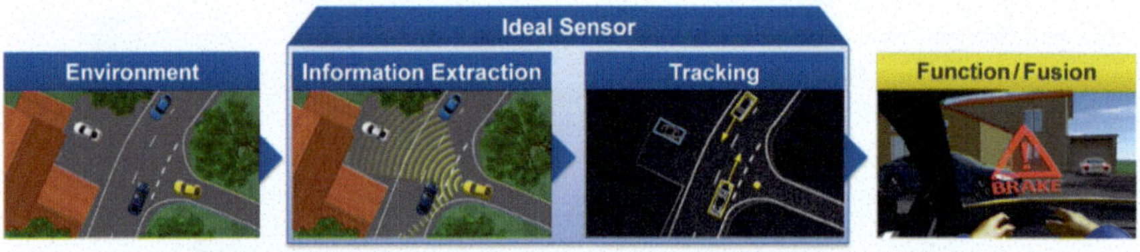

Fig. 6.16 Ideal sensor (CarMaker sensor models)

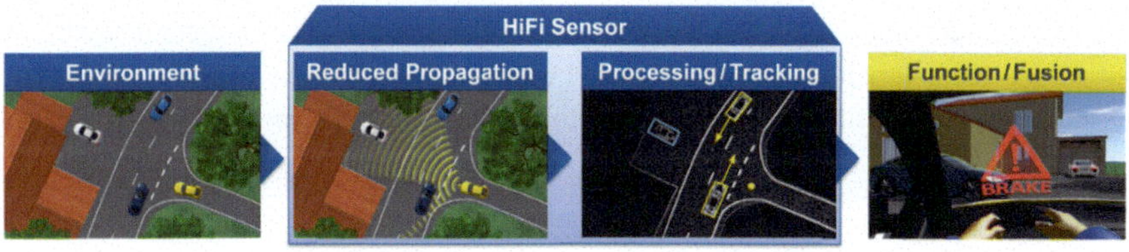

Fig. 6.17 Hi-Fi sensors (CarMaker sensor models)

calculations to get a custom physical sensor model. For instance, the free space sensor can be extended with an intensity-based detection to model a Lidar sensor.

The Hi-Fi (High Fidelity) sensors are designed for function development and testing. They make it possible to answer the question whether the function under development works in general. Hi-Fi sensors consider the effects specific to the basic technology caused by propagation of the respective signal in the environment, as well as by processing algorithms. They also provide the information in a form appropriate to the function interface (Fig. 6.17) [4].

The RSI (Raw Signal Interface) are made for component development and testing. They focus on detailed physical effects by propagation of the respective signal in the environment. RSI do not offer a processing/tracking component giving the user the freedom to model all electronic and software components himself. This allows the detailed investigation for the real component under development without the need to disclose any know-how (Fig. 6.18) [4].

The remaining sensors are relevant for the development of ADAS. Depending on the application case, each sensor delivers different information regarding the test environment based on different physical models.

The ideal sensors provide ground truth information relevant for environment detection and object tracking. The generated object list is independent of the technology used and these sensors can be parameterized [4].

When it comes to function development and testing, Hi-Fi sensors can be used. The provided object list varies, depending on the technology that is being used. The environment detection is characterized by a reduced propagation, but object tracking is still provided [4].

The RSI follows a different approach: instead of an object list, it provides raw signals, like the input of a camera. This information can be used as input to signal

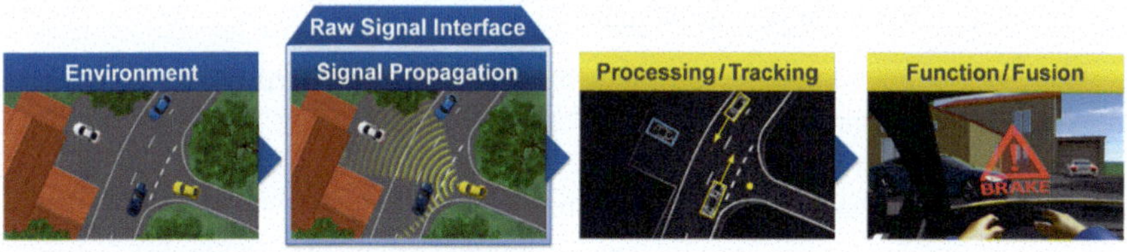

Fig. 6.18 RSI sensors (CarMaker sensor models)

processing and object tracking algorithms. The output considers several levels of physical effects depending on the selected module and lens type [4].

Table 6.1 represent the assignment of the available sensor models for the virtual vehicle model [4].

6.2.3 Algorithm of Simulation Process

The algorithm of the modeling and development process of the virtual model for the autonomous vehicle is structured in steps. These steps have as a desideratum the creation of a new project for the modeling of a virtual vehicle, respectively a new version for an existing virtual model [1].

CarMaker refers to data sets, it does not use the term vehicle model. This is because technically, CarMaker possesses only one vehicle model. This model has defined and fixed masses, DOF (Degrees Of Freedom), spring/damper elements etc. The data set contains the values necessary to parameterize the model, for the weight of the masses, the stiffness of the springs etc. [7].

The control of the autonomous vehicle takes place in the context of the identification of some information sources, respectively of some sensory and perception processes that include acts of orientation towards the environment [8, 9]:

(1) Information and perception source:

   (a) Visual display systems in the vehicle: display instruments, input devices, the possibility of retrieving information provided by the autonomous driving system.
   (b) Main acoustic information systems: acoustic element of the navigation system, acoustic alarm devices.
   (c) Secondary acoustic information systems: radio, telephone, acoustic information transmitted to the passengers.
   (d) Information transmitted from other traffic participants: vehicles, pedestrians.
   (e) Road characteristics: transverse and longitudinal characteristics, intersections, road width, number of lanes.
   (f) Traffic signs, speed limits.
   (g) Road condition, weather conditions, visibility.

(2) Evaluation based on perception data:

   (a) Longitudinal distances from or between other traffic participants.
   (b) Lateral distances to or between other road users.
   (c) The velocity of the autonomous vehicle and of other road users.
   (d) Anticipation of critical traffic situations.

(3) Decision:

   (a) Selection of actions required to move the autonomous vehicle.

Table 6.1 Overview sensor models (CarMaker sensor models)

Application	Physical model	Sensor	Description
Vehicle dynamics	Ideal	Slip angle	Measures the vehicle's side slip angle (CPU (Central Processing Unit) based)
	Ideal	Inertial	Measures inertial body movements (CPU based)
ADAS	Ideal	Object	Detects objects defined as traffic objects (CPU based)
	Ideal	Free space	Detects free and occupied spaces between objects defined as traffic objects, similar to a stereo camera (CPU based)
	Ideal	Free space plus	Detects free and occupied spaces in the entire environment (GPU (Graphics Processing Unit) based)
	Ideal	Traffic sign	Detects traffic signs along the road (CPU based)
	Ideal	Line	Detects road markings such as lines and guard rails, jersey barriers etc. (CPU based)
	Ideal	Road	Provides road information as electronic horizon (CPU based)
	Ideal	Collision	Detects contacts of the test car with other traffic objects (CPU based)
	Ideal	Object by line	Detects traffic objects moving along the selected road lanes (CPU based)
	High fidelity	Radar	Detects objects defined as traffic objects based on the SNR (Signal-to-Noise Ratio) (CPU based)
	High fidelity	Camera	Detects objects defined as traffic objects, traffic signs and traffic lights (CPU based)
	High fidelity	Global navigation	Simulates GPS satellites and their visibility for the receiver on the vehicle (CPU based)
	Raw signal interface	Radar RSI legacy	Radar Sensor simulating the propagation of electromagnetic waves through the virtual environment Outdated and less powerful radar, sensor legacy model (GPU based)

(continued)

Table 6.1 (continued)

Application	Physical model	Sensor	Description
	Raw signal interface	Radar RSI	Radar Sensor simulating the propagation of electromagnetic waves through the virtual environment Enhanced radar sensor model (GPU based)
	Raw signal interface	Ultrasonic RSI	Ultrasonic sensor simulating the propagation of mechanical sound pressure waves through the virtual environment (GPU based)
	Raw signal interface	Lidar RSI	Lidar sensor simulating the propagation of laser light pulses through the virtual environment (GPU based)
	Raw signal interface	Camera RSI	Camera sensor providing image information using IPGMovie (GPU based)

(b) Selection of actions required to guide the autonomous vehicle.

(4) Autonomous vehicle movement control.

(a) Control of the longitudinal movement of the autonomous vehicle (acceleration, braking).
(b) Control of the lateral movement of the autonomous vehicle (steering).
(c) Control of auxiliary systems (lights, wipers, radios etc.).

Step 1 consists in the methodology of collecting and generating the initial data of the virtual model to be developed based on the intermediate steps necessary for collecting the initial data, respectively based on the operations performed for generating these data.

Step 2 consists in virtual model development by choosing the necessary components for the construction of the systems that make up the proposed model and location, respectively their connection according to the construction scheme defined in step 1.

Step 3 consists in defining the elements of the model by entering the initial data generated in step 1, operation after which the elements of the model become components with defined characteristics.

Step 4 consists in defining the virtual road and the virtual environmental in which the virtual model defined in the previous steps is to be integrated.

Step 5 consists in defining the virtual driver behavior which represents the operation algorithm of the autonomous driving system.

Step 6 consists in defining a traffic model based on the following data: number and types of vehicles in traffic, direction of travel, pedestrians in traffic,

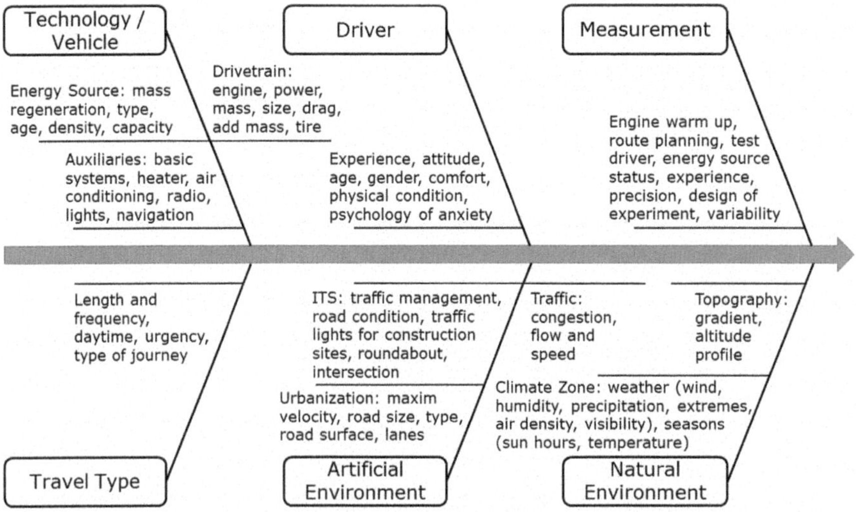

Fig. 6.19 Impact factors on the energy balance for electric vehicle

controlled or uncontrolled traffic, traffic rules, rush hours and traffic congestion, intelligent transport systems etc.

Step 7 consists in establishing the simulation tasks which involves the definition of maneuvers in the movement of the virtual model, maneuvers highlighted when the computer simulations are defined.

Step 8 the characteristics of the simulation process consist in defining the simulation step, the simulation time step, and the number of outputs after every simulation steps.

Step 9 consists of running computer simulations on the virtual model built and validated in the previous steps which can be carried out according to one of the methods related to the duration of the simulation process without time limits "−RT", simulation "RT", respectively accelerated simulation "+RT".

Step 10 consists of viewing and evaluating the results of computer simulations (show results).

All the impact factors involved in calculating the energy balance for an electric vehicle are shown in Fig. 6.19 [10].

6.2.3.1 Initial Data (Step 1)

Step 1 consists of the methodology of collecting and generating the initial data for the virtual model to be developed based on the intermediate steps required for collecting the initial data, respectively based on the operations performed to generate these data (Fig. 6.20).

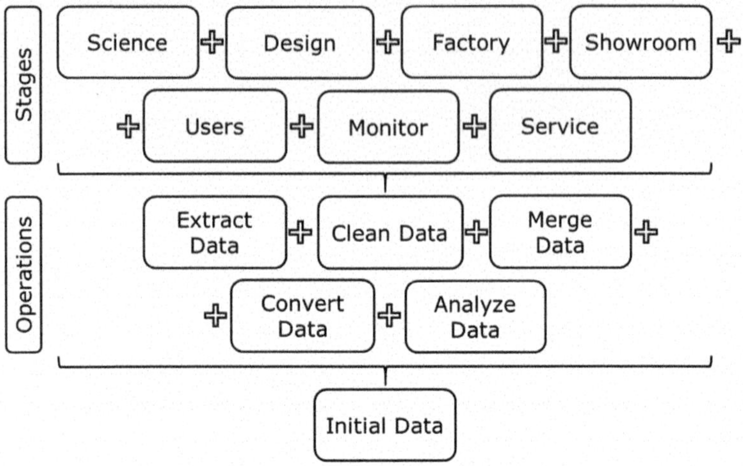

Fig. 6.20 Methodology for collecting and generating initial data

(A) The collection of the initial data for the virtual vehicle model to be developed includes the following intermediate steps:

(1) Science has the role of extracting and collecting data sets based on information about the model considered, data that can be found in scientific papers that have conducted and published studies and research on similar models. Generally, these data are based on in-depth studies, both theoretical, based on mathematical and/or simulation models, and experimental, generally from similar models from accredited laboratories, and these published studies are validated by scientific communities.

(2) Design has the role of collecting data sets on the construction dimensions, respectively on the design elements of the real vehicle considered. These data are based on direct observations, scans, and measurements on the real model of the vehicle considered. Design elements are useful for defining those virtual models for which the data sets cannot be obtained from the manufacturer.

(3) Factory has the role of collecting data sets from the manufacturer that contain the technical details of the systems and components of the vehicle considered. They are the most complete and useful data sets that can be used to define a virtual model and are based on the manufacturer's construction plans and data, respectively on observations and measurements obtained during the production process.

(4) Showroom has the role of collecting data sets on the characteristics of the vehicle considered from the sales tools available at dealerships (Salesforce etc.). In addition to the main constructive features, the sales tools highlight the advantages and disadvantages of the models promoted by the sales teams.

(5) Users has the role of collecting data sets on the vehicle considered based on user experience. This information is especially useful when developing new versions of existing models. To be conclusive, the data sets collected from users must be the result of surveys based on a representative number of respondents.
(6) Monitor has the role of collecting the data sets regarding the considered vehicle from the monitoring of the fleets of similar vehicles existing in operation. The companies that own these fleets have access to operating data monitoring platforms for vehicles, and these data are generally downloaded at the end of a business day, and based on them, statistics and periodic reports are issued on the activities carried out, on the balance sheet and energy efficiency, income, and expenditure balance etc.
(7) Service has the role of collecting the data sets regarding the vehicle considered from the reports of the service centers. The service activities of car dealerships, especially during the warranty period of the models considered, are of major importance in determining the reliability of these vehicles, the frequency of any malfunctions or functional problems that may occur as a result of their operation.

(B) Generating the initial data for the virtual model to be developed includes the following intermediate steps based on the raw data sets that were collected in the previous steps:

(1) Data extract consists in selecting the sets of relevant parameters for defining the components of the developed virtual model. In cases where the data sets have been taken from the time-lapse monitoring platforms (months, years etc.) of the operating data for the vehicles in question, for the selection of initial data sets that will be useful in defining virtual models, it is necessary to extract only certain categories of data (consumption, power, variation of specific parameters etc.), possibly for certain seasonal periods (summer/winter, or certain months of the year etc.), or standard climatic conditions (temperatures, precipitation, other environmental conditions at certain times of the day etc.).
(2) Clean data consists in filtering the data extracted by any deviations and values that exceed the allowed standard limits. Boxplot (Microsoft) statistical analyzes or various genetic algorithms integrated in DE (Design Explorer) applications are designed to filter the extracted data, providing workflows for designing DoE (Design of Experiment) type experiments, necessary for the analysis and optimization of the data sets to be used as initial data in defining the virtual model for the vehicle considered.
(3) Merge data consists of combining data sets in the same category for different modes of operation of the vehicle (consumption, power, variation of parameters etc.), or from different databases, respectively different locations, or time periods, but which have common features.
(4) Convert data consists in converting or transforming the data collected in the formats accepted by the simulation application to define the components

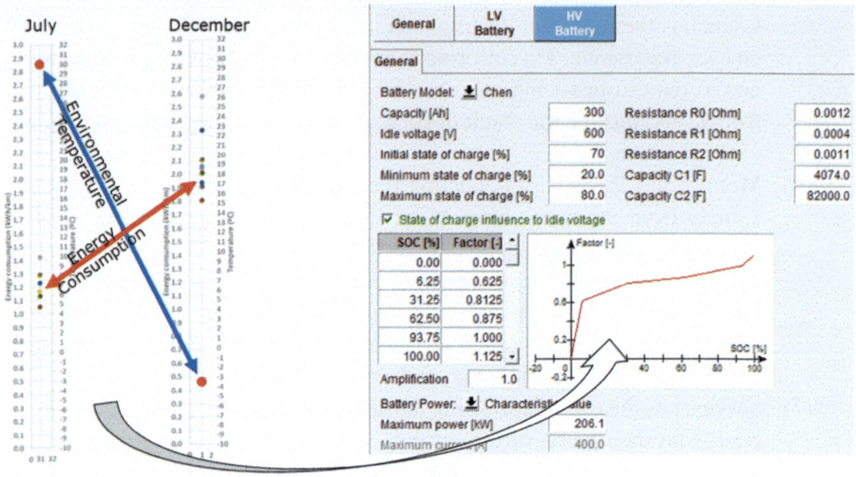

Fig. 6.21 Analysis, validation, and certification of initial data

of the virtual model (pressures, temperatures, flows, volumes etc. in the format accepted by the simulation application).

(5) Analyze data consists of the analysis of the data sets that will be used to define the components of the virtual model. The analysis of the obtained data has to validate and certify that the data sets are compliant and can be used as initial data in defining a virtual model, similar to the real model of the vehicle considered (Fig. 6.21).

6.2.3.2 Virtual Model Development (Step 2)

Step 2 consists in virtual model development by choosing the necessary components for the construction of the systems that make up the proposed model and connecting them according to the constructive architecture defined in step 1—Virtual model in the format provided by the simulation application, which are called components. These have undefined technical characteristics (Fig. 6.22).

A virtual vehicle model consists of systems (main and auxiliary) as they are found in the structure of the real model. These systems consist of interconnected components. The main systems of a virtual model of an autonomous vehicle with electric propulsion are the following:

(1) Electric machine consisting of traction motor (electric motor), power inverter (AC/DC inverter), electric battery (battery).
(2) Driveline consisting of the mechanical systems responsible for transmitting the kinematic chain to the wheels: steering system, braking system, wheels etc.
(3) Battery thermal management system.

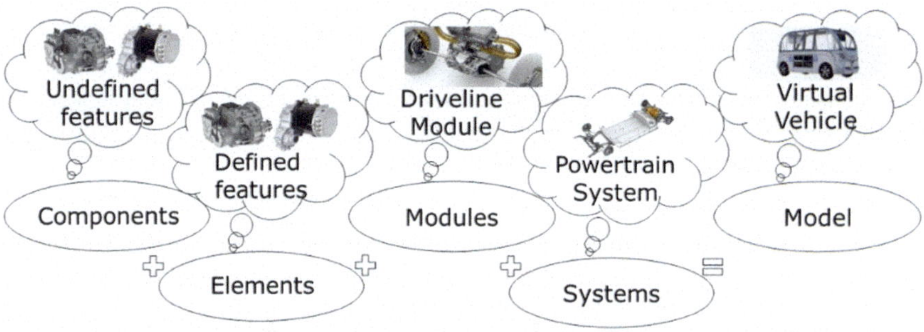

Fig. 6.22 Virtual model development methodology

(4) Auxiliary systems consisting of HVAC (Heating Ventilation and Air Conditioning), DC/DC converter and electrical systems etc.
(5) Autonomous driving systems, sensor systems etc.

6.2.3.3 Defining the Components of the Model (Step 3)

Step 3 consists of defining the components of the model by entering the initial data generated in step 1, operation after which the components of the model become elements with defined constructive characteristics.

There are situations in which the elements of the virtual model are grouped into modules for a clear highlight of their operation within the virtual model. For example, the elements of motion transmission from the propulsion element (but without this element) to the wheels form the driveline module. Another example: the electric motor together with the power inverter and the electric battery form the electric machine module.

The grouping of two or more modules forms the systems of the virtual model. For example, the driveline module together with the electric machine module form the powertrain. All interconnected virtual model systems form the virtual vehicle (see Fig. 6.22).

The virtual model developed based on the components, elements, modules, and systems defined using the initial data is characterized by the following categories of parameters found in the simulation process [1]:

(1) IP* (Initial Parameters) that are defined by the user. These can be defined according to the values of the initial data (for example: power consumption of the electric motor/electric consumers) or can be modified by the user according to the requirements of the simulation tasks.
(2) EP (External Parameters) which are the characteristic parameters of the environmental conditions (temperature, pressure, ambient humidity etc.) or in different points (zones) of the virtual model (temperature of the electric battery modules, temperatures of various modules or systems of the virtual model etc.). The peculiarity of the external parameters (which can also be applied to the real model)

is that the user can intervene through a process to change their value (by establishing a mode of operation of the electric battery thermal management system, the user can control the temperature of the battery elements, or under laboratory testing the actual vehicle may run in a user-conditioned environment at different values of ambient temperature, pressure, and humidity).

(3) CP (Calculate Parameters) which are the parameters calculated by the mathematical apparatus of the virtual model based on the initial data consisting of input parameters and external parameters (electricity consumption at 100 km under certain conditions of the simulation process (specific environmental and climatic conditions, elevation profile of the road)), the amount of electricity recovered by regenerative braking under certain simulation conditions (specific environmental and climatic conditions, elevation profile of the road). The user can intervene by changing the conditions of the simulation process to change the value of these parameters (by customizing the specific environmental and climatic conditions, or by changing the elevation profile of the road).

(4) RP (Resulting Parameters) which are the parameters resulting from the simulation process. The user cannot directly intervene in changing these parameters (electric motor power).

6.2.3.4 Virtual Road and Environment (Step 4)

Step 4 consists in defining the virtual road and the virtual environment in which the virtual model defined in the previous steps is to be integrated. The digital road on which the virtual vehicle will travel during computer simulations can be defined by geographical coordinates (latitude, longitude, and altitude).

It is important to use a digitized route at maximum resolution in the simulation application, similar to the route implemented in the autonomous driving system, considering that the interpretation of this route and the travelling will be similar for both the virtual vehicle and the real vehicle.

In addition to the geographical coordinates of the route, the following set of parameters regarding the route will be defined in the simulation application: altitude, inclination, friction coefficient right/left, direction of travel (right/left), number of lanes, tram lines, bicycle lanes, pedestrian crosswalks, and shelters, turns, speed limits, restrictions etc. Road characteristics data can be retrieved from various sources and contains sets of information about (e.g.):

(1) Road condition between km … and km …
(2) Cause: weather conditions, sector roadworks, accidents etc.
(3) Measures taken speed limits (between km … and km …), or traffic restrictions (one-way traffic between km … and km …).
(4) Traffic exceptions: only certain categories of vehicles can drive (up to 3.5 tons), only drive between the hours of …

The environment utility in the CarMaker application allows to define the atmospheric conditions: ambient temperature, humidity, air density, atmospheric pressure, wind velocity, wind direction, horizontal visibility, dew point temperature at any time. The environment parameters contain settings that define the animation scenes in the IPGMovie utility: visual range in fog, sun elevation, sun azimuth, date, time (Table 6.2):

(i) Reference temperature specifies the environmental air temperature.
(ii) Air density specifies the environmental air density at mean sea level.
(iii) Air pressure specifies the environmental air pressure at sea level.
(iv) Air humidity specifies the environmental air humidity.
(v) Solar radiation specifies the environmental solar radiation.
(vi) Rain rate specifies the amount of rain per hour.
(vii) Visual range in fog specifies the visibility in a foggy environment.
(viii) Sun position specifies the sun positioning.
(ix) Sun azimuth specifies the sun positioning and the cardinal direction angle (azimuth) of the sun.
(x) Sun elevation specifies the sun positioning and the height angle (elevation) of the sun.
(xi) Temperature specifies the environmental air temperature.
(xii) Elevation defines the temperature offset depending on the elevation as a linear function.
(xiii) Depending on time of day defines a temperature offset depending on the time of day.

An important factor in defining the algorithm of the simulation process is the definition of the steps of implementation of the virtual model for the autonomous vehicle on the road and in the virtual environment. The following tools are available for this in CarMaker [3, 5, 8].

Maneuver (see Fig. 6.3) on a vehicle in CarMaker consists of a set of instructions that represent the control action that drives and coordinates the movement of the vehicle. Each set of maneuvers consists of one or more steps called mini-maneuvers that are performed successively.

The maneuver represents the control action that is meant to lead to certain vehicle motion but does not set the actual vehicle motion directly. Usually, the actual vehicle motion follows the desired vehicle motion, but sometimes the desired motion can be beyond the vehicle limits and thus is not reached. A maneuver consists of one or several maneuver steps called mini-maneuvers that are executed one after another (Table 6.3).

The first entry in the list of minimaneuvers "====Global Settings/Preparation====" has a special meaning. It sets the initial conditions at simulation start and some global settings that are valid during all minimaneuvers of this TestRun. To add a step to the end of the list, select the entry "====END====" and then click on the button New. Minimaneuvers can also be copied (Copy) and pasted (Insert), deleted (Delete) or imported from another TestRun (Import).

When right-clicking inside the minimaneuvers list, there will also be an option clear, used to delete all steps at once. Minimaneuvers are composed of the following general specifications [3, 10]:

Table 6.2 Environment

Atmosphere		
Reference temperature (°C)		20.00
Air density (kg/m^3)		1.200
Air pressure (bar)		1.000
Air humidity (%)		60.00
Solar radiation (W/m^2)		400.0
Rain rate (mm/h)		0.000
Visual range in fog (m)		10,000
Sun position		Angles
Sun azimuth (°)		45.00
Sun elevation (°)		45.00
Temperature		
Environment model		Generic
Elevation (km)		– 0.0065
Depending on time of day		
Time (h)	Offset (K)	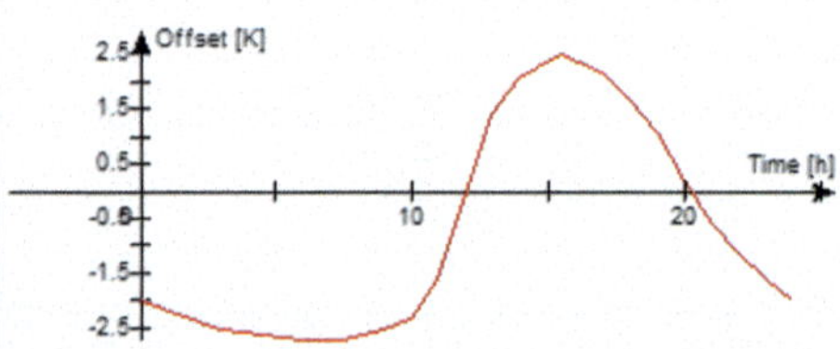
0.000	– 2.000	
3.000	– 2.500	
6.000	– 2.700	
9.000	– 2.500	
10.00	– 2.300	
11.00	– 1.600	
12.00	0.000	
13.00	1.400	
14.00	2.100	
15.50	2.500	
17.00	2.200	
18.00	1.700	
19.00	1.100	
20.00	0.200	
21.00	– 0.600	
22.00	– 1.100	
23.00	– 1.600	
24.00	– 2.000	
Amplification		1.000

Table 6.3 Maneuver

Maneuver					
No	Start	Duration	Long	Lat	Label
====Global Settings/Preparation====					
0	0.000	999.0	25.00		
1	999.0				

Specification of maneuver step				
Label				
Description				
End condition				
Duration (time/distance)	999.0	s	–	m
Longitudinal dynamics				
Mode	IPGDriver			
Speed (km/h)	25.00			
Driver parameter (driver)				
Lateral dynamics				
Mode	IPGDriver			
Track offset (m)	0.000			
Driver parameter (driver)				

The duration of these minimaneuvers defines when this minimaneuver ends. Each minimaneuver needs to have an end condition that defines its duration. If this condition is fulfilled the maneuver control switches to the next minimaneuver or ends the simulation, if the current minimaneuver is the last one.

Longitudinal dynamics actions describe the vehicle motion control in longitudinal direction and thus have an influence on the accelerator control and the brake control.

Lateral dynamics actions describe the vehicle motion control in lateral direction and thus have an influence on the steering wheel position respectively torque.

Additional actions e.g., turning the engine on or off, driver assistance systems or lights.

(A) The longitudinal dynamics section specifies how the maneuver controls the accelerator pedal to control the longitudinal vehicle motion [4]:

   (1) IPGDriver is used for fully automated speed control (closed loop control) that shows a behavior like a human driver. This driver model accelerates and decelerates the vehicle according to its dynamic limits just as a human driver would do. This option is also used to follow a measured, distance-based speed profile read from an external file.
   (2) Speed profiles follow measured, time-based speed profile read from an external file. Optional parameter defines the speed the driver tries to reach, if it is possible with the current route shape and if there are no speed limits

that impose driving at lower speeds. If no value is defined, IPGDriver uses the cruising speed defined in the IPGDriver dialog as maximum speed.

(3) Speed control simulates a speed controller. The speed control option cannot brake the vehicle: once the specified speed is reached, it will be kept without considering the curves of the road.
(4) Speed transition is used to provide a target velocity to the vehicle and to transition from the current velocity to the target velocity as a function of time, distance, or acceleration. Depending on the type of transition chosen, different end conditions are considered for the minimaneuver.
(5) Manual control of the pedal positions. This is an open loop maneuver. The pedal position is not adapted according to any desired vehicle speed.
(6) Stop vehicle – IPGDriver will use it to stop the vehicle. IPGDriver will press the brake pedal in a human-like way and not with a linear function as the manual option would do. This option can be used for stopping during forward driving as well as for stopping during backwards driving.
(7) Drive backwards – IPGDriver will use it to drive backwards. The vehicle should be standing still before the beginning of this minimaneuver.
(8) IPGDriver + user driver enables custom driver model developed by the user can be activated and parametrized. The infofile containing the parameters for the user driver can be selected by clicking on the folder icon. This selection automatically activates the user driver and causes the IPGDriver to be overwritten.
(9) Not specified option selects a default longitudinal controller for the ego vehicle, which internally uses the driver lite model. This option identifies the vehicle's velocity at the start of the minimaneuver step and specifies it as the target velocity for the driver lite model.

(B) The lateral dynamics section specifies how the maneuver control acts on the steering wheel to control the lateral vehicle motion [4]:

(1) IPGDriver is used for controlling the vehicle's position (Closed Loop Control). This model follows the route just as a human driver would do. The track offset defines the driver's deviation from its static desired course. A track offset of zero means that the driver follows its static desired course (that is like the route). Positive values make IPGDriver steer the vehicle to the left and negative values make it steer to the right.
(2) Sinus option applies a regular sinusoidal input to the steering wheel. This is an open loop maneuver. That means the steering wheel input is not adapted according to the vehicle's position or to its yaw movement.
(3) Sinus sweep option applies an ascending or descending sinusoidal input to the steering wheel. This is an open loop maneuver. That means the steering wheel input is not adapted according to the vehicles position or to its yaw movement.
(4) Steer step option applies a steady ramp input to the steering wheel, up to a defined angle, respectively a torque limit. Using a short duration for the ramp, it corresponds to a classical steer step.

(5) Follow course option activates a primitive steering wheel angle (respectively the torque) controller. It is a controller which follows the center line of the road. It does not consider any course changes like pylons.
(6) Lane change option under the lateral dynamics section of the maneuver window enables the user to define a lane change action by directly specifying a new lane as the target lane. On top of this, an additional target offset can also be specified, which is then applied with respect to the new lane's center.
(7) Lane offset option enables the user to define a target offset which is then applied with respect to the current lane's center. The duration of the maneuver must be defined by specifying either a time limit or a distance limit (but not both). If no other end conditions are specified, the minimaneuver ends as soon as the target offset has been reached.
(8) Not specified option identifies the actual lateral offset to the route at the start of the minimaneuver and sets this as the target for the default lateral controller. Like the not specified option in longitudinal dynamics, if the vehicle is in a transient state at the start of the minimaneuver, then this has an influence on the target offset passed on to the controller.

Track offset parameter defines the deviation of the vehicle from the travel course. A zero value for track offset means that the vehicle follows the course on the implemented route without deviations from it. A positive value for track offset causes a deviation of the trajectory to the left, and a negative value causes a deviation of the trajectory to the right.

Minimaneuver commands allows to add special actions and calculation formulas in the form of script commands, such as [3, 4]:

(1) Switching the driver assistance systems on/off.
(2) Calculate additional characteristic values such as brake distance or maximum speed reached.
(3) Abort the simulation automatically if a critical situation comes up.
(4) Switch the vehicle's operation states to turn the ignition on and off.
(5) Jump to another minimaneuver if a specific condition is met.
(6) Turn the vehicle lights on/off.
(7) Apply external forces and torques to the vehicle.
(8) Trigger the result storage and start the postprocessing.
(9) Operate the failsafe tester on HiL (Hardware-in-the-Loop).
(10) Write messages to session log.

Input from file … mechanism (Fig. 6.23) is another possibility to specify the starting conditions. A start velocity and maximum throttle position can be specified. As opposed to the starting conditions in the maneuver dialog, in this case the vehicle is accelerated in the preparation phase of the simulation. At simulation start, the vehicle is already settled at the given initial conditions which prevents disturbances in the signal sequence.

Input from file covers three use cases:

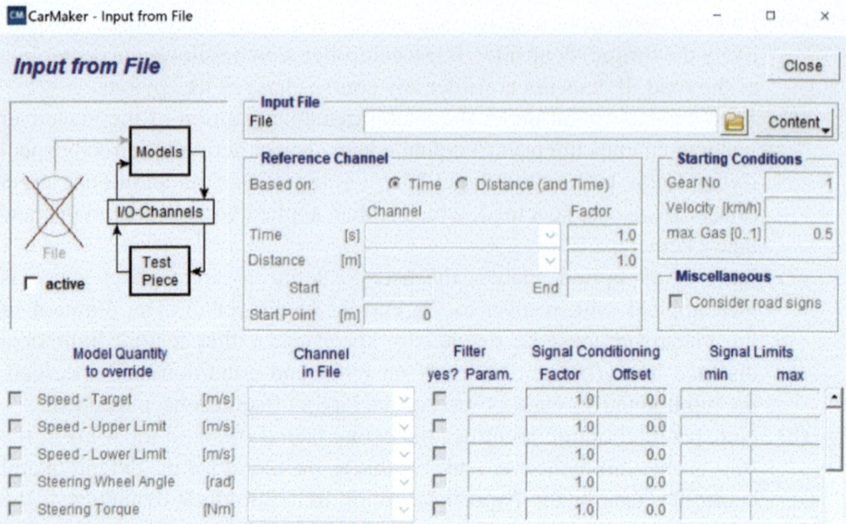

Fig. 6.23 Input from file

(1) Overwriting maneuver control outputs by measurement data. Input from file can be used for overwriting the pedal positions, the gear selection, the steering wheel input etc. provided by the conventional maneuver control by measurement data. The measurements replace the driver's actions that is forwarded to the vehicle model. The vehicle's movement then is a result of the vehicle model calculation based on this input.
(2) Following a measured, distance-based speed profile. Input from file enables to impose a target vehicle speed to IPGDriver as a function of the driven distance. This speed profile is closely linked to the vehicle's position on the route.
(3) Following a measured, time-based speed profile. Following a time-based speed profile is a different task then following a distance-based speed profile. It needs to be executed by a specific version of IPGDriver that is optimized for this task.

To activate it, the speed profile for the longitudinal dynamic's settings of the maneuver is needed to be used. The available channels overwrite the corresponding user accessible quantity of the maneuver control directly. The following channels are in Table 6.4 [4].

6.2.3.5 Virtual Driver Behavior (Step 5)

Step 5 consists in defining the virtual driver behavior which is the algorithm for the operation of the autonomous driving system. It can be defined using different approaches to the driving style of a human driver (preventive driving, dynamic driving etc.). All of these driving behavior patterns can be defined based on the characteristics

Table 6.4 Input from file channel

Model quantity	Unit	Channel in file	Signal factor	Conditioning offset
Speed target	m/s		1.000	0.000
Speed upper limit	m/s		1.000	0.000
Speed lower limit	m/s		1.000	0.000
Steering torque	N m		1.000	0.000
Gas pedal	0…1		1.000	0.000
Brake pedal	0…1		1.000	0.000
Clutch pedal	0…1		1.000	0.000
Gear no	−1…n		1.000	0.000
Selector control	−9…1		1.000	0.000
Parking brake	0…1		1.000	0.000
Brake pressure	Bar		1.000	0.000
User define 1	–		1.000	0.000
User define 2	–		1.000	0.000
User define 3	–		1.000	0.000
User define 4	–		1.000	0.000

of a human driver's reaction speed. The set of parameters that define these behaviors contain data about: starting launch speed, braking speed, maximum braking force, reaction speed when changing direction in the event of an obstacle, data on the modification and adaptation of the driving behavior (acceleration pedal (on/off), shifting according to velocity (up/down), shifting time velocity, shifting according to speed (up/down), shifting time speed, shifting skip gear (up/down), clutch pedal (on/off), time of clutch release, maximum brake force, brake pedal at standstill) etc.

The IPGDriver utility in the CarMaker application makes it easy to add a human driver's control actions to simulate the operation of the virtual vehicle. These actions include [4]:

(1) Choice of driving course within the lane borders (corner cutting).
(2) Steering.
(3) Choice of the driving speed according to the course and the vehicle behavior.
(4) Influence on the speed over the accelerator and brake pedal, as well as over the clutch pedal and gear selection.
(5) Regard traffic objects and follow.

The virtual model for the autonomous vehicle, the virtual driver, the virtual environment in which the virtual model runs and the interaction between all these entities (Fig. 6.24) form the VVE (Virtual Vehicle Environment) system. The CarMaker IPGDriver utility is a component of the VVE virtual system, which allows the definition of the simulation conditions and the behavior of a real driver, setting input parameters and values of these parameters that are essential for the system and it

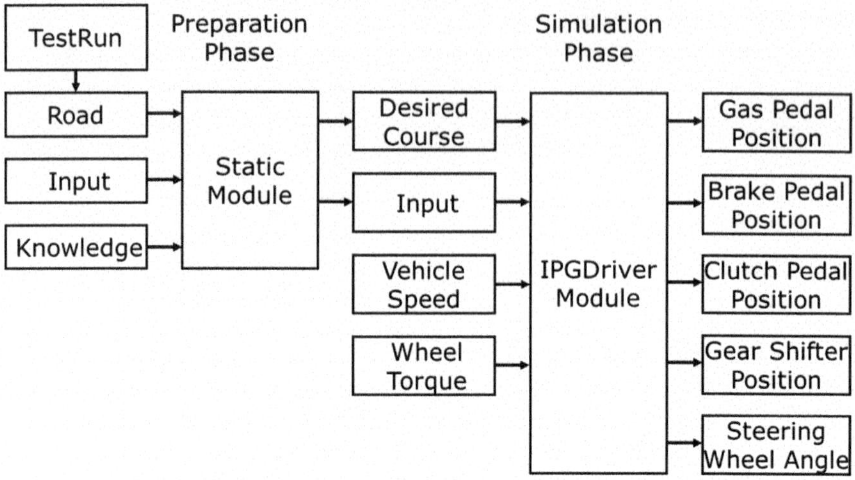

Fig. 6.24 IPGDriver input and output parameters

provides the mathematical device for calculating the values of the output parameters corresponding to the operation of the system in a response loop [8].

The input parameters required to perform the simulation are as follows:

(1) Desired course—that contains the data needed to describe the trajectory along a two-dimensional surface. This is done by specifying the x and y coordinates along the centerline, by defining the corresponding width of the lanes.
(2) Desired speed—to start the simulation, the speed profile is set for the entire route, which considers the following factors: desired cruising speed, deceleration before entering turns, acceleration after turning etc. Taking these factors into account, IPGDriver will maintain the desired speed throughout the simulation process, intervening even when speed corrections are required due to events that occur during the travel.
(3) Vehicle state—that contains information about the current movement of the vehicle, speed, acceleration, side slip angle etc.

The IPGDriver utility calculates the values of the output parameters corresponding to the operation of the powertrain, steering and braking systems so that they ensure a safe movement of the autonomous vehicle. The output parameters required to perform the simulation are as follows:

(1) Steering correction—acts on the angle and torque corresponding to the actuation of the steering system to ensure the current movement of the autonomous vehicle.
(2) Speed correction—acts on the speed of the vehicle to ensure the safe movement of the autonomous vehicle.

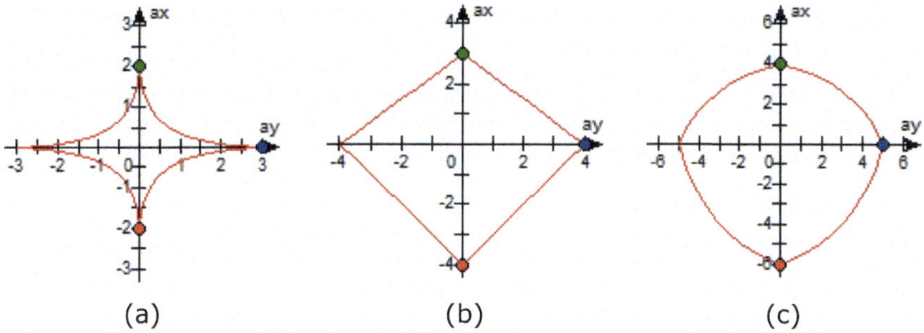

Fig. 6.25 IPGDriver behavior models: **a** defensive, **b** normal, **c** aggressive

(3) Evaluation of vehicle dynamics—determines the current state of motion of the autonomous vehicle based on momentary decisions that a driver would make in a real situation.
(4) Machine Learning—is developed by IPGDriver to gain basic knowledge of the vehicle it controls in order to improve the performance of the autonomous driving system and to allow actions to be performed within the dynamic limits of the vehicle.

Driver is an application that determines the direction of travel and the variation of the speed profile for the set route. Driver allows to define the behavior of a human driver with direct actions on the profile of the virtual vehicle (actions on the steering, braking, acceleration systems).

From the configuration menu of the IPGDriver utility can preselect three different behavior modes of the autonomous driving system, as follows: a defensive driver (parameter set defensive), a normal driver (parameter set normal) and a driver aggressive (set aggressive set) (Fig. 6.25).

The following driver models are available in CarMaker [3, 4, 8]:

(1) User parameterized driver that automatically adjusts the optimal values in the guidance of the vehicle by performing predefined maneuvers.
(2) Racing driver automatically adjusts the maximum values in the guidance of the vehicle by performing predefined sequences of maneuvers.

The default maneuvers performed by the driver system are:

(a) Establish a route within the limits of the driving lane.
(b) Determining the optimal travel speed depending on the route and the vehicle behavior.
(c) Keeping the direction of travel related to static and dynamic objects in traffic, respectively to the road profile (turns, slopes etc.). The corner cutting coefficient allows the virtual vehicle to move on the established driving lane. The value 0 of the corner cutting coefficient is the fact that the vehicle runs exactly in the middle of the driving lane, and the value 1 represents the fact that the vehicle uses the entire width of the driving lane to keep its course static.

(d) Energy efficient driving influences energy consumption. The value 0 of the energy efficient driving coefficient characterizes a hurried driver who permanently changes the velocity, with multiple successions of acceleration and braking. The value 1 of the energy efficient driving coefficient characterizes a cautious driver who ensures a movement of the vehicle with low energy consumption.

IPGDriver offers the possibility to select two main types of presets regarding the behavior of the virtual driver:

(1) Standard that can define a normal, defensive, respectively an aggressive driving style.
(2) Extended, which can define a driving style, defines an energy efficient driver or a stressed driver.

Each of these parameters can be adjusted by means of three coefficients: dynamics which defines the limit of the dynamic stability of the vehicle, energy efficiency, which directly influences the energy consumption in the driving style, respectively nervousness which simulates an aggressive style of a driver. The standard parameters used in configuring the desired user model for the IPGDriver utility are (Fig. 6.26; Table 6.5) [8]:

(i) Cruising speed, the specified speed is the target that the virtual driver tries to reach. It will of course control the vehicle, even brake, if necessary, to keep it

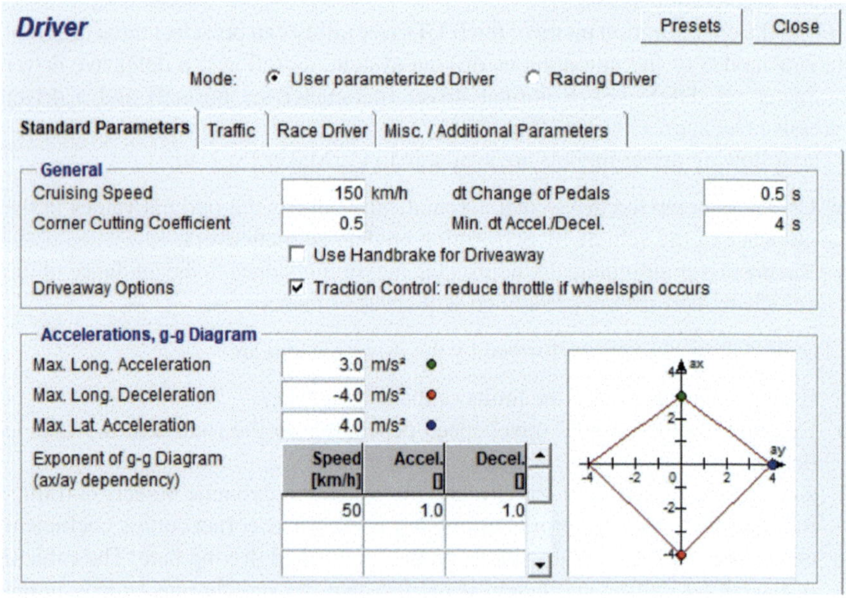

Fig. 6.26 IPGDriver standard parameters

Table 6.5 User parameterized driver

Standard parameters					
Cruising speed (km/h)	25.00				
Corner cutting coefficient	0.500				
Change of pedals (s)	0.500				
Min. acceleration/deceleration (s)	4.000	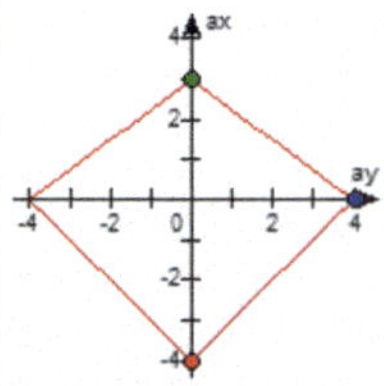			
Traction control (reduce throttle)					
Accelerations, (g-g) diagram					
Max. long. acceleration (m/s^2)	3.000				
Max. long. deceleration (m/s^2)	– 4.000				
Max. lateral acceleration (m/s^2)	4.000				
Exponent of (g-g) diagram (ax/ay dependency)	Speed	Accelerat.	Decelerat.		
	50.00	1.000	1.000		
Declutching/gear shifting					
Time of shifting (s)	1.000				
Gear	Speed	Minim	Maxim	Idle up	Acc down
1	(rpm)	1500	4000	2000	3000

on the road according to the limits he learned about this vehicle. But he will always try to accelerate to reach this speed.

(ii) Corner cutting coefficient, the parameter allows the virtual driver to move off the given driving lane. The values of the corner cutting coefficient range from zero to infinite. Value 0 means the vehicle virtual drives exactly in the middle of the driving lane, value 1 means the driver uses the whole width of the driving lane to calculate its static desired course.

(iii) Change of pedals is the time required by the virtual driver to move its command from the gas to the brake pedal. If IPGDriver switches from the gas to the brake it waits this duration (starting after the throttle value equals zero) before allowing the brake value to change.

(iv) Minimum acceleration/deceleration, it is the time the virtual driver is cutting speed peaks in the static desired speed because of too short acceleration and deceleration sequences.

(v) Maximum longitudinal acceleration, maximum longitudinal deceleration, maximum lateral acceleration, specify the maximum absolute acceleration (longitudinal or lateral). Only one value is necessary for the lateral acceleration, but two are required for the longitudinal acceleration and deceleration. The value for the lateral acceleration needs to be positive, but the value for the longitudinal acceleration can be positive or negative.

(vi) Exponent of g-g diagram (ax/ay dependency) defines the dependency between the lateral and the longitudinal acceleration and between the lateral and the longitudinal deceleration. The exponents can be observed on the g-g diagram

Table 6.6 Traffic in driver parameter

Traffic			
Consider traffic			
Following leading vehicle			
Minim time gap (s)	1.800		
Maxim time gap (s)	5.000		
Minim distance (m)	6.000		
Maxim distance (m)	250.0		
Energy efficient driving (0…1)	0.750		
Misc./additional parameters			
Parameters	Longitudinal	Lateral	
Tolerated deviation	0.000 km/h	0.000 m	
Reaction time	0.000 s	0.000 s	
Maxim steering wheel angle		630.0	°
Maxim steering wheel velocity		500.0	°/s
Maxim steering wheel acceleration		3000	$°/s^2$
Respect speed limits			
Respect stop markers			
Respect traffic lights			

(red curves) linking the maximum longitudinal and lateral accelerations. The virtual driver will drive so that the accelerations applied to the vehicle remain in the area defined by the exponents and the maximum limits.

Traffic parameters used in configuring the desired user model for the IPGDriver utility are as follows (Table 6.6) [8]:

(i) Minimum time gap/maximum time gap variables are speed dependent distances for realistic driving. IPGDriver will try to keep the time gap to the leading vehicle within these limits.
(ii) Minimum distance defines the minimum distance to a parking traffic object. Once the road gap is smaller than this value, IPGDriver will bring the vehicle to a standstill.
(iii) Maxim distance specified function of the absolute limitation for the follow traffic. Once the road gap to the leading vehicle is larger than this value, IPGDriver switches over to the normal driving mode.
(iv) Energy efficient driving value from zero stands for a very hurried, agile driver who permanently changes its velocity and gap to the leading vehicle. The economic driver with a coefficient set to 1 rather adopts to the driving behavior of the leading car. Thus, the average velocity is lower than the more aggressive driver, but there are also fewer stopping phases. The fuel consumption is also influenced by the nervousness factor of the driver presets. A more nervous

driver results in higher fuel consumption when following traffic. All in all, the difference in fuel consumption between the different driver characteristics can be about 30%.

(v) Maxim steering wheel angle specifies at which point the driver will stop to turn the steering wheel, even if the vehicle is still not in the desired direction.
(vi) Maxim steering wheel velocity defines the maximum steering wheel velocity that can be achieved.
(vii) Maxim steering wheel acceleration defines the maximum steering wheel acceleration that can be reached.

The adaptation of the driver to the control of the autonomous vehicle according to the traffic situations is done with the help of the driver adaption utility:

(i) Vehicle limits calculates the maximum longitudinal acceleration, deceleration, and lateral acceleration of the vehicle.
(ii) Controller dynamics calculates the time preview used by the driver to predict his actions during a TestRun.
(iii) Engine speeds for shifting calculates the optimal engine speeds for shifting.
(iv) Road friction the coefficient of friction of the tire-road-contact point that will be used for the driver adaption can be defined. If no value is entered, the default value 1.0 is used.

6.2.3.6 Virtual Traffic Data (Step 6)

Step 6 consists of defining a traffic data model based on the following data: number and types of vehicles in traffic, direction of travel, pedestrians in traffic, controlled or uncontrolled traffic, traffic rules, rush hours and congestion in traffic, intelligent transport systems etc. (Fig. 6.27).

Car traffic, especially in large, congested cities, is a very important feature in terms of road safety. Virtual traffic can be defined based on the following data: number and types of vehicles in traffic, direction of travel, pedestrians in traffic, controlled or uncontrolled traffic, traffic rules, rush hours and traffic congestion, intelligent transport systems etc.

Traffic in the CarMaker application defines existing objects on the route (pedestrians, vehicles etc.), static or dynamic. The longitudinal maneuvers of the virtual model of the autonomous vehicle will be defined separately for each object in traffic [3]. The movement of objects in traffic is done for defined values of travel velocities when autonomous driving is simulated. Alternatively, objects in traffic can be defined according to free-movement maneuvers (Fig. 6.28).

General attributes of traffic objects (Table 6.7) can be parameterized:

(i) Object mode can be made between a movable object with predefined motion and two types of stationary objects.
(ii) Name for the object is used to refer to this traffic object.

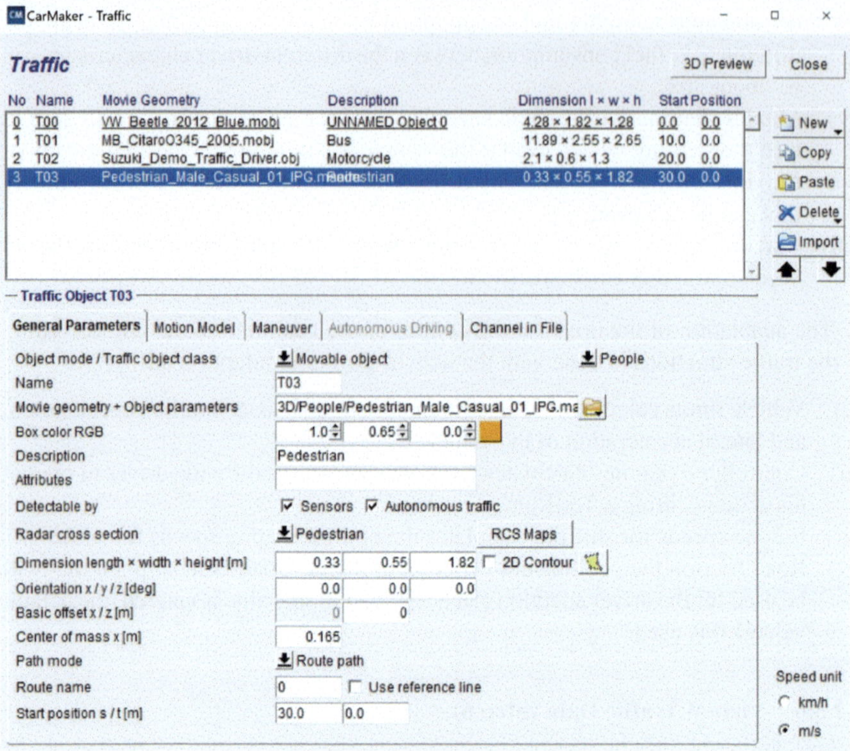

Fig. 6.27 Traffic

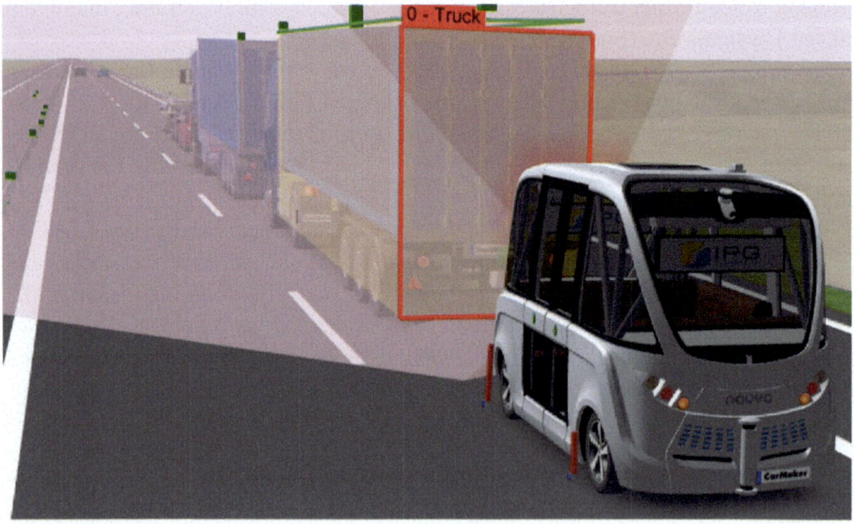

Fig. 6.28 Autonomous shuttle bus in traffic

Table 6.7 Traffic

No	Name	Movie geometry	Description	Dimension l/w/h	Start
0	T00	Navya_Arma	Object 0	4.74 × 2.10 × 2.40	10.00

Traffic object T00					
General parameters					
Object mode/traffic object	Movable object	Car			
Name	T00				
Movie geometry, object parameters	3D/Vehicles/Navya_Arma_2016.obj				
Box color RGB	1.000	0.000	0.000		
Description	Object 0				
Detectable by sensors, autonomous traffic					
Radar cross section: car-RCS maps					
Dimension (l/w/h) (m)	4.740	2.100	2.400		
Orientation (x/y/z) (°)	0.000	0.000	0.000		
Basic offset (x/z) (m)	0.000	0.190			
Center of mass (x) (m)	2.400				
Path mode	Route path				
Route name	0.000				
Start position (s/t) (m)	10.00	0.000			
Motion model					
Motion model	Single track (4 wheel)				
Overall mass (kg)	2700				
Moment of inertia ($I_{xx}/I_{yy}/I_{zz}$) (kg m^2)	2500	10,000	10,000		
Overhang (front/rear) (m)	0.930	0.930			
Front cornering stiffness (N/rad)	1.5e5				
Rear cornering stiffness (N/rad)	1.5e5				
Roll stiffness rate (N m/rad)	1.0e6				
Roll damping rate (N ms/rad)	1.0e6				
Pitch stiffness rate (N m/rad)	1.0e6				
Pitch damping rate (N ms/rad)	1.0e6				
Maximum steer angle (°)	40.00				
Maneuver-maneuver					
Update rate (Hz)	200.0				
Start	Long	Long sup	Lat	Lat sup	Hidden
===Global Settings/Start Conditions===					
Start velocity (km/h)	25.00				
Action at end	Freeze position				
Autonomous driving					

(continued)

Table 6.7 (continued)

Cruising and following model	Generic		
Update rate (Hz)	20.00		
Preview distance (m)		200.0	
Standstill distance (m)		3.000	
Maximum acceleration (m/s^2)		2.000	
Comfort deceleration (m/s^2)		– 2.000	
Proportional gain speed control (–)		0.410	
Target time gap (–)		1.500	
Adaptation time of distance (s)		4.000	
Adaptation time of low relative speed (s)		2.600	
Adaptation time of high relative speed (s)		1.300	
Reference relative speed (m/s)		1.000	
Maximum deceleration (m/s^2)		– 10.00	
Curve driving			
Maximum lateral acceleration (m/s^2)		4.000	
Lane change			
Factor of cautiousness (–)		0.500	
Longitudinal vehicle dynamics			
Maximum acceleration (m/s^2)		0.500	
Maximum velocity (m/s)		25.00	
Maximum driving power (kW)		15.00	
Response time PT1-filter (s)		0.150	
Channel in file	Channel	Factor	Offset
Longitudinal velocity (m/s)		1.000	0.000
Longitudinal position (m)		1.000	0.000
Lateral position (m)		1.000	0.000

(iii) Movie geometry, object parameters, or movie geometry and import object parameters, or select movie geometry, or import object parameters can be selected.
(iv) Box color RGB (Red–Green–Blue) defines varying color for the traffic objects.
(v) Description adds the additional information about the traffic object.
(vi) Attributes option assigns up to 10 attributes for traffic object, which are evaluated by the sensor model.
(vii) Detectable assigned the traffic object should be detectable for object sensors and free space. Radar cross section is defined as the measure of how detectable the object is when subjected to a radar signal.
(viii) Dimension (l/w/h) defines the overall dimensions of a traffic object as simulated in the following order: length, width, height.

(ix) Orientation (x/y/z) attribute allows an initial orientation to be specified to the traffic object with reference to the road.
(x) Basic offset (x/z) is used to define the distance of the traffic object from the road surface.
(xi) Center of mass (x) defines the center of mass location.
(xii) Path mode initializes the motion planning for the traffic object. Route path traffic defines the object's motion by a route set in the scenario editor (fix sequence of paths). Dynamic path defines the traffic object's motion for any individual path sequence.
(xiii) Route name defines the route name or ID.
(xiv) Start position (s/t) defines the traffic object in route/path coordinates: (s) longitudinal direction along the center line and (t) lateral deviation from the reference line or the absolute lane.
(xv) Overall mass defines the entire mass for the traffic object.
(xvi) Moment of inertia (lxx/lyy/lzz) defines the moment of inertias of the traffic object.
(xvii) Overhang (front/rear) defines the distance between the center point of the front axle and the front edge of the traffic object as first parameter, followed by the distance between the center point of the rear axle and the rear edge of the traffic object.
(xviii) Front cornering stiffness corresponds to the front tires.
(xix) Rear cornering stiffness corresponds to the rear tires.
(xx) Roll stiffness rate is the coefficient for roll stiffness for the dynamic and specific roll model.
(xxi) Roll damping rate is the coefficient for roll damping for the dynamic and specific roll model.
(xxii) Pitch stiffness rate is the stiffness coefficient for the dynamic model.
(xxiii) Pitch damping rate is the coefficient for the dynamic pitch model.
(xxiv) Maximum steer angle specifies the maximum steering angle of the traffic object.
(xxv) Start velocity defines the start velocity of the traffic object.
(xxvi) Start condition for the execution of the maneuver list can be defined using RT expressions.
(xxvii) Action at end defines the behavior of the traffic object when all minimaneuvers are completed.

6.2.3.7 Simulation Tasks (Step 7)

Step 7 consists in establishing the simulation tasks which involve defining maneuvers in the movement of the virtual model, maneuvers highlighted in the definition of computer simulations.

The virtual model simulation methodology for the considered vehicle is carried out according to the following steps:

(1) Selection of simulation typologies corresponding to a functional cycle of the virtual model.
(2) Scenarios consisting of control parameters of the simulation process.
(3) Simulation series for various values of the functional parameters.
(4) Running computer simulations to determine the performance parameters, analysis, and interpretation of results.

Since different objects of investigation are examined when using CarMaker, there are also different approaches taken when implementing the VVE [3]:

(A) RT approach when the VVE is simulated on a computer that is running a RT operating system. The RT operating system enables the execution of the VVE to be deterministic, and the VVE will run on a timeline that corresponds to the timeline of the real world. If the VVE is not able to meet the criteria for RT simulation, or in other words, fails to meet the specified timing deadlines, an error or warning will occur.
(B) Office when the VVE is simulated on a usual computer. There is no way to ensure a deterministic behavior, since the running VVE can be blocked or otherwise delayed by applications that run with a higher priority, or for some other reason depending on the way the operating system scheduling has been implemented. With an office simulation, the VVE may run with a timeline that is faster, slower, or equal to the RT.

6.2.3.8 Characteristics of the Simulation Process (Step 8)

Characteristics of the simulation process (step 8) consist in defining the simulation step, the simulation time step, and the number of runs through the same simulation (output after each simulation steps).

Because the simulation application offers the possibility to evaluate, calculate and save all the parameters that characterize each element of the virtual model and considering that selecting all these parameters would mean a very long duration of the simulation process, respectively a large collection of results, a selective collection of results is required.

The steps taken to develop a virtual model for the vehicle considered are based on the V diagram, a diagram that describes the development cycle from a virtual model to a real model by representing the built-in control applications, respectively by tracking the evolution over time of the main steps of development (Fig. 6.29) [1].

The steps on the V diagram for the development of a virtual vehicle model from the generation phase of the values for the initial data sets, to the implementation phase of the prototype developed on a real vehicle are presented in Table 6.8 [1].

The MiL concept includes all the functionality and performance requirements of the virtual vehicle to be developed. It is defined by the constructive characteristics, respectively by the functional parameters of the main systems. The analysis and evaluation methodology used to define the systems to be developed at the block diagram level has the following characteristics:

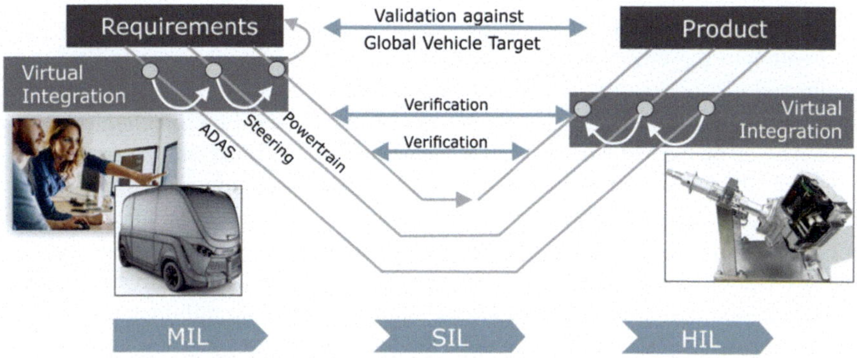

Fig. 6.29 Design/development process steps

Table 6.8 The steps on the V diagram

Vehicle system	MiL	SiL	PiL	HiL			Real vehicle
				EiL	ViL[1]	ViL[2]	
Code/ECU	**Virtual**	*Real*	*Real*	*Real*	*Real*	*Real*	*Real*
ECU (system)	**Virtual**	**Virtual**	*Real*	*Real*	*Real*	*Real*	*Real*
Engine/motor	**Virtual**	**Virtual**	**Virtual**	*Real*	*Real*	*Real*	*Real*
Vehicle	**Virtual**	**Virtual**	**Virtual**	**Virtual**	*Real*	*Real*	*Real*
Environmental	**Virtual**	**Virtual**	**Virtual**	**Virtual**	**V**/*R*	**V**/*R*	*Real*
Vehicle dynamics	**Virtual**	**Virtual**	**Virtual**	**Virtual**	**Virtual**	*Real*	*Real*
Driver	**Virtual**	**Virtual**	**Virtual**	**Virtual**	**Virtual**	**V**/*R*	*Real*
Behavior	**Virtual**	**Virtual**	**Virtual**	**Virtual**	**Virtual**	**V**/*R*	*Real*
Traffic	**Virtual**	**Virtual**	**Virtual**	**Virtual**	**Virtual**	**Virtual**	*Real*

Where MiL is the Model-in-the-Loop concept, SiL is the Software-in-the-Loop concept, PiL is the Processor-in-the-Loop concept, HiL is the Hardware-in-the-Loop concept divided into the following stages: EiL is Electric/motor-in-the-Loop, ViL[1] is Vehicle-in-the-Loop static (prototype is tested on a testbed without dynamic elements), ViL[2] is Vehicle-in-the-Loop dynamic (prototype is tested on a roller stand with dynamic elements)

(1) It provides the ability to implement any calculation formula for the selected data.
(2) It provides the ability to edit the developed model at any step of the development process directly by the user.
(3) It provides a direct link between the basic models used as a reference for the data collection and the models to be developed.
(4) It offers the possibility to simultaneously process an unlimited number of files with the collected data.
(5) It provides the selection and allows the adjustment of the parameters (manual or automatic).

The MiL concept allows to define the initial data, the input parameters, and the creation of the virtual working environment, respectively the system boundary conditions.

The SiL concept aims to develop the virtual model based on the characteristics defined based on the MiL concept and to integrate it into a virtual simulation environment using the software application dedicated to the development and interconnection of embedded systems (CarMaker). The virtual model is ready to be integrated into an offline computer simulation process and contains all the constructive-functional characteristics of a real vehicle. The concept SiL allows the use of software features to develop, optimize and validate a virtual vehicle model with minimal costs and in a short time period. The main features of the SiL concept are as follows [1]:

(1) Minimum two interconnected systems (powertrain-driveline).
(2) RT simulation process.
(3) The SiL model must control real HiL equipment.
(4) Response loop of interconnected systems must exist (X-in-the-Loop).

The SiL concept involves the development of a prototype of the considered vehicle, which based on computer simulations in an offline virtual environment to test the basic concept of the model through the physical connection to a real system. For example, the virtual prototype of an embedded system can be connected and tested on an electric motor or an embedded system that equips the real vehicle, based on the algorithms that were the basis of the original model and that were developed and simulated in the software applications.

The rapid controller prototyping allows testing and establishing control strategies by RT computer simulation with real I/O (Input/Output) devices [1]. The rapid controller prototyping differs from the HiL concept in that the control strategy is simulated in RT and the reference model is real. The rapid controller prototyping is a typical method used to develop and test control strategies.

The PiL concept is the transition from the design and simulation of the virtual vehicle to the first component of the real system to be implemented on a real vehicle model. The component is an ECU of any real vehicle system, for which PiL defines characteristic parameter sets based on RT computer simulations performed on the virtual model developed and implemented in the response loop (in-the-Loop).

The HiL concept allows control prototypes to be tested at a level of fidelity that cannot be achieved through the computer simulations performed in the SiL stage. HiL processes run in a shorter time than SiL simulations for the same control prototype due to the fact that the real mode operating regime involves a single phase of the defined process, without any series of cases to go through for validation. The HiL concept has the following conditionings [1]:

(1) At least one of the system components is real and at least one of the system components is simulated but works in RT.
(2) Usually the control system (software and hardware) is a real component, used in series production.
(3) The controlled process (consisting of sensors, process, actuators), can include both real and simulated components.

6.2.3.9 Run Simulation (Step 9)

Step 9 consists of running computer simulations on the virtual model built and validated in the previous steps which can be carried out according to one of the methods related to the duration of the simulation process without time limits "−RT", simulation "RT" and accelerated simulation "+RT".

(1) Simulation without time limits "−RT" in case of investigations on the behavior of the virtual model, for the design and implementation of control components or functional processes.
(2) "RT" the simulation process takes place in RT.
(3) Accelerated simulation "+RT" the simulation process is carried out at a higher speed than RT, for model optimizations, development of implementation strategies, planning etc.

Performing a simulation in CarMaker requires similar definitions to when carrying out a real test drive. The type of vehicle, tires, driver, test track and a maneuver that the driver is to perform, need to be defined. CarMaker provides several predefined models for each of these requirements. A combination of these models and settings form what is called a TestRun.

CarMaker is based on fixed virtual models whose properties can be varied. This means that the number of bodies and the DOF between them are already defined and the user does not need to model these on his own.

The models are already defined, but they still need to be parameterized according to their environment. For this, data set is manually implemented or loaded for the models. Parameterizing includes selecting a vehicle, selecting, or designing a road, defining a type of driver, and defining a maneuver. After all these components are set, CarMaker has the information necessary to control the VVE and simulate. All these settings are stored in a file used by the VVE during simulation. This file, which can be saved, loaded, or edited, is the TestRun definition. Loading and executing this file results in the simulation of that specific test. The TestRun files have the format of a text file and can be viewed with an editor [11].

6.2.3.10 Show Results (Step 10)

Step 10 consists of viewing and evaluating the computer simulations results (Show Results) (Fig. 6.30).

There are several ways to analyze the data that is generated during a CarMaker simulation. IPGControl tool was developed for viewing and analyzing purposes and can also be used standalone. IPGControl offers the functionalities of an online result management application. Diagrams can be displayed directly during the simulation (Fig. 6.31). By loading external result files, the user also has the possibility to display the results of previous TestRun. This is called offline result management [11, 12].

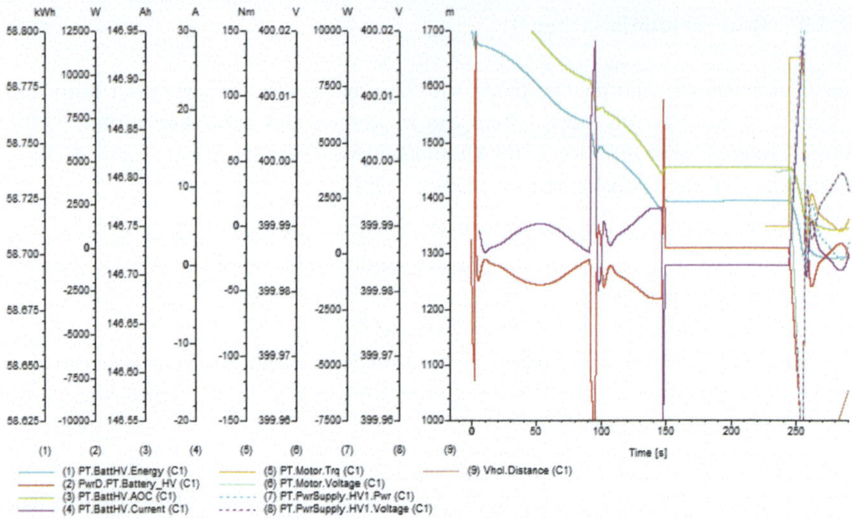

Fig. 6.30 Computer simulations results (show results)

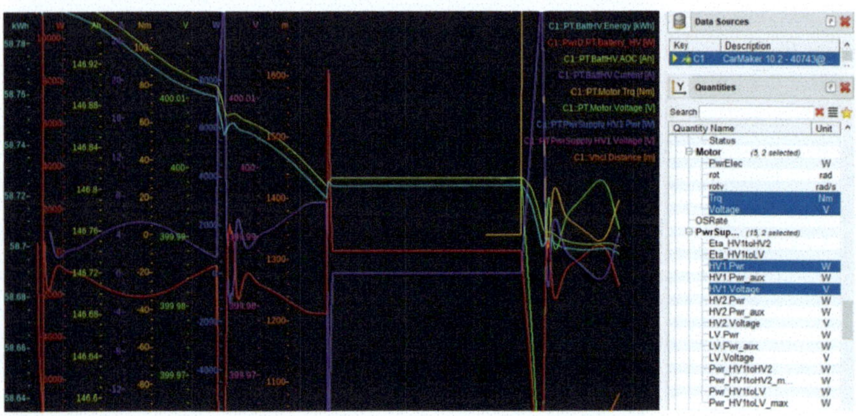

Fig. 6.31 Diagrams during the simulation

6.2.4 Constructive Architecture of the Virtual Configuration

CarMaker consists of two components: CIT (CarMaker Interface Toolbox) and the VVE in which the developed vehicle runs (Fig. 6.32) [12–14]. The virtual vehicle model consists of systems, subsystems and elements which are grouped, interconnected, and defined according to the structure presented below [3, 8].

CIT can be viewed as several individual tools that are used to manage the VVE. The tools can be classified as [3]:

Fig. 6.32 Virtual Vehicle Environment

(1) Control and direct access tools that control the actions performed by the simulation and allow certain parts of the simulation to be directly controlled by the user.
(2) Parameterization tools serve to specify the parameters that will be used in the VVE.
(3) Analysis and visualization tools that allow the data to be viewed and analyzed either during or after a simulation.
(4) File management tools that create, save, and modify files that define simulations, settings, output data etc.

The virtual vehicle model developed in the CarMaker simulation application consists of systems and subsystems that are grouped and defined according to the structure shown in Table 6.9 [8].

6.2.4.1 Assembly

The vehicle data set generator allows the user to create an entire vehicle data set based on only little basic information regarding the vehicle:

(1) In case the user does not have access to detailed information about the vehicle that is intended for the simulation.
(2) If merely an average vehicle of a certain class is necessary to test other modules, such as driver assistance systems.

Table 6.9 Vehicle data set (CarMaker)

Assembly	Body	Suspensions	Steering	Tire	Brake	Powertrain	Sensors
Configuration	Structure	Spring	Gear ratio	All axles	Model	General	Slip angle
Engine mount	Outer shell	Secondary spring	Mechanical	Front axle		Drive source	Inertial
Body	Aerodynamics	Damper	Power assist	Rear axle		Driveline	Object
Chassis		Buffer				Power supply	Free space
Powertrain		Stabilizer				Control unit	Traffic sign
Trim load		Kinematics					Line
Sensor mounting		Compliance					Road
		Wheel bearing					Collision
		External force					Object by line
							Global navigation
							Radar
							Camera
							Radar RSI
							Ultrasonic RSI
							Lidar RSI
							Camera RSI

(3) As a starting point for a vehicle data set parameterization that is to serve as a basis for further detailing.

To generate an autonomous shuttle bus, the appropriate model for a delivery van (Vehicle Data Set Van.car) was selected. The vehicle's graphics are generated based on an image file (Fig. 6.33) that faithfully reproduces the body of the generated model (Vehicle Graphics Van.tcl).

Vehicle data set defines a name for the generated vehicle. Vehicle graphics define the selected vehicle with a picture file. The picture file is generated, and it is a .tcl file. Vehicle class selection of a vehicle class influences the basic attributes of the generated vehicle. A predefined parameter set that corresponds to an average vehicle of this class is loaded and can be refined by altering some of the vehicle data set generator parameters [4]:

(i) Vehicle data set parameters defines a name for the generated vehicle.
(ii) Vehicle graphics field when activated it generates the selected vehicle.

Fig. 6.33 Virtual model graphics

(iii) Vehicle class selection influences the basic attributes of the generated vehicle. A predefined parameter set that corresponds to an average vehicle of this class is loaded and can be refined by altering some of the vehicle data set generator parameters.

The vehicle parameters are specified from the initial data in Table 6.10.

(i) Unloaded weight vehicles tare weight including all components excepting additional loads.
(ii) Vehicle length distance between the front-most and the rear-most point of the vehicle.
(iii) Vehicle width distance between the left-most point and the right-most point of the vehicle.
(iv) Vehicle height distance between the top-most point of the vehicle and the ground.
(v) Wheelbase longitudinal distance between the center of the front and rear axle.
(vi) Track width front lateral distance between tire centers at front axle.
(vii) Track width rear lateral distance between tire centers at rear axle.
(viii) Rear overhang distance between the rear axle and the rear-most point on the body.

Advanced settings for the initial parameters are presented in Table 6.11.

(i) Vehicle body where the generated vehicle can have a rigid or flexible body model which allows bending and torsion of the main body.
(ii) Steering system configuration can be optimized by defining the turning circle diameter and the maximum steering angle.
(iii) Driving axle generated vehicle data set can either be front driven, rear driven, or all wheels driven.

Table 6.10 Vehicle data set generator (basic settings) (Some of the values that define the virtual model may differ from real values, they have been determined via experimental studies or mathematical calculus)

Basic settings			
Generate			
Vehicle data set	Van.car		
Vehicle graphics	Van.tcl		
Vehicle class	Delivery van		
Vehicle parameters			
Unloaded weight		2600	kg
Vehicle length		4755	mm
Vehicle width		2100	mm
Vehicle height		2640	mm
Wheelbase		3393	mm
Track width front		1800	mm
Track width rear		1800	mm
Rear overhang		945.0	mm
Tire size		195/65R15	
Generate tire data set			
Tire data set	RT_195_65R15		
Tire type	Car		
Tire size	195/65R15		
Inflation pressure	2.50		
Coefficient of friction	1.00		
Coefficient of stiffness	1.00		

(iv) Engine characteristics of the generated vehicle, the nominal power and the maximum engine torque can be configured along with the including corresponding engine rotation speeds.
(v) Axle suspension allows adaption of the ride frequency of the front and rear springs and dampers.
(vi) Aerodynamics drag coefficient can be adjusted. It defines the drag coefficient acting at the frontal face of the vehicle along the x-axis.

The main configuration elements of the virtual vehicle model refer to the location of the engine (engine (motor) mount), the body of the vehicle (body), the body structure (chassis), respectively the powertrain, so as shown in Table 6.12 [3, 11].

Body model includes a separate engine body that is elastically mounted on the main body (assembly). The position can be defined, as well as the location of the joint between engine and the main body. The joint can be changed to a flexible mounting consisting of spring damper elements which can be characterized in the lower part of the dialog:

(i) Hitch, position of the hitch relative to the design frame FrD.
(ii) Jack, the four positions of the jacking points of the vehicle.

Table 6.11 Vehicle data set generator (advanced settings) (Some of the values that define the virtual model may differ from real values, they have been determined via experimental studies or mathematical calculus)

Advanced settings		
Vehicle parameters		
Vehicle body	Flexible	
Driving axle	Rear driven	
Gearbox	Automatic	
Steering		
Turning circle	9.000	m
Max steering angle	600.0	°
Engine		
Nominal power	15.00	kW
Rotation speed	5000	rpm
Maximum torque	130.0	N m
Rotation speed	8000	rpm
Transmission		
Number of gears	1	–
Spread of gears	1	–
Axle suspension		
Front ride frequency	1.300	Hz
Rear ride frequency	1.500	Hz
Aerodynamics		
Drag coefficient	0.400	–

(iii) Virtual force the point of attack of a possibly defined virtual force.

Chassis including their center of gravity and inertias. The definition takes place in the design position, according to the frame defined in the tab assembly configuration:

(i) Wheels unsprung spinning masses.
(ii) Wheel carriers unsprung masses that do not rotate.

For each chassis body the following parameters can be specified:

(i) Position of the center of gravity of each mass. The location of each body is parameterized in the design state.
(ii) Mass of each body.
(iii) Moments of inertia of the body mass can be defined in the axis orientation of Fr1. The position of the center of gravity of the mass considered is taken as origin.

Powertrain including the reference frame, position, orientation, and mounting frame of the optional powertrain masses are defined. Electric motors can also be mounted to the wheel frames Fr2FL, Fr2FR, Fr2RL, Fr2RR.

In the sensor mountings tab, the mounting of all sensors in the vehicle is defined; this includes position, orientation, and attachment to a reference frame for each

Table 6.12 Vehicle data set configuration (Some of the values that define the virtual model may differ from real values, they have been determined via experimental studies or mathematical calculus)

Assembly							
Configuration							
Reference frame	Position (x/y/z) (m)						
Vehicle (Fr1)	0.000	0.000	0.000				
Overall mass (kg)			1770				
Engine (motor) mount							
Position (x/y/z) (m)	0.000	0.000	0.000				
Orientation (x/y/z) (m)	0.000	0.000	0.000				
Mounting	Rigid						
Body							
Hitch position (x/y/z) (m)	– 0.100	0.000	0.400				
Hitch system	--not specified--						
Jack FL	3.904	0.845	0.312				
Jack RL	3.904	– 0.845	0.312				
Jack RL	1.449	0.846	0.312				
Jack RR	1.449	– 0.846	0.312				
Virtual force	3.050	0.000	0.789				
Chassis	Position (x/y/z) (m)			Mass	Inertia (x/y/z) (kg m^3)		
Wheel carrier FL	3.820	0.845	0.312	39.32	0.365	0.365	0.365
Wheel carrier FR	3.820	– 0.845	0.312	39.32	0.365	0.365	0.365

(continued)

Table 6.12 (continued)

Wheel carrier RL	0.980	0.846	0.312	37.91	0.285	0.285	0.285
Wheel carrier RR	0.980	– 0.846	0.312	37.91	0.285	0.285	0.285
Wheel FL	3.820	0.845	0.312	29.54	0.788	1.576	0.788
Wheel FR	3.820	– 0.845	0.312	29.54	0.788	1.576	0.788
Wheel RL	0.980	0.846	0.312	26.84	0.768	1.535	0.768
Wheel RR	0.980	– 0.846	0.312	26.84	0.768	1.535	0.768

Powertrain								
Drive source 0	Position (x/y/z) (m)				Orientation (x/y/z) (m)			
Engine	FrD	0.00	0.00	0.00	0.00	0.00	0.00	FrEng
Gearbox	FrD	0.00	0.00	0.00	0.00	0.00	0.00	FrEng
Electric motor	FrD	0.00	0.00	0.00	0.00	0.00	0.00	Fr1A
Gearbox M	FrD	0.00	0.00	0.00	0.00	0.00	0.00	Fr1A
Tank	FrD	0.00	0.00	0.00	0.00	0.00	0.00	Fr1A
Drive source 1	Position (x/y/z) (m)				Orientation (x/y/z) (m)			
Electric motor	FrD	0.00	0.00	0.00	0.00	0.00	0.00	Fr1A

(continued)

Table 6.12 (continued)

Gearbox M	FrD	0.00	0.00	0.00	0.00	0.00	0.00	Fr1A
Drive source 2	Position (x/y/z) (m)				Orientation (x/y/z) (m)			
Electric motor	FrD	0.00	0.00	0.00	0.00	0.00	0.00	Fr1A
Gearbox M	FrD	0.00	0.00	0.00	0.00	0.00	0.00	Fr1A
Drive source 3	Position (x/y/z) (m)				Orientation (x/y/z) (m)			
Electric motor	FrD	0.00	0.00	0.00	0.00	0.00	0.00	Fr1A
Gearbox M	FrD	0.00	0.00	0.00	0.00	0.00	0.00	Fr1A
Power supply	Position (x/y/z) (m)				Orientation (x/y/z) (m)			
LV Battery 0	FrD	0.00	0.00	0.00	0.00	0.00	0.00	Fr1A
LV Battery 1	FrD	0.00	0.00	0.00	0.00	0.00	0.00	Fr1A
HV Battery 0	FrD	0.00	0.00	0.00	0.00	0.00	0.00	Fr1A

(continued)

Table 6.12 (continued)

HV Battery 1	FrD	0.00	0.00	0.00	0.00	0.00	0.00	Fr1A
Sensor mounting								
Slip angle								
Name			Sensor name					
Parametrization			Parameter name					
Position (x/y/z) (m)			2.700		0.000		0.900	
Body mounting			Frame Fr1A					

(continued)

Table 6.12 (continued)

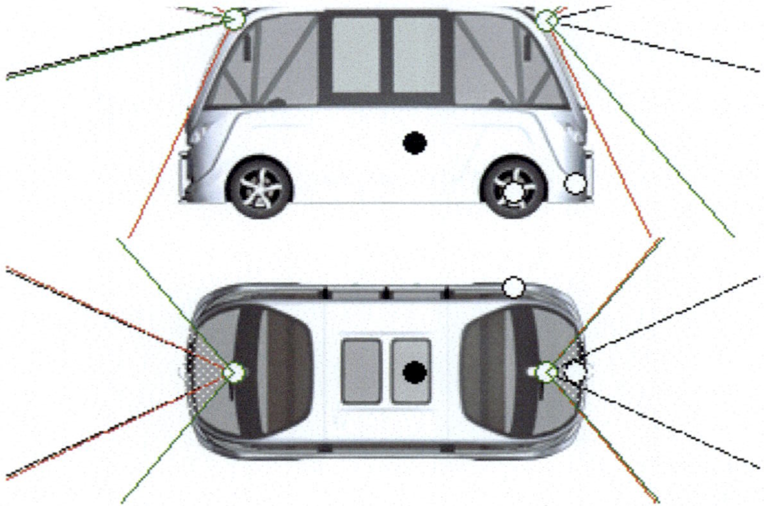

Inertial			
Name	Sensor name		
Parametrization	Parameter name		
Position (x/y/z) (m)	3.800	1.000	0.300
Orientation (x/y/z) (°)	0.000	0.000	0.000
Body mounting	Frame Fr1A		

(continued)

Table 6.12 (continued)

Calculation class	Global		

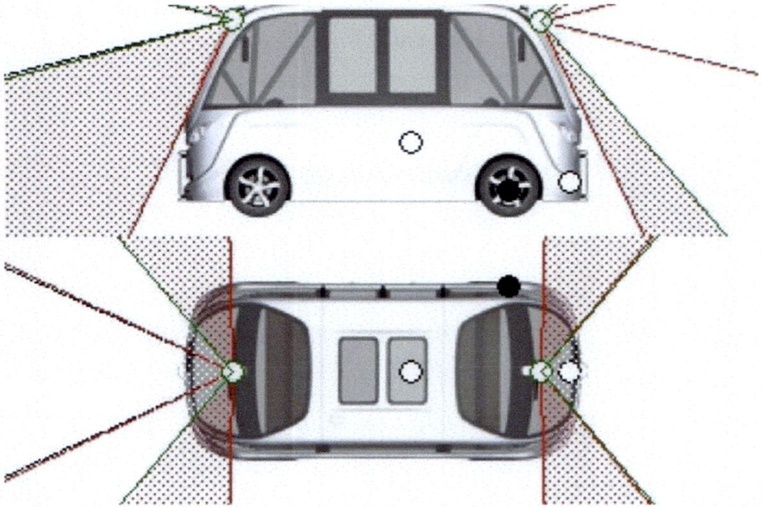

Object			
Name	Sensor name		
Parametrization	Parameter name		
Position (x/y/z) (m)	4.200	0.000	2.300
Orientation (x/y/z) (°)	0.000	0.000	0.000

(continued)

Table 6.12 (continued)

Body mounting	Frame Fr1A		
Visualization	Color		

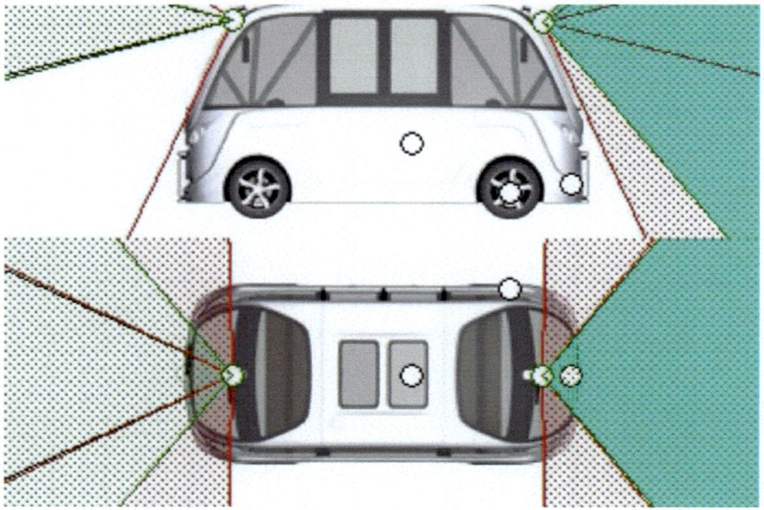

Position (x/y/z) (m)	0.620	0.000	2.300
Orientation (x/y/z) (°)	0.000	0.000	– 180.0
Body mounting	Frame Fr1A		
Visualization	Color		

(continued)

Table 6.12 (continued)

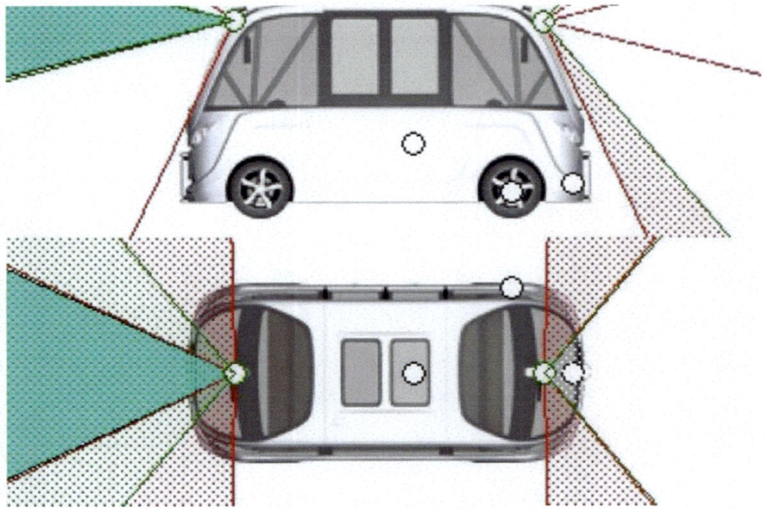

Free space			
Name	Sensor name		
Parametrization	Parameter name		
Position (x/y/z) (m)	4.200	0.000	2.300
Orientation (x/y/z) (°)	0.000	0.000	0.000
Body mounting	Frame Fr1A		

(continued)

Table 6.12 (continued)

Visualization	Color		

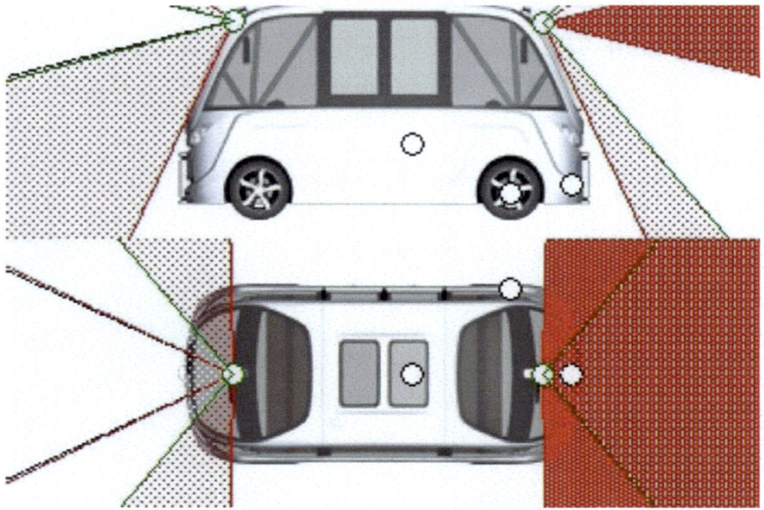

Position (x/y/z) (m)	0.620	0.000	2.300
Orientation (x/y/z) (°)	0.000	0.000	– 180.0
Body mounting	Frame Fr1A		
Visualization	Color		

(continued)

Table 6.12 (continued)

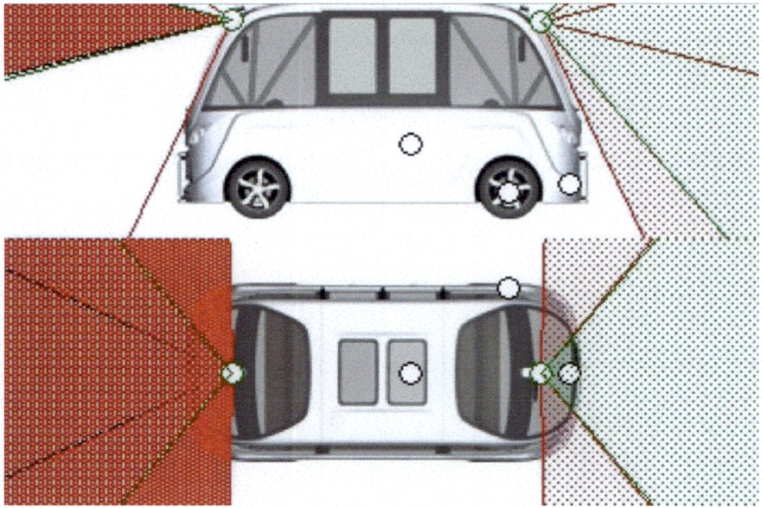

Traffic sign			
Name	Sensor name		
Parametrization	Parameter name		
Position (x/y/z) (m)	0.620	0.000	2.300
Orientation (x/y/z) (°)	0.000	0.000	– 180.0
Body mounting	Frame Fr1A		

(continued)

Table 6.12 (continued)

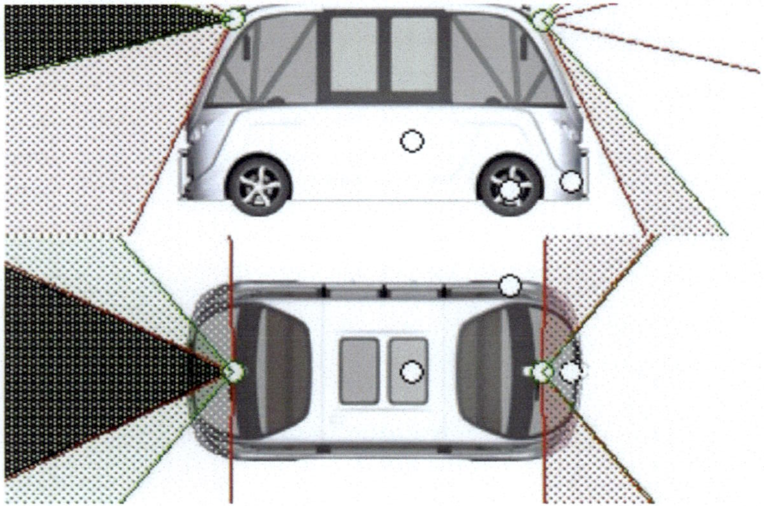

Position (x/y/z) (m)	4.200	0.000	2.300
Orientation (x/y/z) (°)	0.000	0.000	0.000
Body mounting	Frame Fr1A		

(continued)

Table 6.12 (continued)

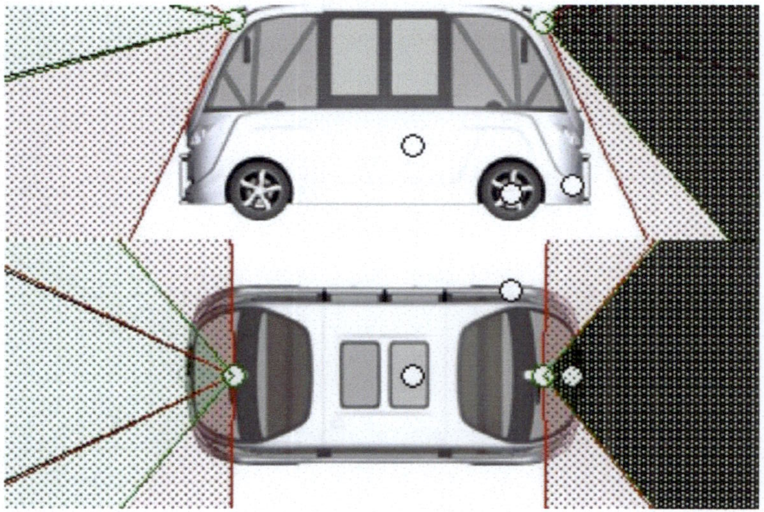

Collision	
Name	Sensor name
Parametrization	Parameter name

(continued)

Table 6.12 (continued)

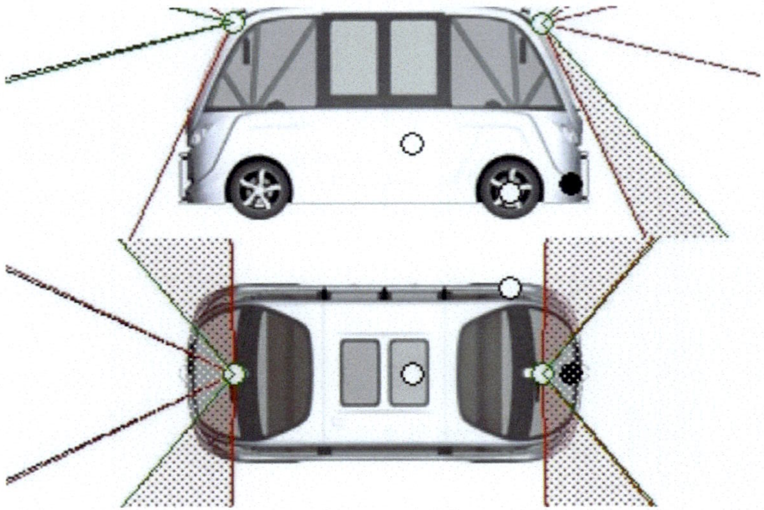

Camera			
Name	Sensor name		
Parametrization	Parameter name		
Position (x/y/z) (m)	0.620	0.000	2.300
Orientation (x/y/z) (°)	0.000	15.00	– 180.0
Body mounting	Frame Fr1A		

(continued)

Table 6.12 (continued)

Visualization	Color		

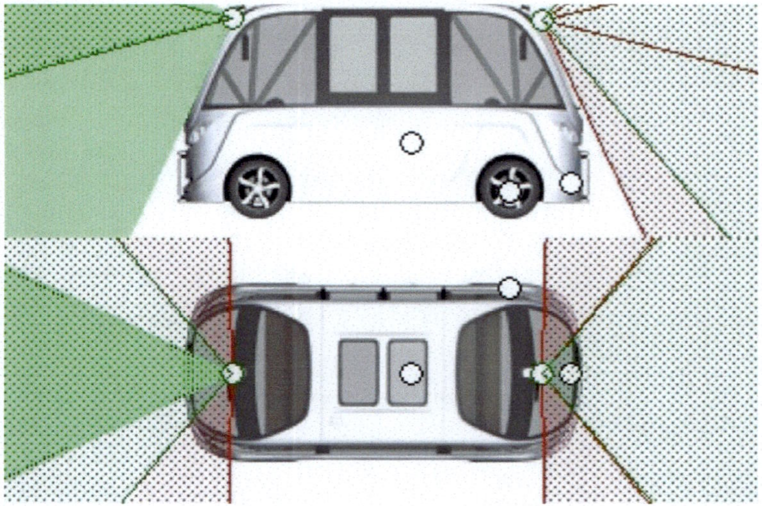

Position (x/y/z) (m)	4.200	0.000	2.300
Orientation (x/y/z) (°)	0.000	15.00	0.000
Body mounting	Frame Fr1A		
Visualization	Color		

(continued)

Table 6.12 (continued)

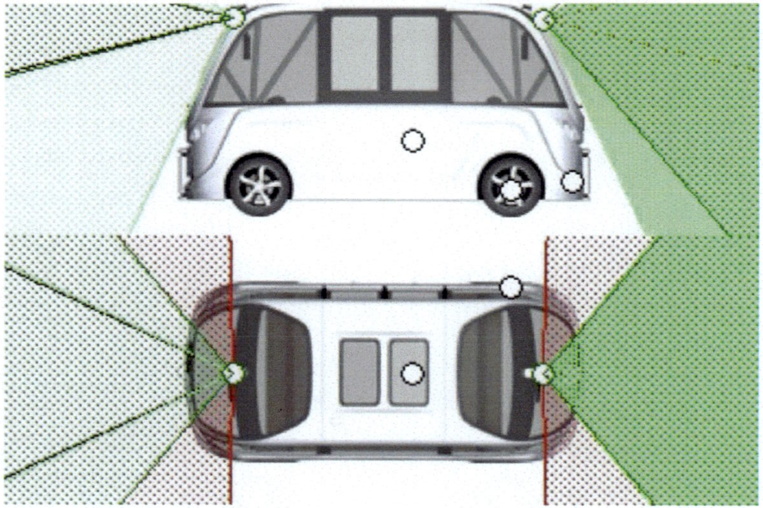

Radar			
Name	Sensor name		
Parametrization	Parameter name		
Position (x/y/z) (m)	4.600	0.000	1.200
Orientation (x/y/z) (°)	0.000	0.000	0.000
Body mounting	Frame Fr1A		

(continued)

Table 6.12 (continued)

Visualization	Color		

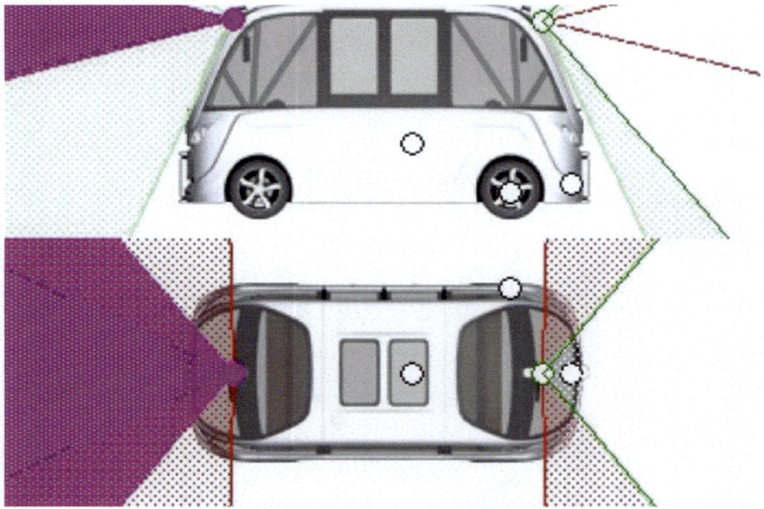

Position (x/y/z) (m)	4.200	0.000	2.300
Orientation (x/y/z) (°)	0.000	0.000	0.000
Body mounting	Frame Fr1A		
Visualization	Color		

(continued)

Table 6.12 (continued)

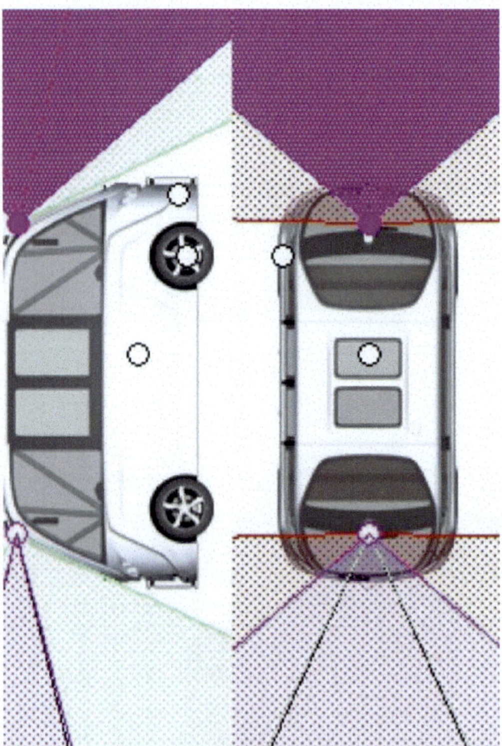

sensor. The parameters in greater detail will be set for each sensor class in the sensors tab [4]:

(i) Each sensor mounting can be linked to a parameter set from the sensors tab via the dropdown menu. By clicking the horizontal arrow button next to it, one jumps to the related parameters in the sensors tab directly.
(ii) Position coordinates (x/y/z) indicate the position of the sensor mounting on the vehicle. The sensor position is visualized in the graphical representation of the vehicle.
(iii) Body mounting coordinates are always determined in the vehicle-based coordinate system Fr1A or Fr1B.
(iv) Orientation angles (x/y/z) specify the rotation of the sensor frame in the vehicle design frame (FrameD). This specifies the viewing direction of the sensor; the rotation order is z-y-x.
(v) Visualization uses the color of the sensor beam for visualization purposes in IPGMovie without the effect on the simulation results.

6.2.4.2 Body

Vehicle body allows the CarMaker application to size the body of the vehicle (flexible) as consisting of two parts (body A, body B). Body A includes all spring masses at the front (excluding the engine) and body B includes all spring masses at the rear (excluding the engine).

The mass and inertia of the main bodies are parameterized in the Body field and will be distributed automatically and equally between both bodies (half for body A, half for body B).

The stiffness will be parameterized in two directions of rotation x for torsion, respectively y for bending.

Aerodynamics applies additional variable forces to the body of the vehicle depending on its velocity.

The reference area of the vehicle is the front area, including the surface of the tires and the underside of the body.

Reference length calculates the torque applied to the center of pressure due to aerodynamic effects.

The body consists of the following elements: structure, body (outer shell) and aerodynamics of the vehicle, whose parameters are shown in Table 6.13 [3, 4, 11]:

(i) Position of the center of gravity of the main body in the three fields x, y, and z of the body tab. The location of each body, especially of the main body, is parameterized in the design state. The design state can be a virtual state, or the static state. The position of the wheels is defined for the design state, too.
(ii) Mass defines the weight influence of the body.
(iii) Moments of inertia of this body mass (I_{xx}, I_{yy} and I_{zz}) can be defined in the axis orientation of frame 1. The position of the center of gravity of the mass regarded is taken as origin.

Table 6.13 Vehicle body (Some of the values that define the virtual model may differ from real values, they have been determined via experimental studies or mathematical calculus)

Structure							
Vehicle body							Flexible
Manual body split							
	Position (x/y/z) (m)			Mass (kg)	Inertia (x/y/z) (kg m^3)		
Body A	3.100	0.000	0.86	852.0	465.0	1735	1742
Body B	2.430	0.000	0.60	650.0	235.0	750.0	800.0
Joint A–B	3.100	0.000	0.86				
Mode	Characteristic value						
	Rotation X (torsion)				Rotation Y (bending)		
Stiffness (N m/°)	5000				5000		
Amplification (–)	1.000				1.000		
Damping (N ms/°)	100.0				100.0		
Amplification (–)	1.000				1.000		
Outer shell							
3D Object	3D/Vehicles/Navya_Arma_2016.mobj						
Picture	Navya_Arma_2016.png						

(continued)

Table 6.13 (continued)

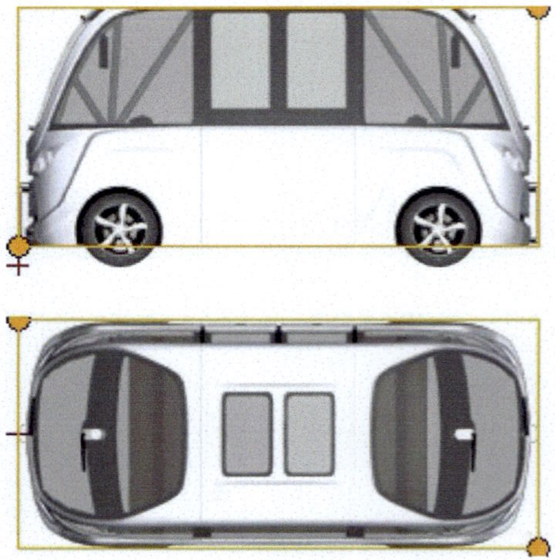

Bounding box	Position (x/y/z) (m)		
Rear lower left point	0.000	1.060	0.180
Front upper right point	4.720	– 1.060	2.410
Aerodynamics			
Interface	Position (x/y/z) (m)		
Marker	4.373	0.000	0.864

(continued)

Table 6.13 (continued)

Center of pressure	2.474			0.000	0.000	
Model	1D Look-Up table					
Reference area (m^2)	4.750	Crosswind build-up				
Reference length (m)	4.750	Kind of build-up				Step
Aerodynamics characteristics (aerodynamic coefficients)						
tau (°)	Wind angle					
cD (–)	Aerodynamical drag force (x-axis)					
cS (–)	Aerodynamical side force (y-axis)					
cL (–)	Aerodynamical lift force (z-axis)					
cRM (–)	Aerodynamical roll moment (around x-axis)					
cPM (–)	Aerodynamical pitch moment (around y-axis)					
cYM (–)	Aerodynamical yaw moment (around z-axis)					
tau (°)	cD (–)	cS (–)	cL (–)	cRM (–)	cPM (–)	cYM (–)
– 180.0	– 0.615	– 0.000	0.149	0.000	0.002	0.000
– 160.0	– 1.034	– 0.376	0.239	– 0.037	– 0.100	0.119
– 140.0	– 1.133	– 0.950	0.293	– 0.176	– 0.210	0.245
– 120.0	– 0.825	– 1.429	0.307	– 0.359	– 0.207	0.240
– 100.0	– 0.269	– 1.523	0.283	– 0.436	– 0.077	0.089
– 90.00	0.000	– 1.415	0.258	– 0.411	0.000	– 0.000
– 80.00	0.254	– 1.440	0.283	– 0.412	0.073	– 0.084
– 70.00	0.498	– 1.368	0.299	– 0.373	0.136	– 0.157

(continued)

Table 6.13 (continued)

− 60.00	0.699	− 1.211	0.307	− 0.304	0.176	− 0.204
− 50.00	0.833	− 0.993	0.305	− 0.220	0.185	− 0.215
− 40.00	0.885	− 0.743	0.293	− 0.138	0.164	− 0.191
− 30.00	0.854	− 0.493	0.271	− 0.071	0.123	− 0.144
− 25.00	0.811	− 0.378	0.256	− 0.046	0.098	− 0.115
− 20.00	0.751	− 0.273	0.239	− 0.027	0.073	− 0.086
− 15.00	0.678	− 0.182	0.22	− 0.013	0.049	− 0.059
− 10.00	0.594	− 0.105	0.198	− 0.005	0.028	− 0.035
− 5.000	0.504	− 0.044	0.174	− 0.001	0.011	− 0.015
0.000	0.410	0.000	0.149	− 0.000	− 0.002	0.000
5.000	0.504	0.044	0.174	0.001	0.011	0.015
10.00	0.594	0.105	0.198	0.005	0.028	0.035
15.00	0.678	0.182	0.22	0.013	0.049	0.059
20.00	0.751	0.273	0.239	0.027	0.073	0.086
25.00	0.811	0.378	0.256	0.046	0.098	0.115
30.00	0.854	0.493	0.271	0.071	0.123	0.144
40.00	0.885	0.743	0.293	0.138	0.164	0.191
50.00	0.833	0.993	0.305	0.220	0.185	0.215
60.00	0.699	1.211	0.307	0.304	0.176	0.204
70.00	0.498	1.368	0.299	0.373	0.136	0.157
80.00	0.254	1.440	0.283	0.412	0.073	0.084

(continued)

Table 6.13 (continued)

90.00	0.000	1.415	0.258	0.411	0.000	0.000
100.0	− 0.269	1.523	0.283	0.436	− 0.077	− 0.089
120.0	− 0.825	1.429	0.307	0.359	− 0.207	− 0.240
140.0	− 1.133	0.950	0.293	0.176	− 0.210	− 0.245
160.0	− 1.034	0.376	0.239	0.037	− 0.100	− 0.119
180.0	− 0.615	0.000	0.149	− 0.000	0.002	− 0.000

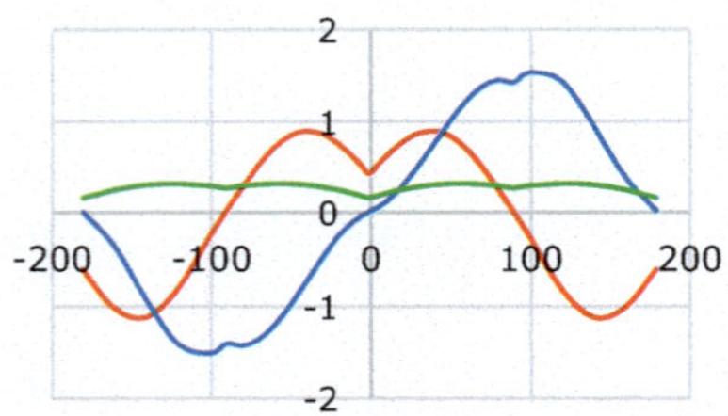

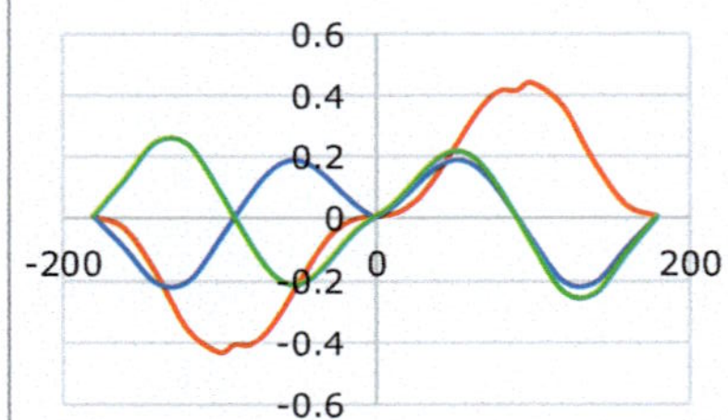

Information to the entire vehicle			
Vehicle overall mass (kg)	1770		
Vehicle overall center of gravity (x/y/z) (m)	2.755	0.000	0.684
Vehicle overall inertia tensor (kg m^2)	965.0	0.000	− 101.0
	0.000	3296	0.000
	− 101.0	0.000	3472
Estimate gyration radius ($I = M * R^2$) (m)	0.738	1.365	1.400

(continued)

Table 6.13 (continued)

Axle load (front/rear) (kg)	1106	663.0	
Wheelbase (m)	2.840		
Distance CoG to front axle/rear axle (m)	1.065	− 1.775	
Engine mount overall mass (kg)	0.000	0.000	0.000
Engine mount center of gravity (x/y/z) (m)	0.000	0.000	0.000
Engine mount overall inertia tensor (kg m^2)	0.000	0.000	0.000
Bodies			

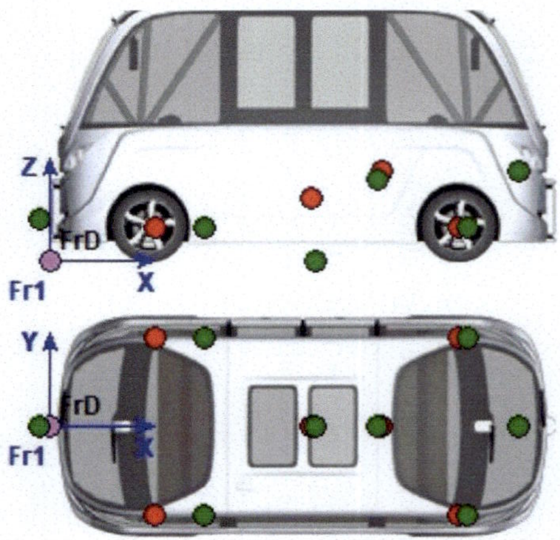

Reference
Bodies
Marker
Powertrain
Trim load

The flexible model allows parameterization and connection between the front and rear bodies (bodies A and B). The interface only parameterizes the overall body weight or the weight of each body A and B separately (by activating the manual division of the body) joint A–B [4]:

(i) Body A the mass and inertia of the main bodies are parameterized in the field of body and then automatically and equally distributed between both bodies.
(ii) Body A+B mass and inertia of the main bodies are parameterized separately for each of the bodies A and B.
(iii) Joint A–B defines the position of the joint between both bodies.
(iv) Stiffness/damping mode of the joint can be linear (single parameter). The torsional stiffness has to be parameterized in two directions: X (torsion) and Y (bending). Typically, those values can be gained from real measurements.
(v) Stiffness amplification parameter enables to modify the stiffness very fast and easily, either to compare two simulation results or to convert the values to the units required by CarMaker.
(vi) Damping uses a single linear coefficient, necessary for each degree of freedom. This value is quite difficult to obtain and has a lower effect than the stiffness.
(vii) Damping amplification parameter enables to modify the damping very fast and easily, either to compare two simulation results or to convert the values to the units required by CarMaker.

The object file used by IPGMovie can be selected under 3D object (in obj, kmz, dae or 3ds format). File definition guidelines:

(i) Only polygons and triangles, no splines, or free forms.
(ii) No more than 50,000 nodes.
(iii) Generate the normal for all points.
(iv) Colors generate the corresponding material file.

The aerodynamic model applies additional variable forces to the vehicle body according to the vehicle speed [4]:

(i) Aero marker parameter when the vehicle is gripped by side wind and the wind starts to take effect only from a certain point on. Ahead of this point, the vehicle body does not offer enough contact surface for the wind to take effect.
(ii) Center of pressure points are applied for the aerodynamic forces, including the side wind forces.
(iii) Model selected is Look-Up Table 1D.
(iv) Reference area is the frontal area, tires, and underbody parts.
(v) Reference length is used to calculate the torques applied to the center of pressure due to aerodynamic effects.
(vi) Aerodynamic coefficients used by the model Look-Up Table 1D have the following meanings from Table 6.13.

6.2.4.3 Suspensions

Suspensions is a mechanism that connects the wheels and the body of the vehicle. Suspensions evenly transmits the forces acting on the vehicle to the road surface and at the same time, it isolates the vehicle from the forces that appear from the road surface, thus improving its comfort and maneuverability. Suspensions consists of the elements: spring, secondary spring, damper, buffer, stabilizer, kinematics, wheel bearing [3].

The spring's characteristics are the same for both sides of the vehicle. The spring is modeled as a component that generates a force when it is compressed/stretched. The spring is seen as a "stand alone" component with properties can be measured independently (Table 6.14) [4]:

(i) Stiffness parameter is a coefficient or a table according to the selected parameter. A positive variation means that the spring is compressed.
(ii) Amplification parameter enables to scale the spring stiffness very quickly for test purposes instead of modifying the whole table.
(iii) Length IO is the relaxed or unstretched length of the spring, the distance between the lower and upper attachment point of the spring.

Table 6.14 Suspension (spring) (Some of the values that define the virtual model may differ from real values, they have been determined via experimental studies or mathematical calculus)

Spring			
Front			
Model	Hookean spring		
Stiffness mode	1D Look-Up table		
Stiffness (N/m)	Compress (m)	Force (N)	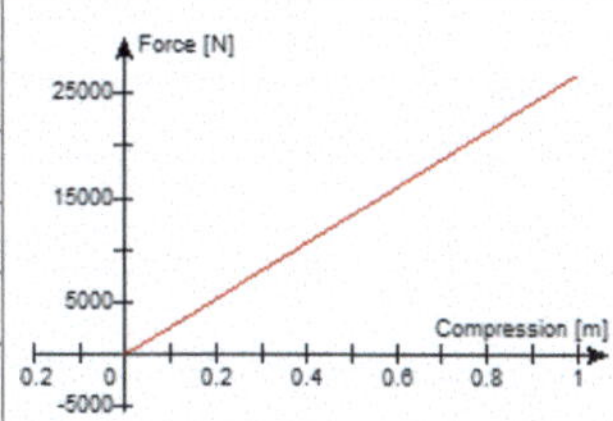
	0.000	0.000	
	0.100	2500	
	1.000	25,000	
Amplification (–)	1.000		
Length IO (m)	0.110		
Rear			
Model	Hookean spring		
Stiffness mode	1D Look-Up table		
Stiffness (N/m)	Compress (m)	Force (N)	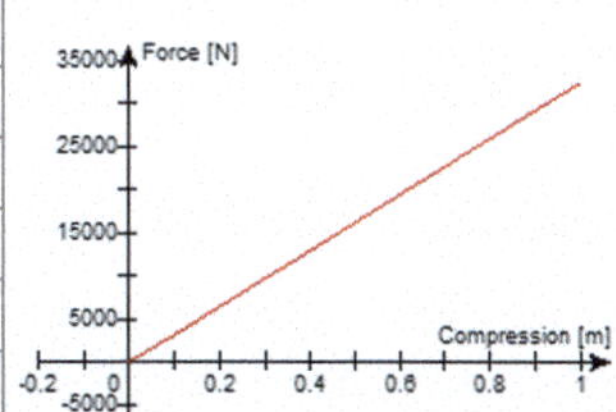
	0.000	0.000	
	0.100	3000	
	1.000	30,000	
Amplification (–)	1.000		
Length IO (m)	0.060		

Damper characteristics are same for both sides of the vehicle. The damper is modeled as a component that generates a force when being compressed/deflected (reaction to the change of velocity) (Table 6.15) [4]:

Table 6.15 Suspension (damper) (Some of the values that define the virtual model may differ from real values, they have been determined via experimental studies or mathematical calculus)

Damper			
Front			
Mode	Newtonian damper		
Damping mode	1D Look-Up table		
Damping (N s/m) push	Velocity (m/s)	Force (N)	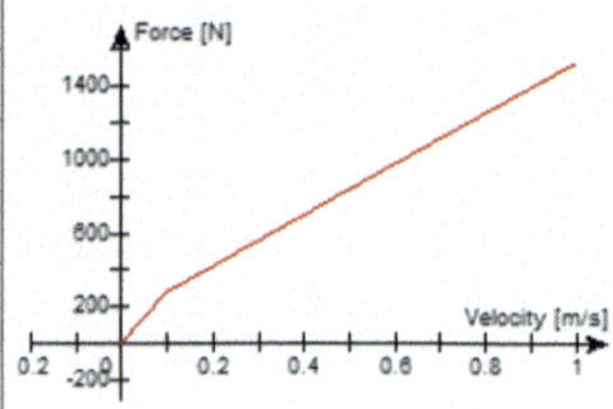
	0.000	0.000	
	0.100	300	
	1.000	2000	
Amplification (–)	1.000		
Damping (N s/m) pull	Velocity (m/s)	Force (N)	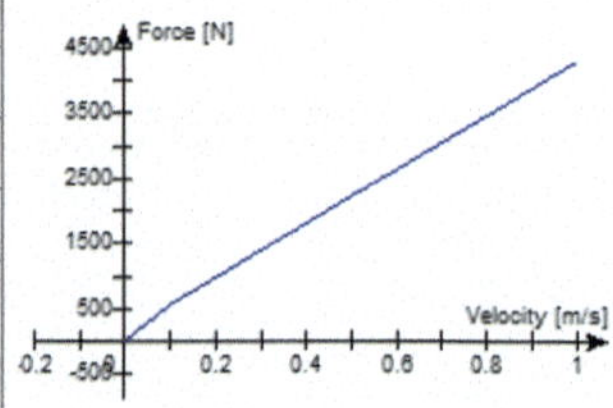
	0.000	0.000	
	0.100	500	
	1.000	4000	
Amplification (–)	1.000		
Rear			
Mode	Newtonian damper		
Damping mode	1D Look-Up table		
Damping (N s/m) push	Velocity (m/s)	Force (N)	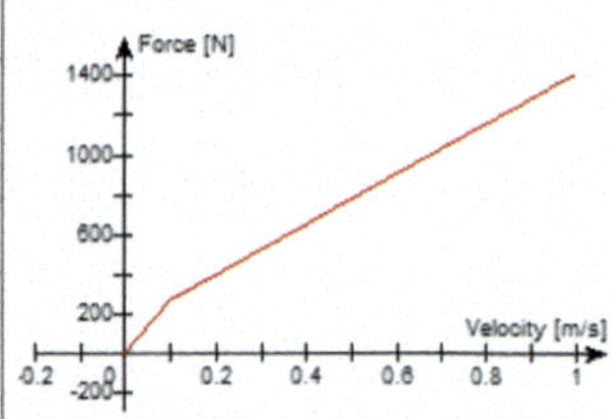
	0.000	0.000	
	0.100	300	
	1.000	1500	
Amplification (–)	1.000		
Damping (N s/m) pull	Velocity (m/s)	Force (N)	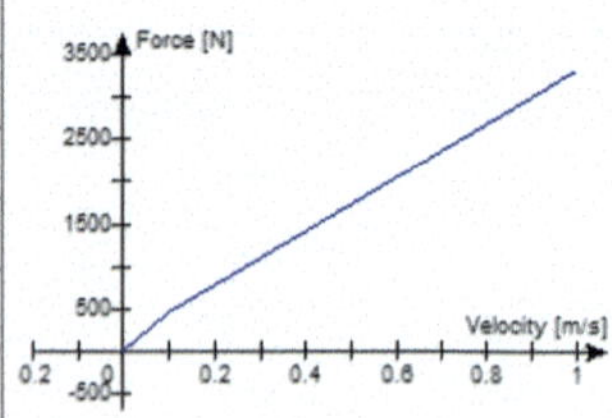
	0.000	0.000	
	0.100	500	
	1.000	3500	
Amplification (–)	1.000		

(i) Damper characteristic is specified by a table (mode Look-Up table 1D).
(ii) Damping characteristic is defined by a table of values according to the selected mode. The characteristics of the damper are split into two domains: push and pull. The push domain corresponds to a positive damper speed, which means that the damper is being compressed.
(iii) Amplification parameter enables to scale the damper characteristics very quickly instead of modifying the whole table.
(iv) Damper top mount is an additional spring element with Stiffness and damper element with a damping factor, added to the upper damper mounting point.

Buffer characteristics are the same for both sides of the vehicle. The buffer is modeled as a spring which is activated only at a defined wheel travel. The buffers are used to limit the wheel travel in one direction or in both directions, up and down. If the wheel travelled far enough to hit one of the buffers, the buffer acts like an additional spring (Table 6.16) [4]:

(i) Model name Hookean spring defines pull/push characteristic.
(ii) Mode buffer stiffness Look-Up table 1D is defined in a table of values.
(iii) Buffer stiffness is defined by a table of values according to the selected mode. The characteristics of the buffer are split into two domains: push and pull, where the push corresponds to a positive wheel travel.
(iv) Amplification parameter enables to scale the buffer stiffness very quickly instead of modifying the whole table.
(v) Length tz0 is the wheel travel in the vertical direction from which the buffer is activated, calculated from the position in the configuration.

The stabilizer/anti-roll bar is a force element, which generates the stabilizer force depending on its right and left side deflection length, deflection velocity and a control signal for each side. While the deflection lengths and velocities are passed from the suspension kinematics module, the control signals are passed from the suspension control unit.

Stabilizer specified the properties of the stabilizer element (Table 6.17):

(i) Mode stiffness of the stabilizer defined in a table (Look-Up table 1D).
(ii) Stiffness of the stabilizer can be defined to the selected parameter:

    (1) Stabilizer as a virtual spring, the movement of the stabilizer end rod.
    (2) Stabilizer is a real torsional component.

(iii) Amplification enables to scale the stiffness of the stabilizer very quickly instead of modifying the parameter itself.

Kinematics model also includes the length variation of the spring, damper, and buffer against the same variables (wheel travel and steering rack movement). The kinematics must be specified for the front and rear axle separately (Table 6.18).

Model Linear 2 DOF characteristics can be used for both front and rear axles: front axle, the kinematics will depend on the wheel travel and the steering rack

Table 6.16 Suspension (buffer) (Some of the values that define the virtual model may differ from real values, they have been determined via experimental studies or mathematical calculus)

Buffer		
Front		
Model	Hookean spring	
Stiffness mode	1D Look-Up table	
Stiffness (N/m) push	Compress (m)	Force (N)
	0.000	0.000
	0.005	35.00
	0.015	165.0
	0.025	430.0
	0.035	960.0
	0.045	1960
	0.055	3800
	0.065	7135
	0.075	13,000
Amplification (–)	1.000	
Length (tz0) (m)	0.030	

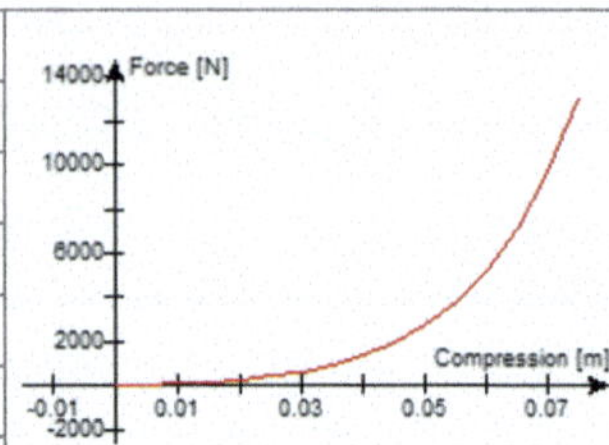

Stiffness (N/m) pull	Compress (m)	Force (N)
	0.000	0.000
	0.005	160.0
	0.010	650.0
	0.015	2000
	0.020	5000
	0.025	13,000
Amplification (–)	1.000	
Length (tz0) (m)	– 0.099	

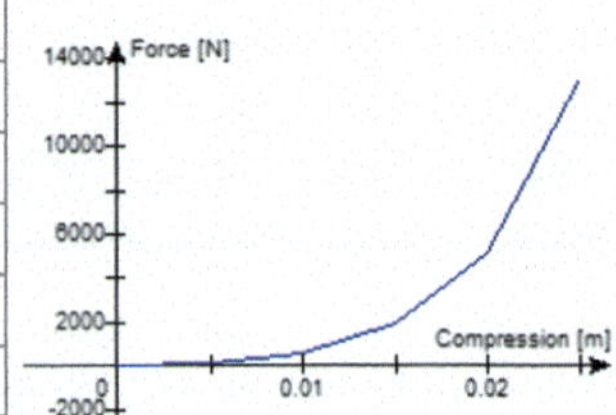

Rear		
Model	Hookean spring	
Stiffness mode	1D Look-Up table	
Stiffness (N/m) push	Compress (m)	Force (N)
	0.000	0.000
	0.005	15.00
	0.015	70.00
	0.025	165.0
	0.035	335.0
	0.045	620.0
	0.055	1100

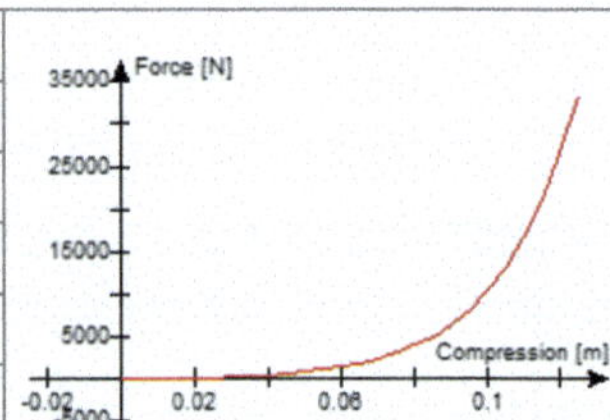

(continued)

Table 6.16 (continued)

	0.065	1880	
	0.075	3100	
	0.085	5150	
	0.095	8300	
	0.105	13,300	
	0.115	21,100	
	0.125	33,200	
Amplification (–)	1.000		
Length (tz0) (m)	0.130		
Stiffness (N/m) pull	Compress (m)	Force (N)	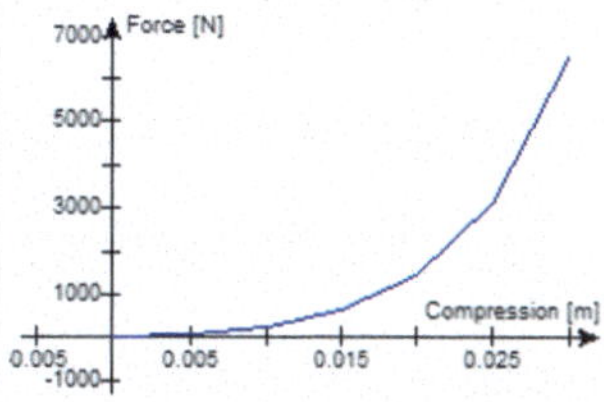
	0.000	0.000	
	0.005	70.00	
	0.010	240.0	
	0.015	600.0	
	0.020	1500	
	0.025	3100	
	0.030	6550	
Amplification (–)	1.000		
Length (tz0) (m)	– 0.130		

displacement and rear axle, the kinematics will depend on the wheel travel and the movement of the opposite wheel (rigid):

(i) Linear two-dimensional kinematics (front). The kinematics depend on the wheel travel and the steering rack displacement.
(ii) Linear two-dimensional kinematics (rear). The kinematics depend on the wheel travel and the movement of the opposite wheel.

The Linear 2 DOF model can be used for both front and rear axles:

(i) Front axle where the kinematics will depend on the wheel travel and the steering rack displacement.
(ii) Rear axle where the kinematics will depend on the wheel travel and the movement of the opposite wheel (rigid or semi-rigid axle).

6.2.4.4 Steering

The steering system consists of a mechanical module that defines the ratio between the steering wheel angle (steering wheel torque) and the movement of the rack. The Pfeffer steering model is a power steering unit that refers to the friction losses in the steering system. Steering system is defined by a static model whose parameters are shown in Table 6.19 [3].

Table 6.17 Suspension (stabilizer) (Some of the values that define the virtual model may differ from real values, they have been determined via experimental studies or mathematical calculus)

Stabilizer			
Front			
Model	Hookean spring		
Stiffness mode	1D Look-Up table		
Stiffness (N/m)	Compress (m)	Force (N)	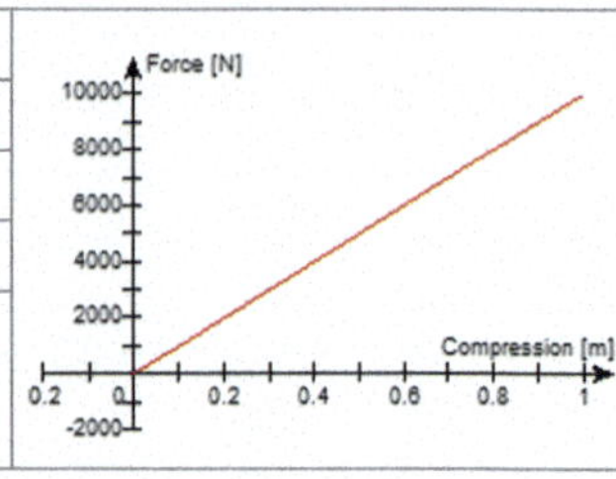
	0.000	0.000	
	0.100	1000	
	1.000	10,000	
Amplification (–)	1.000		
Rear			
Model	Hookean spring		
Stiffness mode	1D Look-Up table		
Stiffness (N/m)	Compress (m)	Force (N)	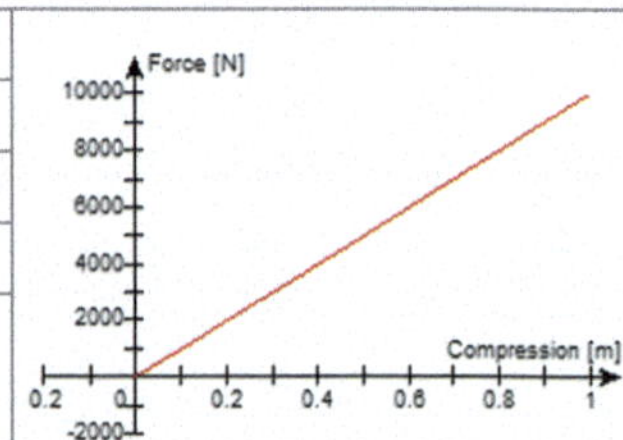
	0.000	0.000	
	0.100	1000	
	1.000	10,000	
Amplification (–)	1.000		

The static steer ratio steering model expects the steering angle and its derivative as inputs. It includes the calculation of the theoretical steering torque corresponding to the steer angle. Characteristic value ratio defined by a constant coefficient of steering wheel angle (rad) to steering rack travel in.

6.2.4.5 Tires

The tires fitted to the virtual vehicle are identical for all four wheels and are defined according to the example in Table 6.20 [3].

6.2.4.6 Brake

Brake system is divided into the command-and-control unit of the braking system and the braking system. The brake system command and control unit manages the

Table 6.18 Suspension (kinematics) (Some of the values that define the virtual model may differ from real values, they have been determined via experimental studies or mathematical calculus)

Kinematics				
Front model	Linear 2 DOF			
	Static (m)	Compr. (m/m)	Oppos. (m/m)	Steer (m/m)
Translation (tx)	0.000	0.028	0.000	– 0.341
Translation (ty)	0.000	– 0.077	0.000	0.122
Translation (tz)	0.000	1.000	0.000	– 0.027
	Static (rad)	Compr. (rad/m)	Oppos. (rad/m)	Steer (rad/m)
Rotation (rx)	0.010	0.628	0.000	– 0.855
Rotation (ry)	0.000	– 0.620	0.000	– 1.293
Rotation (rz)	– 0.003	0.105	0.000	5.000
	Static (m)	Compr. (m/m)	Oppos. (m/m)	Steer (m/m)
Deflection ISpring	0.000	– 1.000	0.000	0.000
Deflection IDamp	0.000	– 1.000	0.000	0.000
Deflection IBuf	0.000	– 1.000	0.000	0.000
Deflection IStabi	0.000	– 1.000	0.000	0.000
Rear model	Linear 2 DOF			
	Static (m)	Compr. (m/m)	Oppos. (m/m)	Steer (m/m)
Translation (tx)	0.000	0.000	0.000	0.000
Translation (ty)	0.000	0.000	0.000	0.000
Translation (tz)	0.000	1.000	0.000	0.000
	Static (rad)	Compr. (rad/m)	Oppos. (rad/m)	Steer (rad/m)
Rotation (rx)	0.000	0.591	– 0.591	0.000
Rotation (ry)	0.000	0.000	0.000	0.000
Rotation (rz)	0.000	0.000	0.000	0.000
	Static (m)	Compr. (m/m)	Oppos. (m/m)	Steer (m/m)
Deflection ISpring	0.000	– 0.900	– 0.100	0.000
Deflection IDamp	0.000	– 0.900	– 0.100	0.000
Deflection IBuf	0.000	– 0.900	– 0.100	0.000
Deflection IStabi	0.000	1.000	0.000	0.000

Table 6.19 Steering (Some of the values that define the virtual model may differ from real values, they have been determined via experimental studies or mathematical calculus)

Steering gear ratio	
Steering model	Static steer ratio (GenAngle)
Mode	Characteristic value
Rack travel to steering pinion angle (rad/m)	80

Table 6.20 Tires (Some of the values that define the virtual model may differ from real values, they have been determined via experimental studies or mathematical calculus)

All axles	
Front axle	
Tire front left	Examples/RT_215_60R17_p3.75
Tire front right	Examples/RT_215_60R17_p3.75
Rear axle	
Tire front left	Examples/RT_215_60R17_p3.75
Tire front right	Examples/RT_215_60R17_p3.75

operation of the hydraulic system. For the autonomous vehicle, which is an electrically powered vehicle, it distributes the braking load between the electric motors of the engine and the hydraulic brake to ensure a reliable regenerative braking.

Regenerative brake serial defines the distribution of the braking energy to the electric motors (used as generators for recovering the braking energy) and the hydraulic braking system, so that the electric motors take over as much of the resulting braking energy as possible (for a most efficient regenerative braking). The hydraulic braking system is only used if the electric motors do not have the capacity to take over all the resulting braking energy. Share recuperative braking torque defines the percental maximum value for which the hydraulic braking system is activated to take over some of the regenerative braking energy that cannot be regenerated by the electric generators.

The parameters of the braking system are presented in Table 6.21 [3].

Hydraulic basic controller model defines the controller for hydraulic brake systems and supports several recuperative brake modes. It does not contain any brake assistance system as ABS or ESP.

Regenerative brake mode defines how the target braking torque is distributed between electric motors (used as generator for energy recuperation) and hydraulic brake system. For model serial the electric motors apply as much of the demanded brake torque as possible (regenerative braking). The hydraulic brake system is used only if the electric motors are not able to satisfy the brake torque demand:

(i) Brake torque controller P-value/I-value parameter define the proportional (P) and integral (I) of the PI controller that controls the hydraulic brake system valves signals.
(ii) Share recuperative braking torque factor enables to set the distribution of regenerative and hydraulic braking torque.
(iii) Brake torque recuperative margin—if the difference between the target brake torque and the maximum possible regenerative brake torques exceeds this margin value, then the hydraulic brake system is activated to support the regenerative brake torques.

Pressure distribution model are the ways to control the brake pedal force:

(iv) Pedal actuation to pMC (pressure in Master Cylinder) parameter is used only by the mode pedal actuation. It is the ratio of brake pedal position to the corresponding pressure in the master cylinder.

Table 6.21 Brake (Some of the values that define the virtual model may differ from real values, they have been determined via experimental studies or mathematical calculus)

General			
Model	Hydraulic		
Brake torque amplification axle (–)		Front	Rear
Left		1.000	1.000
Right		1.000	1.000
Control			
Model	Hydraulic basic controller		
Regenerative brake	Serial		
Brake torque controller P-value (–)		1.0E−2	
Brake torque controller I-value (–)		1.0E−2	
Share recuperative braking torque (–)		0.500	
Brake torque recuperative margin (N m)		10.00	
System			
Model	Pressure distribution		
Mode	Pedal force		
Pedal actuation to pMC (bar)		150.0	
Pedal actuation to pedal force (N)		500.0	
Pedal force to pMC (bar/N)		0.400	
Response time (s)		0.005	
Build-up time (s)		0.080	
Pressure to brake torque (N m/bar)		Front	Rear
Left		50.00	25.00
Right		50.00	25.00
Parkbrake torque at wheel (N m)		Front	Rear
Left		0.000	2500
Right		0.000	2500

(v) Pedal actuation to pedal force parameter is only used by the mode pedal force. It is the ratio of the force the driver applies to the brake pedal and the corresponding pedal movement.
(vi) Pedal force to pMC parameter is used only by the mode pedal force. It is the ratio between the force applied to the brake pedal and the corresponding pressure in the master cylinder.
(vii) Response time is the period from pressing the pedal until the brake pressure begins to build up.
(viii) Build-up time is the period from the beginning of the building of brake pressure until it reaches 75% of the desired value.
(ix) Pressure to brake torque parameters describe the ratios between the pressure in the master cylinder and the brake torque to each wheel.

(x) Parkbrake torque at wheel describes the brake torque applied to each wheel if the park brake is activated.

6.2.4.7 Powertrain

Powertrain system allows the configuration of the constructive architecture of the powertrain. For the virtual model of the autonomous vehicle a powertrain with 1 electric motor will be selected.

The virtual model for the electric motor is (1D Look-Up table), includes the characteristics for the torque of both the motor and of the electric generator. The characteristic parameters of the electric motor are [4]:

(i) Inertia of the electric motor sums up all rotating and spinning components of the motor. For I_{xx} it is assumed that the x-axis is along the transmission components.
(ii) Ratio parameter describes a constant transmission ratio between the electric motor output shaft and the driven shaft that it is connected to.
(iii) Build-up time specifies the build-up time of the load signal.
(iv) Voltage level specifies to which electric circuit of the PS the electric motor is connected to. The electric power consumed or generated by the starter motor is automatically considered in the power balance of the selected circuit.

The electric motor is seen as a torque source producing its output depending on the throttle load signal and on its speed. In motor mode (conversion of electric power into mechanic power) the sign of the rotation speed and torque are equal. In generator mode (conversion of mechanic power into electric power) the rotation speed and torque have different signs.

Engine torque is defined by a model of characteristic value in electric motor mode and in electric generator mode, the values being specified for the following parameters: motor/electric generator power, maximum torque for motor/generator, maximum engine speed.

The efficiency of the motor/electric generator is defined by the model (2D Look-Up table), and it depends on the normalized speed (actual speed/maximum speed) and the normalized torque (current torque/maximum torque) of the motor/electric generator mechanism.

The driveline is defined by a rear drive model with no external torque (external torque off) applied to the rear wheels that are not driven (for the current constructive version).

The PS manages the supply voltages provided by the LV source consisting of a battery, respectively by the HV source consisting of a battery and a HV circuit (DC/DC converter). Maximum power of the converter is the maximum value that can be transferred between the LV circuit and the HV circuit (DC/DC converter from HV to LV). The Chen battery model consists of an electric battery model combined with a characteristic curve that represents the influence of the SOC on the electrical voltage. Initial SOC defines battery charge at start of the simulation (%).

For the virtual model, a constant load is defined on each electrical circuit that considers the electrical power consumed (or generated) by the auxiliary devices (auxiliary consumer on LV+ auxiliary consumer on HV). Positive consumption values decrease the electricity available in the circuit, and negative values increase the electricity available in the circuit.

Battery management according to the control strategy determines the SOC of electric batteries by defining the values for the following parameters for both the LV and the HV power source: maximum power of generator, initial battery charge in percentage (target SOC of LV/HV battery), control parameter of SOC (SOC control I-Value), electric consumer control parameter (consumer control I-Value).

The command-and-control units comprise the following devices that manage the operation of the powertrain: the global command and control unit powertrain control provides calculated values for the algorithms of the controllers of the electronic control unit of the electric motor MCU, respectively of the control unit of the electric battery BCU*.

The role of the BCU* electronic battery control unit is to calculate the characteristic values of the battery that cannot be measured directly, for example the SOC (%). The virtual model for BCU* is (LV + HV), it models a LV circuit and a HV circuit.

The powertrain consists of the following elements: drive source, driveline, PS, and control units [3].

Electrical powertrains can contain 1–4 electric motors that transfer driving torque to the wheels. The distribution of drive torques is calculated as a function of the final drive output torque of an electric motor (Table 6.22).

Electric motor model is 1D Look-Up table. This model includes motor and generator characteristics (Table 6.23):

(i) Inertia sums up all rotating and spinning components of the motor.
(ii) Ratio parameter describes a constant transmission ratio between the electric motor output shaft and the driven shaft.
(iii) Build-up time specifies the build-up time of the load signal.
(iv) Friction coefficient defines the internal motor friction torque that acts as a drag torque if the motor load is zero.
(v) Voltage level specifies to which electric circuit of the PS the electric motor is connected to. The electric power consumed or generated by the starter motor is automatically considered in the power balance of the selected circuit.

Torque motor/generator model using the electric motor model mapping, the electric motors torque characteristics are defined by two curves:

(A) Maximum motor torque (available if the load signal is 1).
(B) Maximum generator torque (available if the load signal is – 1).

Characteristic value—the torque characteristic is described by three (motor) or four (generator) characteristic values:

Table 6.22 Powertrain (general) (Some of the values that define the virtual model may differ from real values, they have been determined via experimental studies or mathematical calculus)

General	
Model	Electrical
Configuration number of electric motors	1 Electric motor
General information	

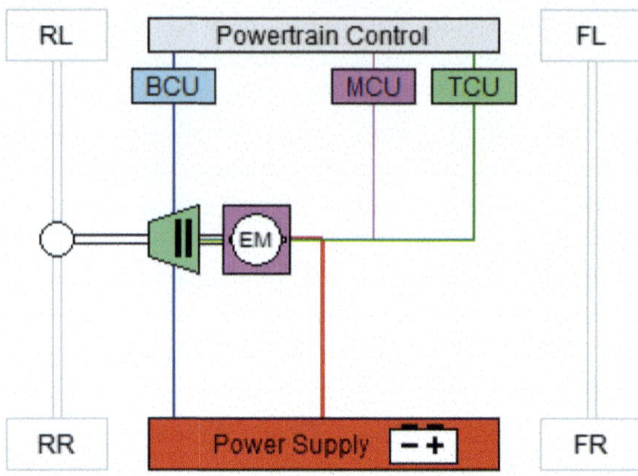

Powertrain	Electrical
GearboxM	No gearbox
Motor	1D Look-Up table
Driveline	Rear drive
Powertrain control	Electrical
TCU	–
MCU	Basic
BCU (=BCU*)	LV + HV
Power supply	LV + HV
Integration substeps	5

(I) Mechanical power—the electric motors' mechanical power in motor/generator mode influences the curvature of the motor/generator torque curve.

(II) Maximum torque—the overall maximum motor/generator torque corresponds to the torque available for small motor rotation speeds.

(III) Maximum rotational speed—the parameter defines the maximum allowed motor rotation speed for motor/generator mode.

(IV) Rotational speed maximum torque is available for generator mode.

Amplification parameter enables to scale the torque characteristics very quickly instead of modifying the entire table.

Table 6.23 Powertrain (drive source) (Some of the values that define the virtual model may differ from real values, they have been determined via experimental studies or mathematical calculus)

Drive source	
Gearbox model: no gearbox	
Electric motor	
General	
Motor model	1D Look-Up table
Inertia (kg m^2)	0.002
Ratio (–)	1.000
Build-up time (s)	0.010
Friction coefficient	3.0e−4
Voltage level	HV1
Torque	
Motor model	Characteristic value
Mechanical power (kW)	15.00
Maximum torque (N m)	130.0
Maximum speed (rpm)	8000

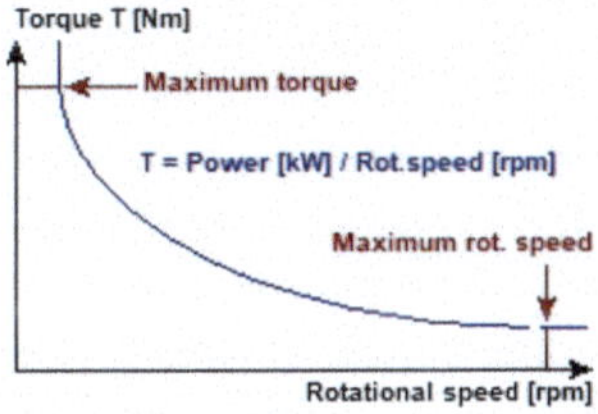

Generator model	Characteristic value
Mechanical power (kW)	20.00
Maximum torque (N m)	250.0
Maximum rotational speed (rpm)	8000
Rotational speed max torque (rpm)	500.0

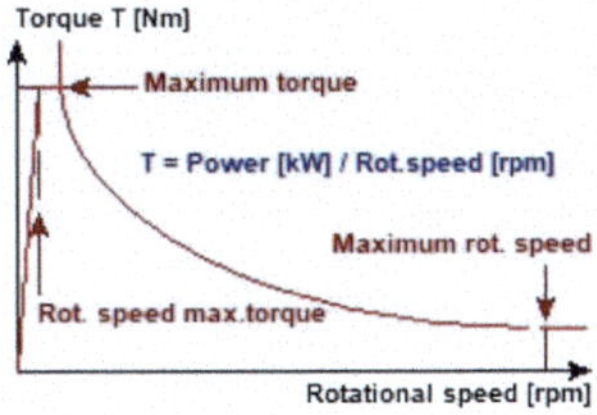

Efficiency

(continued)

Table 6.23 (continued)

Motor model		2D Look-Up table			
Rot. speed norm (–)	Torque norm (–)	Eta (–)	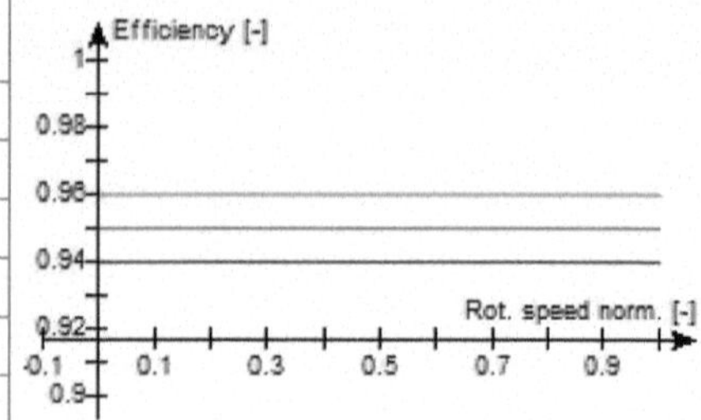		
0.000	0.000	0.940			
0.000	0.500	0.950			
0.000	1.000	0.960			
1.000	0.000	0.940			
1.000	0.500	0.950			
1.000	1.000	0.960			
Amplification		1.000			
Generator model		Characteristic value			
Efficiency (–)		0.950			
Bodies					
Electric motor					
Mass (kg)			0.000		
Inertia $I_{xx}/I_{yy}/I_{zz}$ (kg m^2)			0.000	0.000	0.000

Efficiency parameter corresponds to the electric motor's overall efficiency in motor/generator mode.

Driveline models connect the drive sources to the wheels. The driveline models consist of shafts and differentials and realize different torque distributions. Driveline model is rear drive mode. The differential model that is used here transfers the torque from the input shaft to each wheel, or from a mid-differential to each axle. On the other hand, the speeds coming from both output shafts are converted to a mid-speed imposed to the input shaft. The model is described by the following general parameters (Table 6.24) [4]:

(i) Coupling models modify the torque distribution to the left and right wheel.
(ii) Inertia in includes all rotating and spinning components at the input side of the differential whose inertia has not been considered by another parameter.

Table 6.24 Powertrain (driveline) (Some of the values that define the virtual model may differ from real values, they have been determined via experimental studies or mathematical calculus)

Driveline			
Model	Rear drive		
External torque off			
Rear axle			
Differential			
Inertia in (kg m^2)	0.001	Transmission (–)	4.000
Inertia out (kg m^2)	0.001	Efficiency (–)	1.000
Inertia cage (kg m^2)	0.000		

(iii) Inertia out includes all rotating and spinning differential output shaft components.
(iv) Inertia cage includes the inertia of the cage and of all rotating and spinning components that are in it.
(v) Transmission parameter is the transmission ratio from differential input shaft to cage.
(vi) Efficiency parameter sets the state efficiency of the differential. Values must be between 0…1.

The PS consists of the vehicle's electric circuits, the batteries, and auxiliary consumers. PS configurations consist of one LV circuit with battery and one HV circuit with battery (LV + HV1) (Table 6.25) [4]:

(i) LV PS consisting of a LV circuit with battery only.
(ii) LV + HV1 PS consisting of a LV circuit and one HV circuit (with battery).
(iii) Auxiliary consumers on LV/HV define a constant load on the electric circuit to consider the electric power consumed by auxiliary devices.
(iv) Maximum converter power from HV to LV sets the maximum electric power that can be transferred between LV/HV circuit.
(v) Efficiency model from HV to LV describes the efficiency of the energy transfer as being characteristic value.
(vi) Battery model of Chen consists of an electric battery model combined with a characteristic curve to consider the influence of SOC on the voltage. The battery's equivalent circuit consists of an ideal voltage source, two RC-circuits and a single resistance that are connected in series.
(vii) Capacity sets the overall capacity of the battery.
(viii) Idle voltage parameter is the basic voltage of the ideal source.
(ix) Initial SOC defines the state of charge of the batteries at the start of the simulation.
(x) Minimum SOC—SOC should not fall below this limit because this would damage the battery.
(xi) Maximum SOC—SOC should not exceed this limit because this would damage the battery.
(xii) Resistance R0/R1/R2 set the resistances R0, R1 and R2.
(xiii) Capacity C1/C2 set the capacities C1 and C2.
(xiv) SOC influence on idle voltage sets the influence of the battery's SOC on its idle voltage. It is defined as a factor on U0.
(xv) Amplification factor applies to all values in the column factor if SOC influence on idle voltage is selected.
(xvi) Battery power limits the battery charging and discharging power at a time (characteristic value), limited by a maximum power.
(xvii) Mass sets the weight of the battery.
(xviii) Inertia of these body masses can be defined in the axis orientation of frame FrD.

Table 6.25 Powertrain (PS) (Some of the values that define the virtual model may differ from real values, they have been determined via experimental studies or mathematical calculus)

Power supply			
Model	LV + HV1		
Auxiliary consumer on LV (W)		0.000	
Auxiliary consumer on HV (W)		0.000	
Converter HV LV			
DC/DC converter from HV to LV			
Maximum converter power (kW)		0.000	
Efficiency model	Characteristic value		
Efficiency (–)		0.950	
LV battery			
Battery model	Chen		
Capacity (Ah)	10	Resistance R0 (Ω)	0.002
Idle voltage (V)	12	Resistance R1 (Ω)	0.001
Initial SOC (%)	75	Resistance R2 (Ω)	0.001
Minimum SOC (%)	10	Capacity C1 (F)	4000
Maximum SOC (%)	90	Capacity C2 (F)	8000
SOC influence on idle voltage			
SOC (%)	Factor (–)	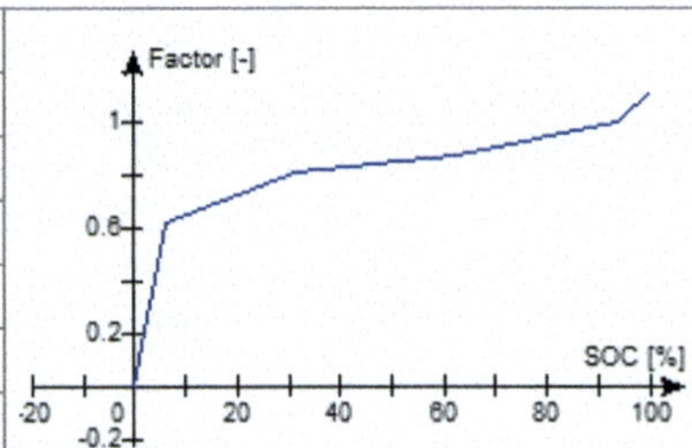	
0.000	0.000		
6.000	0.600		
30.00	0.800		
60.00	0.900		
90.00	1.000		
100.0	1.125		
Amplification	1.000		
Battery power	Characteristic value		
Maximum power (kW)	2.000		
Bodies			
LV battery			
Mass (kg)		0.000	
Inertia $I_{xx}/I_{yy}/I_{zz}$ (kg m^2)	0.000	0.000	0.000
HV battery			
Battery model	Chen		
Capacity (Ah)	200.0	Resistance R0 (Ω)	0.002
Idle voltage (V)	400.0	Resistance R1 (Ω)	0.001
Initial SOC (%)	75.00	Resistance R2 (Ω)	0.001

(continued)

Table 6.25 (continued)

Minimum SOC (%)	20.00	Capacity C1 (F)	4000
Maximum SOC (%)	80.00	Capacity C2 (F)	8000
SOC (%)	Factor (–)		
0.000	0.000		
6.000	0.600		
30.00	0.800		
60.00	0.900		
90.00	1.000		
100.0	1.125		
Amplification	1.000		
Battery power	Characteristic value		
Maximum power (kW)	100.0		
Bodies			
HV battery			
Mass (kg)	0.000		
Inertia $I_{xx}/I_{yy}/I_{zz}$ (kg m^2)	0.000	0.000	0.000

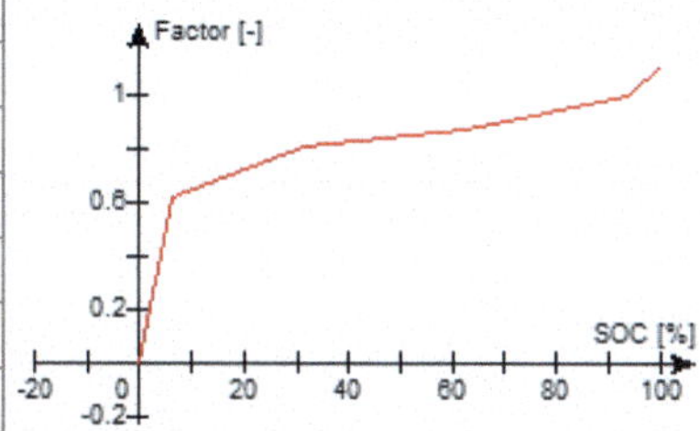

Control unit's models correspond to the software part of the powertrain elements. Their task is to provide control signals to the electromechanical powertrain components and to calculate/measure characteristic values that describe the current state of these components (Table 6.26):

(I) MCU (Motor Control Unit)—electric motors.
(II) BCU* (Battery Control Unit)—PS including batteries.

Powertrain control is a global control unit whose main task consists in providing target values for the controller algorithms of the other control units. Its main tasks are regulation of the vehicle operation states:

(i) Mechanical energy management translating the virtual driver actions into targets for the electro mechanic components.
(ii) Electric energy management regulating the SOC of the batteries and the energy transfer between different electric circuits.
(iii) Providing information about maximum and current regenerative braking torque to the brake control unit to organize energy recuperation.

Battery management parameters for the LV and the HV circuit:

(iv) Maximum power of generator corresponds to the maximum admissible generator power for recharging the battery in LV/HV circuit.
(v) Target SOC of LV/HV battery is used as target value for the controller algorithm of the battery management.

Table 6.26 Powertrain (control unit) (Some of the values that define the virtual model may differ from real values, they have been determined via experimental studies or mathematical calculus)

Control unit		
Powertrain control		
General		
Model	Electrical	
Start engine with button		
Operation state machine		
Model	Generic	
Gas		
Mode	Characteristic Value	
Maximum torque at full gas (N m)		180.0
Minimum torque at zero gas (N m)		– 10.00
Battery		
Battery management		
LV battery		
Maximum power of generator (kW)		2.000
Target SOC of LV battery (%)		70.00
SOC control I-value (–)		50.00
Consumer control I-value (–)		1.000
HV battery		
Maximum power of generator (kW)		2.000
Target SOC of HV battery (%)		50.00
SOC control I-value (–)		50.00
Consumer control I-value (–)		1.000
Regeneration		
Regeneration brake		
SOC limit (%)		75.00
SOC limit range (%)		5.000
Speed limit (km/h)		2.000
Torque limit (N m)		2.000
Regeneration drags		
SOC limit (%)		75.00
SOC limit range (%)		5.000
Speed limit (km/h)		5.000
Torque limit (N m)		2.000
MCU model	Basic	
Electric motor		
Torque controller P-value (–)		0.050
Torque controller I-value (–)		100.0

(continued)

Table 6.26 (continued)

Control unit		
Speed controller P-value (–)		0.020
Speed controller I-value (–)		5.0e–5
BCU* (=BCU)		
General		
Model	LV + HV1	

(vi) SOC control I-Value parameter scales the influence of the difference between target and current battery SOC (integral element) on the additional generator torque demand.

(vii) Consumer control I-Value parameter scales the influence of the auxiliary power consumption on the additional generator torque demand.

MCU model basic parametrization contains the following elements [4]:

(viii) Torque controller P/I-value parameters correspond to the proportional (P-value) and integral (I-value) of the torque speed controller which is mainly a PI-controller. The torque controller becomes active if the powertrain control strategy provides a target motor torque and transforms it into a motor load signal ($-1\ldots1$) depending on the current motor torque. A negative load signal means generator mode.

(ix) Speed controller P/I-value parameters correspond to the proportional (P-value) and integral (I-value) of the rotation speed controller which is mainly a PI-controller. The speed controller becomes active if the powertrain control strategy provides a target motor rotation speed and transforms it into a motor load signal ($-1\ldots1$) depending on the current motor rotation speed. A negative load signal means generator mode.

The main task of the BCU* is the calculation of characteristic battery values that cannot be measured directly SOC.

6.2.4.8 Sensors

Sensors system allows modeling and definition of different types of sensors. Slip angle sensors and inertial sensors represent the behavior of the virtual vehicle and are mainly used to analyze the dynamics of the vehicle. The other sensors are relevant for generating the command-and-control signals of ADAS. The sensors that equip the virtual model of the autonomous vehicle are divided into virtual sensors and real RSI sensors.

The virtual sensors allow the command-and-control of the developed vehicle model, and the CarMaker application allows the use of the following sensors: slip angle, inertial, object, free space, traffic sign, line, road, collision, object by line, global navigation, radar, camera. These sensors exist only as equipment of the virtual model and have the role of extracting the existing information in the virtual simulation environment and delivering it to the user.

Real RSI sensors provide raw information about the environment in which the virtual vehicle is evolving, leaving it to the user to choose which of this information should be filtered and processed based on functions implemented by the user. The existing RSI sensors in the CarMaker application are RSI radar, RSI ultrasonic, RSI Lidar and RSI camera. These sensors detect virtual traffic, static and dynamic objects, and 3D surfaces in the virtual environment (Fig. 6.34).

Slip Angle Sensor

Slip angle sensor measures the angle of lateral slip between the direction in which the wheel is directed and the direction in which the virtual vehicle is moving. The following options configure the slip angle sensor (Table 6.27) [3, 4]:

(i1) Slip angle sensor defines the identification of the sensor quantities.

Inertial Sensor

Inertial sensor can be positioned anywhere on the vehicle and is used to determine the position, velocity, and acceleration of the vehicle. The correspondent of the inertial sensor for a real autonomous vehicle is the sensor of the IMU which operates on the basis of a three-axis accelerometer and a three-axis gyroscope, respectively.

Fig. 6.34 Sensors system

Table 6.27 Slip angle sensor (Some of the values that define the virtual model may differ from real values, they have been determined via experimental studies or mathematical calculus)

Slip angle sensor	Parameter name

Table 6.28 Inertial sensor (Some of the values that define the virtual model may differ from real values, they have been determined via experimental studies or mathematical calculus)

Inertial sensor	Parameter name

Table 6.29 Object sensor (Some of the values that define the virtual model may differ from real values, they have been determined via experimental studies or mathematical calculus)

Object sensor 1	Parameter name		
Field of view (h/v) (°)	50.0	28.0	(max 180)
Range (min/max) (m)	20.0		
Update (Hz) cycle offset (–)	60.00	1.000	
Observation radius (m)	500.0		
Target selection	Nearest		
Calculation class	Nearest point		
Object sensor 2	Parameter name		
Field of view (h/v) (°)	50.00	28.00	(max 180)
Range (min/max) (m)	20.0		
Update (Hz) cycle offset (–)	60.00	0.000	
Observation radius (m)	500.0		
Target selection	Nearest		
Calculation class	Nearest point		

The following options configure the inertial sensor (Table 6.28) [3, 4]:

(i1) Inertial sensor defines the identification of the sensor quantities.

The inertial sensor delivers the following output signals:

(o1) Sensor position in global frame.
(o2) Translational and rotational velocity of sensor in body frame and optionally in global frame.
(o3) Translational and rotational acceleration of sensor in body frame and optionally in global frame.

Object Sensor

Object sensor detects objects in traffic and calculates the distance from these objects to the sensor. The object that is located at the smallest distance from the sensor is considered to be the most relevant target object. The correspondent of the object detection sensor for a real autonomous vehicle is the radar sensor that equips the autonomous vehicle.

The following options configure the object sensor (Table 6.29) [3, 4]:

(i1) Object sensor defines the identification of the sensor quantities.

(i2) Observation radius defines an observation area around the vehicle with a constant to avoid the calculation of far-off objects.
(i3) Field of view (h/v) defines the horizontal/vertical span of sensor.
(i4) Range (min/max) defines the longitudinal range of the sensor beam.
(i5) Update defines the frequency with which the environment is scanned.
(i6) Target selection specifies the sensor type nearest (sensor module finds the closest object and calculates the distance from this object to the sensor). For nearest the sensor module finds the closest object and calculates the distance from this object to the sensor. The object with the smallest overall distance is the relevant target.
(i7) Calculation class specifies whether the calculated distance to an object refers to the reference or nearest point. Calculation of relative quantities in reference point (including relative angles) and in nearest point.

The object sensor delivers the following output signals:

(o1) Time stamp for updated sensor signals.
(o2) Number of sensors defined.
(o3) Position and approaching speed of the relevant target's nearest point in polar coordinates and in x/y/z direction in sensor middle frame.

Free Space Sensor

Free space sensor is an extended module of the object detection sensor, whose sensor beam is divided into horizontal and vertical segments. Each segment determines the nearest detected point of the objects in traffic, the corresponding angles, and the approach speed of these objects.

This sensor can scan the virtual environment as a stereo video camera and captures free space and occupied space around the vehicle, which will be used as a maneuvering assistant to avoid objects in traffic.

The free space sensor plus module extends the visual field of the sensor, which is divided into segments, each segment being further divided into other horizontal and vertical equiangular segments, up to the level of points from which the image is formed.

The following options configure the free space sensor (Table 6.30) [3, 4]:

(i1) Free space sensor defines the identification of the sensor quantities.
(i2) Field of view (h/v) defines the horizontal/vertical span of sensor.
(i3) Range min/max defines the longitudinal range of the sensor beam.
(i4) Update defines the frequency with which the environment is scanned by the sensor.
(i5) Segments (h/v) defines the number of horizontal and vertical sensor beam segments.

The free space sensor delivers the following output signals:

Table 6.30 Free space sensor (Some of the values that define the virtual model may differ from real values, they have been determined via experimental studies or mathematical calculus)

Free space sensor 1	Parameter name		
Field of view (h/v) (°)	50.0	28.00	(max 180)
Range (minim/maxim) (m)	30.0		
Update (Hz) cycle offset (–)	30.00	0.000	
Segments (h/v) (°)	200.0	1.000	
Free space sensor 2	Parameter name		
Field of view (h/v) (°)	50.0	28.00	(max 180)
Range (minim/maxim) (m)	100.0		
Update (Hz) cycle offset (–)	30.00	0.000	
Segments (h/v) (°)	200.0	1.000	

(o1) Time stamp for updated sensor signals.
(o2) Number of sensors defined.
(o3) Bearing azimuth of the detected point in sensor middle frame.
(o4) Distance to the nearest detected point of a traffic object in polar coordinates.
(o5) Relative distance to sensor in Cartesian coordinates.
(o6) Approaching speed of the nearest detected point.
(o7) Approaching speed of the nearest detected point in x/y/z direction in sensor middle frame.
(o8) Bearing elevation of the detected point in sensor middle frame.
(o9) Identification number of detected traffic object (integer).

Traffic Sign Sensor

Traffic sign sensor detects preselected traffic signs that fall within the defined range and horizontal/vertical range of the sensor's visibility. The sensor checks if the detected traffic sign is oriented towards the virtual vehicle and sorts all the detected signs in ascending order according to the distance to them. The sensor provides additional information about the detected traffic sign such as identification of the sign, information about a possible additional sign, description of the sign and its values.

The following options configure the traffic sign sensor (Table 6.31) [3, 4]:

(i1) Traffic sign sensor defines the identification of the sensor quantities.
(i2) Field of view (h/v) defines the horizontal/vertical span of sensor.
(i3) Position (x/y/z) indicate the position of the sensor on the vehicle.
(i4) Range min/max defined the longitudinal range of the sensor beam.
(i5) Update defines the frequency with which the environment is scanned by the sensor.
(i6) Calculation of relative position in mounted frame specifies whether the relative sign positions are expressed and sorted in relation to the body frame on which the sensor is mounted.

Table 6.31 Traffic sign sensor (Some of the values that define the virtual model may differ from real values, they have been determined via experimental studies or mathematical calculus)

Traffic sign sensor 1	Parameter name			
Field of view (h/v) (°)	50.00	28.00		(max 180)
Range (min/max) (m)	30.0			
Update (Hz) cycle offset (–)	30.00		0.000	
Calculation of relative position in mounted frame				
Reference point (x/y/z) (m)	0.000	0.000		0.000
Detection of all signs				
Traffic sign sensor 2	Parameter name			
Field of view (h/v) (°)	50.00	28.00		(max 180)
Range (min/max) (m)	30.0			
Update (Hz) cycle offset (–)	30.00		1.000	
Calculation of relative position in mounted frame				
Reference point (x/y/z) (m)	0.000	0.000		0.000

(i7) Reference point (x/y/z) calculates the position of the reference point in the vehicle frame FrD in which the relative sign position.
(i8) Detection of all signs detects that fall within the sensor beam.
(i9) Selection of the types of traffic signs that need to be detected can be chosen. The number of traffic sign selections is limited to ten types.

The traffic sign sensor delivers the following output signals:

(o1) Total number of detected signs.
(o2) Distance to traffic sign position.
(o3) Distance to traffic sign position in sensor frame or mounted frame.
(o4) Traffic sign identification given by road.
(o5) User defined sign attribute values.
(o6) Additional external rotation of sensor, expressed in mounted frame.
(o7) Additional external travel of sensor, expressed in mounted frame.
(o8) Time stamp of updated sensor signals.

Collision Sensor

Collision sensor detects if any part of the virtual vehicle comes in contact with other objects in traffic. This sensor is also used to detect collisions with other vehicles during a parking maneuver.

The following options configure the collision sensor (Table 6.32) [3, 4]:

(i1) Collision sensor defines the identification of the sensor quantities.
(i2) Detection vehicle wheels considers the wheels as individual vehicle bodies.

Table 6.32 Collision sensor (Some of the values that define the virtual model may differ from real values, they have been determined via experimental studies or mathematical calculus)

<table>
<tr><th colspan="3">Collision sensor</th><th colspan="2">Parameter name</th></tr>
<tr><td colspan="3">Envelope mode</td><td colspan="2">2D contour</td></tr>
<tr><td>X (m)</td><td>Y (m)</td><td colspan="3" rowspan="14">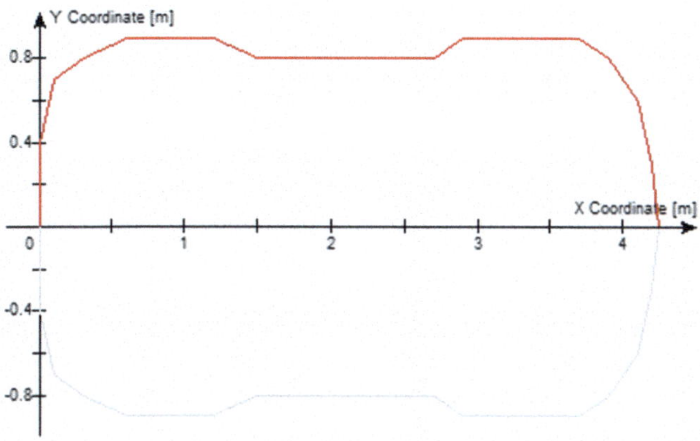
</td></tr>
<tr><td>0.000</td><td>0.400</td></tr>
<tr><td>0.100</td><td>0.700</td></tr>
<tr><td>0.300</td><td>0.800</td></tr>
<tr><td>0.600</td><td>0.900</td></tr>
<tr><td>1.200</td><td>0.900</td></tr>
<tr><td>1.500</td><td>0.800</td></tr>
<tr><td>2.700</td><td>0.800</td></tr>
<tr><td>2.900</td><td>0.900</td></tr>
<tr><td>3.700</td><td>0.900</td></tr>
<tr><td>3.900</td><td>0.800</td></tr>
<tr><td>4.100</td><td>0.600</td></tr>
<tr><td>4.200</td><td>0.300</td></tr>
<tr><td>4.250</td><td>0.000</td></tr>
<tr><td colspan="2" rowspan="2">Mirroring</td><td>Basic offset (x/z) (m)</td><td>0.000</td><td>0.200</td></tr>
<tr><td>Contour height (m)</td><td>1.200</td><td></td></tr>
</table>

(i3) Envelope mode defines the outer skin of the test car as seen by the collision sensor module: 2D contour, when the vehicle's outer skin is defined by a user defined (x/y) contour that is extruded in (z) direction, or cuboid (outer skin), when the vehicles outer skin is a cuboid with the dimensions defined in the vehicle data set.

(i4) Mirroring defines either an asymmetric contour or mirrors one side of the vehicle.

(i5) Basic offset (x/z) defines to move the vehicle contour in longitudinal (x) or vertical (z) direction.

(i6) Contour height defines the height of the outer contour.

(i7) Detection of upper and lower area during collisions with the upper and lower surface of the cuboid are detected in addition to the four peripheral sides.

The collision sensor delivers the following output signals:

(o1) Counter for number of collisions with the vehicle body of collision partner.

(o2) Counter for number of collisions with the vehicle wheel of collision partner.

Camera Sensor

Camera sensor has detecting lane delimitation lines, detecting the distance to other objects or vehicles in traffic and detecting traffic signs. The following options configure the camera sensor (Table 6.33) [3, 4]:

(i1) Camera sensor defines the identification of the sensor quantities.
(i2) Field of view (h/v) defines the horizontal and vertical direction of the sensor beam.
(i3) Range min/max defines the observation range of the sensor beam.
(i4) Cycle time offset defines the period length in which the environment will be captured, and the corresponding quantities updated. The cycle offset in number of cycles can be used to avoid all sensors being calculated in the same simulation cycle.
(i5) Objects max defines the maximum number of detectable objects for which output quantities are provided.
(i6) Resolution (h/v) defines the resolution of the camera image in horizontal and vertical direction.
(i7) Resolution factor defines the factor which will be multiplied with the camera resolution.
(i8) Threshold sets the detection threshold depending on the distance.
(i9) Distance estimation influences the distance estimation for stereo camera—the following additional parameters will be set:

    (a) Baseline defines the distance between two cameras in meters.
    (b) Disparity error occurring when matching the image of two cameras.
    (c) Focal length specifies the focal length of the camera in pixels.

Environment parameters direct link to parameters > environment. The camera sensor is influenced by the environment temperature, the rain rate, and the visual range in fog.

The camera sensor delivers the following output signals:

(o1) Number of detected objects.
(o2) Detected objects: ID of detected object, object type, confidence, number of visible pixels, minimum bounding rectangle.
(o3) Traffic sign/light: facing flag, light state, sign values.

Radar Sensor

Radar sensor detects obstacles on the virtual vehicle's travel path based on a SNR. The following types of obstacles can be detected: static and dynamic objects in circulation, geometric objects, trees (less than 10 m to the road), guide poles etc. Radar sensor considers the effects due to the overlapping objects in traffic. Real radar sensor uses cellular units to detect objects. If two objects overlap, they will

Table 6.33 Camera sensor (Some of the values that define the virtual model may differ from real values, they have been determined via experimental studies or mathematical calculus)

Camera sensor 1		Parameter name	
Field of view (h/v) (°)		50.00	100.0
Range (min/max) (m)		1.000	100.0
Cycle time (ms)/offset (–)		30.00	0.000
Object maximum (–)		50.00	
Detection			
Resolution (h/v) (px)		1280	960.0
Resolution factor (–)		0.500	
Distance (m)	Threshold (–)		
0.000	0.200		
100.0	0.500		

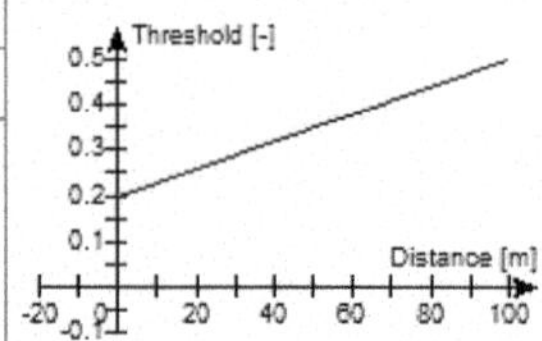

Distance estimation			
Camera type	Stereo		
Baseline (m)	0.120		
Disparity error (px)	0.500		
Focal length (px)	2400		
Camera sensor 2		Parameter name	
Field of view (h/v) (°)		100.00	100.0
Range (min/max) (m)		1.000	100.0
Cycle time (ms)/offset (–)		30.00	0.000
Object maximum (–)		50.00	
Detection			
Resolution (h/v) (px)		1280	960.0
Resolution factor (–)		0.500	
Distance (m)	Threshold (–)		
0.000	0.200		
100.0	0.500		

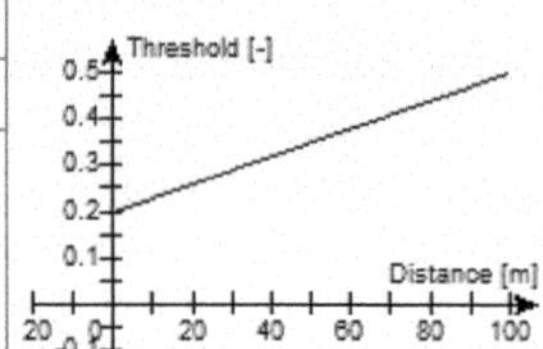

Distance estimation			
Camera type	Stereo		
Baseline (m)	0.120		
Disparity error (px)	0.500		
Focal length (px)	2400		

be placed in the same cell units. In this case, the radar sensor cannot separate the objects, instead it will only detect one.

RCS (Radar Cross Section) maps define different cross section maps for objects detected by the radar sensor in traffic. RCS maps influence the signal strength and the characteristic of the signal reflected by objects in traffic depending on the angle of incidence. These characteristics depend on the geometry and material of the objects in traffic. The antenna feature created by the generator can be viewed as a horizontal section (x, y) labeled azimuth.

Radar sensor is influenced by ambient temperature, atmospheric conditions (rain, snow, fog etc.), parameters that can be defined from the environment parameters menu. The following options are available to configure the radar sensor (Table 6.34) [3, 4]:

(i1) Radar sensor defines the identification of the sensor quantities.

Basic parameters:

(i2) Field of view (h/v) defines the horizontal and vertical span of the sensor beam.
(i3) Range min/max defines the longitudinal range of the sensor beam.
(i4) Cycle time offset defines the period length in which the environment will be captured, and the corresponding quantities updated. The cycle offset in number of cycles can be used to avoid all sensors being calculated in the same simulation cycle.
(i5) Latency factor—that is multiplied with the cycle time in the relative boundary's interval (min, max).

RCS maps define different radar cross section maps for traffic objects. These maps highly influence the signal strength and therefore also the object detection. The maps specify the characteristic of the signal reflected by the traffic object depending on the angle of incidence. This characteristic depends on the object's geometry and material. It must be measured or calculated using dedicated software. Under view RCS data, the RCS maps that are currently available are listed.

Traffic direct link to parameters > traffic.

Environment parameters direct link to parameters > environment. The radar sensor is influenced by the environment temperature, the rain rate, and the visual range in fog.

Additional parameters settings to influence object detection and defines the antenna gain map:

(i6) Antenna gain map defines the sensors radiant emittance in dB.
(i7) Distance accuracy specifies the standard deviation ($\pm$ given distance) for the Gaussian distribution curve that is used for the distance calculation.
(i8) Distance specifies the cell resolution in longitudinal direction.
(i9) Azimuth accuracy calculates the bearing angle of a detected object, representing its position relative to the sensor frame.
(i10) Azimuth specifies the cell resolution in horizontal direction.

Table 6.34 Radar sensor (Some of the values that define the virtual model may differ from real values, they have been determined via experimental studies or mathematical calculus)

Radar sensor 1		Parameter name			
Basic parameters					
Field of view (h/v) (°)		100.00		80.00	
Range (min/max) (m)		0.200		200.0	
Cycle time (ms)/offset (–)		60.00		0.000	
Latency (–)		0.000			
RCS maps		Traffic			
RCS_Car		RCS_Pedestrian		RCS_Truck	
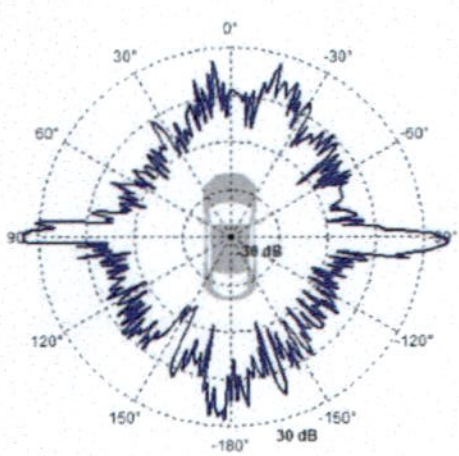		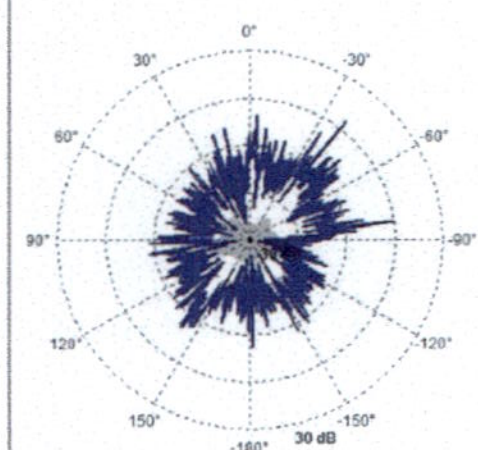		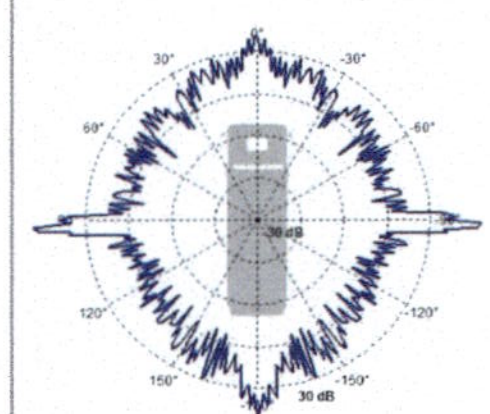	
Data max	38 dB	Data max	18 dB	Data max	48 dB
Data min	– 6 dB	Data min	– 27 dB	Data min	6 dB
Scale plot minim (dB)			– 30.00		
Scale plot maxim (dB)			30.00		
Traffic—CarMaker traffic					
Environment parameters—CarMaker environment					

(continued)

Table 6.34 (continued)

Additional parameters		
Antenna gain map	Radar_Default	
Antenna gain data set		
Antenna gain parameter		
Scanning range h/v (°)	0.000	0.000
Beam width h/v (°)	20.00	15.00
Antenna efficiency (–)	1.000	

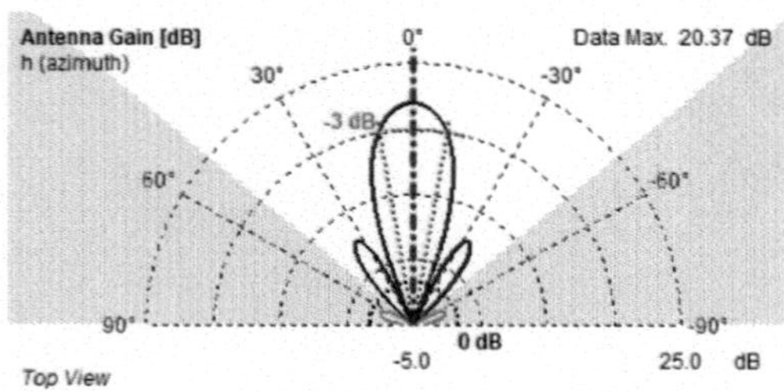

(continued)

Table 6.34 (continued)

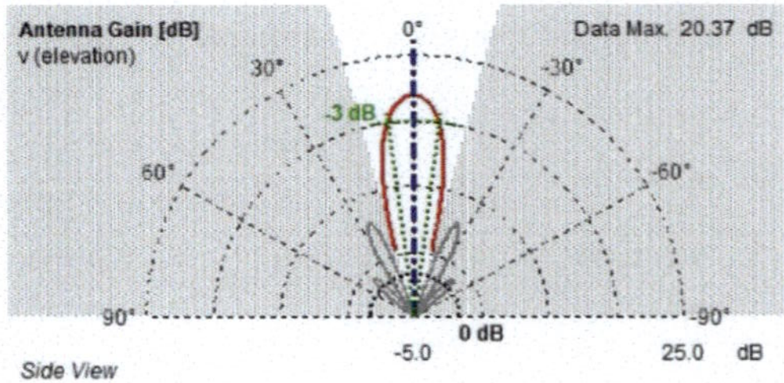

	Accuracy	Resolution
Distance (m)	0.400	1.800
Azimuth (°)	0.100	1.600
Speed (km/h)	0.100	0.400
Separability (–)		1.500
Object max (–)	40.00	
Transmit frequency (GHz)	77.00	
Transmit power (dBm)	14.00	
System losses (dB)	0.000	
Noise bandwidth (Hz)	25,000	
Noise figure (dB)	4.800	
Probability of detection min (–)	0.500	
Probability of false alarm (–)	10e−6	

(continued)

Table 6.34 (continued)

Radar sensor 2		Parameter name			
Basic parameters					
Field of view (h/v) (°)		100.00		100.00	
Range (min/max) (m)		0.200		200.0	
Cycle time (ms)/offset (–)		60.00		1.000	
Latency (–)		0.000			
RCS maps		Traffic			
RCS_Car		RCS_Pedestrian		RCS_Truck	
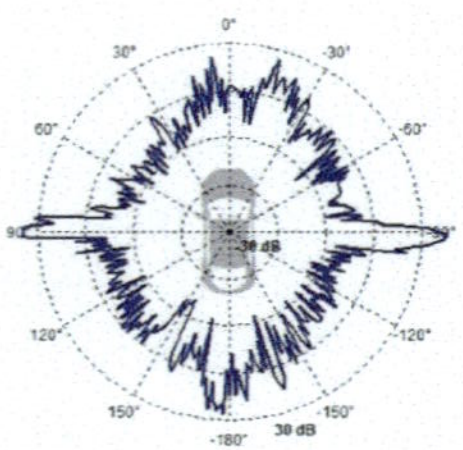		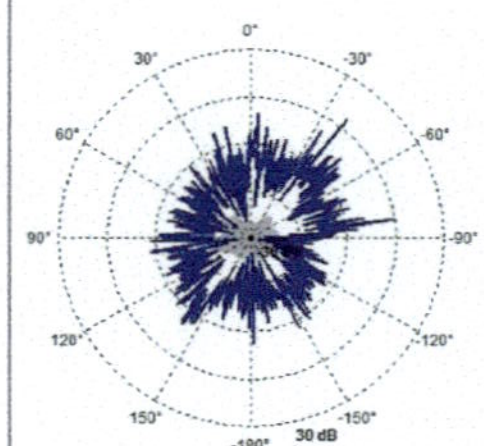		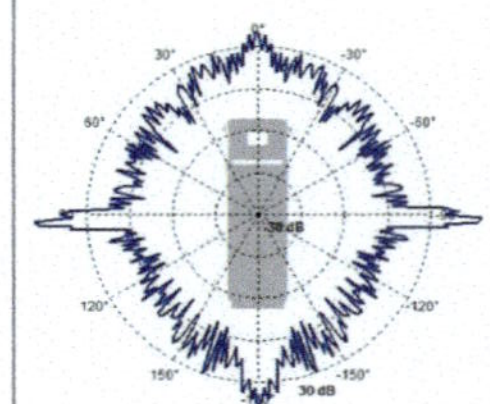	
Data max	38 dB	Data max	18 dB	Data max	48 dB
Data min	– 6 dB	Data min	– 27 dB	Data min	6 dB
Scale plot minim (dB)			– 30.00		
Scale plot maxim (dB)			30.00		
Traffic—CarMaker traffic					
Environment parameters—CarMaker environment					

(continued)

Table 6.34 (continued)

Additional parameters		
Antenna gain map	Radar_Default	
Antenna gain data set		
Antenna gain parameter		
Scanning range h/v (°)	0.000	0.000
Beam width h/v (°)	20.00	15.00
Antenna efficiency (–)	1.000	

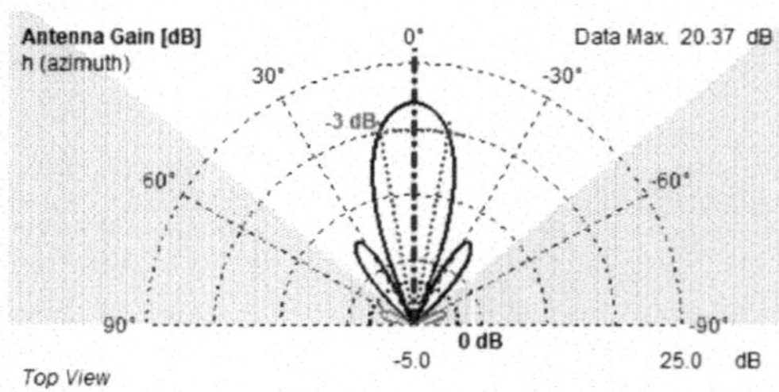

(continued)

Table 6.34 (continued)

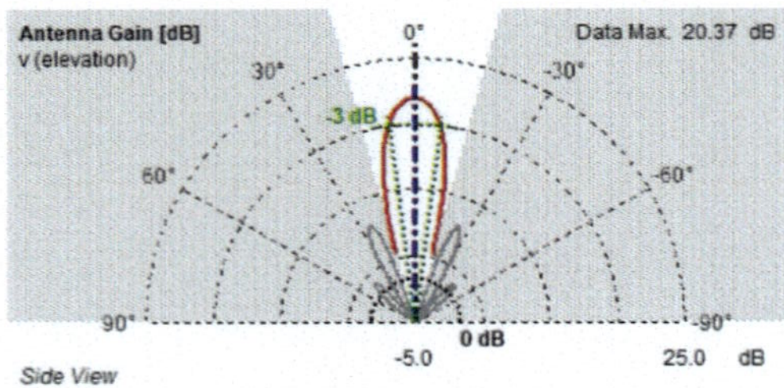

Distance (m)	Accuracy	Resolution
	0.400	1.800
Azimuth (°)	0.100	1.600
Speed (km/h)	0.100	0.400
Separability (–)		1.500
Object max (–)	40.00	
Transmit frequency (GHz)	77.00	
Transmit power (dBm)	14.00	
System losses (dB)	0.000	
Noise bandwidth (Hz)	25,000	
Noise figure (dB)	4.800	
Probability of detection min (–)	0.500	
Probability of false alarm (–)	10e−6	

(i11) Speed accuracy defines the standard deviation (± given velocity) for the Gaussian distribution curve that is used to calculate the velocity.
(i12) Speed resolution defines the minimum speed difference two objects must have to be identified as individual objects.
(i13) Separability adds additional cell units between two cell units that detect two different objects.
(i14) Objects max defines the maximum number of detectable objects for which output quantities are provided.
(i15) Transmit frequency specified the sensor operating frequency.
(i16) Transmit power configures the transmitting power.
(i17) System losses combines all power and system losses in a single factor, including antenna supplies.
(i18) Noise bandwidth is the noise bandwidth of the receiver.
(i19) Noise figure represents the input and output signal to noise ratio.
(i20) Probability of detection min is the threshold for the probability of detection.
(i21) Probability of false alarm is an input to the error function which is used to calculate the probability of detection.
(i22) False positive enables an algorithm giving detection flags for objects that do not exist in the virtual environment.

Radar RSI Sensor

Radar RSI sensors provide raw environmental information. RSI radar is a real sensor that simulates the propagation of electromagnetic waves in the virtual environment. The radar sensor senses the overlapping objects in traffic.

The following options are available to configure the radar RSI sensor (Table 6.35) [3, 4]:

(i1) Radar RSI sensor defines the identification of the sensor quantities.

Basic parameters:

(i2) Field of view (h/v) defines the horizontal and vertical direction of the sensor beam.
(i3) Range min/max defines the observation range of the sensor beam.
(i4) Rays (h/v) represents the discretization of the vertical number of rays within the field of view.
(i5) Channels max defines the maximum number of channels.
(i6) Visualization color is the color of the sensor beam that is used for visualization in IPGMovie.
(i7) Body mounting defines the mounting point of the sensor.

Front end parameters:

(i8) Transmit frequency is used to calculate the coefficient of atmospheric damping and the absolute doppler shift.

Table 6.35 Radar RSI sensor (Some of the values that define the virtual model may differ from real values, they have been determined via experimental studies or mathematical calculus)

Radar RSI sensor	Parameter name		
Basic			
Field of view (h/v) (°)	120.0		20.00
Range (min/max) (m)	0.100		150.0
Rays (h/v) (–)	671.0		130.0
Front end			
Antenna/VRx	RadarRSI_Default		
Antenna gain			
Antenna gain parameter	Beam width		Efficiency
	h (°)	v (°)	
Transmit	10.00	30.00	1.000
Receive	10.00	30.00	1.000
Display data			
Scale			
max (dB)	30.00		
min (dB)	– 30.00		

(continued)

Table 6.35 (continued)

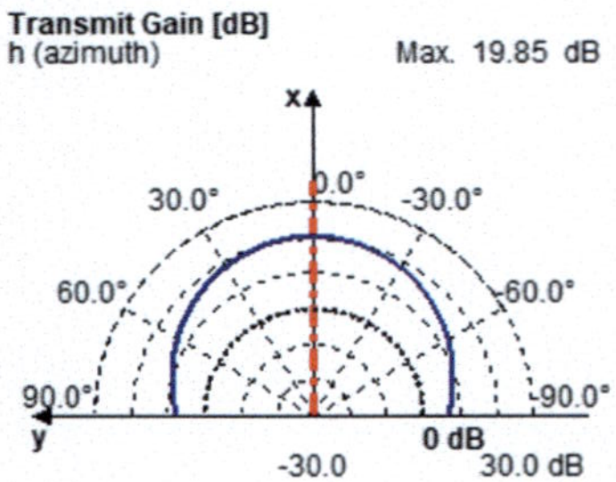	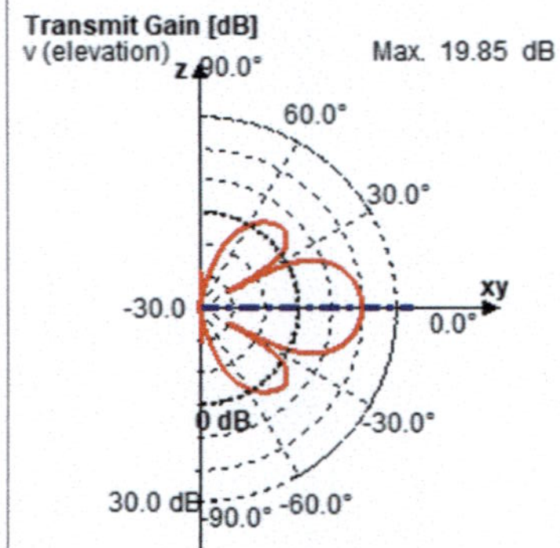
Scale	
max (dB)	30.00
min (dB)	– 30.00

(continued)

Table 6.35 (continued)

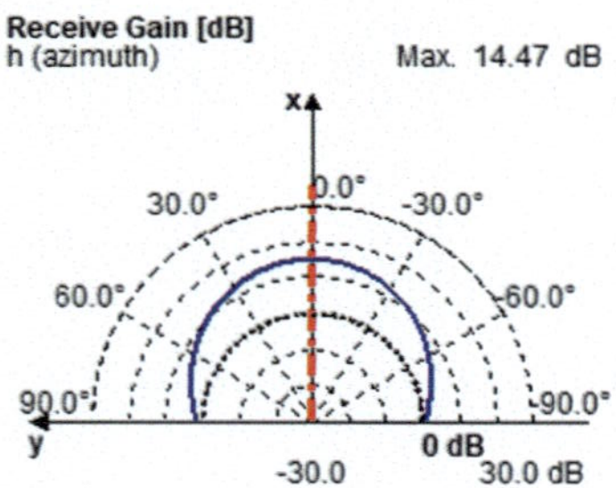

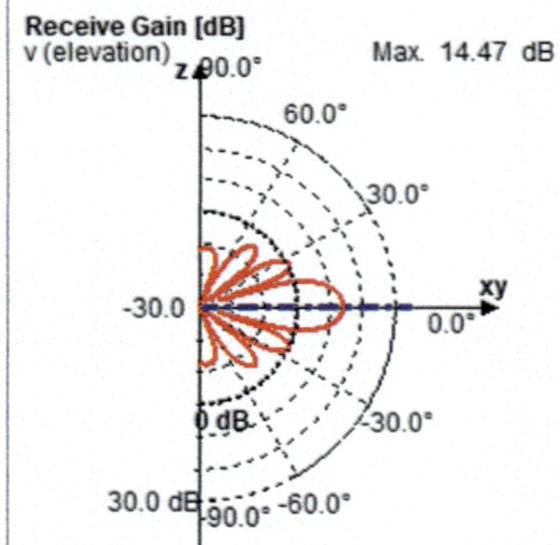

n	y-Pos (m)	z-Pos (m)	Amp (–)	φ (rad)
0	– 0.0156	0	1	0
1	– 0.0136	0	1	0
2	– 0.0117	0	1	0
3	– 0.0097	0	1	0
4	– 0.0078	0	1	0
5	– 0.0058	0	1	0
6	– 0.0039	0	1	0
7	– 0.0019	0	1	0
8	0.0000	0	1	0

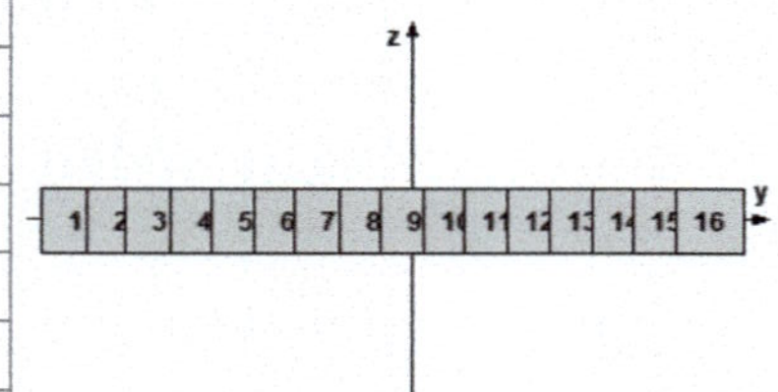

(continued)

Table 6.35 (continued)

9	0.0019	0	1	0
10	0.0039	0	1	0
11	0.0058	0	1	0
12	0.0078	0	1	0
13	0.0097	0	1	0

Transmit freq. (GHz)	77.00	Noise bandwidth (MHz)	200.0
Transmit power (dBm)	30.00	Noise figure (dB)	5.000
System losses (dB)	0.000	Noise temperature (K)	300.0

Noise scaling

Range (m)	Factor (dB)	
0.000	10.00	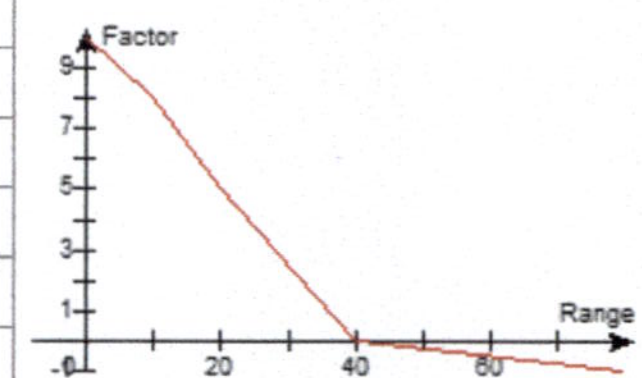
10.00	8.000	
20.00	5.000	
40.00	0.000	
80.00	− 1.000	

Velocity (m/s)	Factor (dB)	
− 20.00	0.000	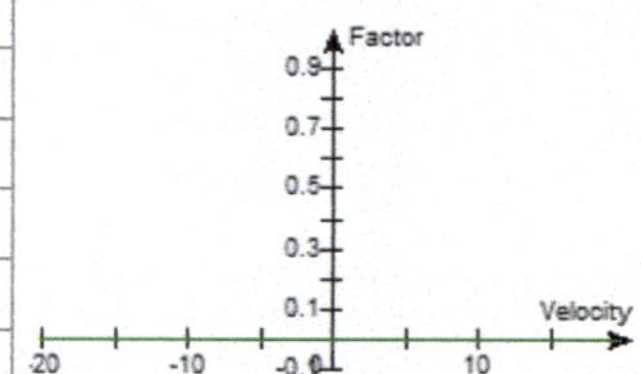
0.000	0.000	
20.00	0.000	
0.000	0.000	
0.000	0.000	

(continued)

Table 6.35 (continued)

Signal processing				
Discretization	Min	Max	Samples	Paddings
Range (m)	0.000	200.0	512.0	0.000
Doppler velocity	– 40.00	40.00	256.0	0.000
Angle (°)	– 90.00	90.00	60.00	100.0
Filtering	Layers	Guards	SnR (dB)	Thres (%)
Range doppler	5.000	2.000	11.00	95.00
Angle	8.000	2.000	12.00	0.000
Peak finder	Layers			
Range doppler	2.000			
Angle	2.000			
Output type	Cartesian			

(i9) Transmit power is used to set the transmit power of the sensor.
(i10) System losses defines the additional losses in the system.
(i11) Noise bandwidth specifies the bandwidth of the noise signals.
(i12) Noise figure is used in the calculation of noise power.
(i13) Noise temperature is used in the calculation of noise power.
(i14) Noise scaling helps to scale the same noise parameterization with respect to range and velocity for different sensors in order to model the sensor-specific characteristics.
(i15) Antenna/VRx data describes the gain map information.
(i16) Antenna gain parameter provides the beam widths in both the horizontal and vertical directions for the transmitter and the receiver gain maps, along with their respective efficiencies.
(i18) Display data the gain maps are visualized as a function of azimuth and elevation angles in the top and the side views.
(i19) VRx parameterization permits parameterized weights and positions in cartesian coordinates for each individual virtual receivers via VRx tab under the Antenna/VRx Data Set window.

Discretization parameters:

(i20) Range defines maximum and minimum limits for the range.
(i21) Doppler velocity defines the value for doppler velocity.
(i22) Angle defines the value for angular processing.

Filtering parameters:

(i24) Range doppler defines the number of layers for the filtering, the number of layers to separate the cell under test and the surrounding cells.
(i25) Angle defines the parameterization for the angular filtering is same as that of the range doppler filtering.

Output parameters:

(o1) Number of detections.
(o2) Time stamp of the raytracing job.
(o3) Detections as point cloud: point coordinates, power, velocity.
(o4) Detections as complex amplitudes of all virtual receivers.

Ultrasonic RSI Sensor

Ultrasonic RSI sensor is a real sensor that simulates the propagation of mechanical sound pressure waves through the virtual environment. This sensor considers the effects of overlapping objects and the effects of physical propagation, such as direct, indirect, or cross-echo, as well as false positive/negative object detection. The following options are available to configure the ultrasonic RSI sensor (Table 6.36) [3, 4]:

(i1) Ultrasonic RSI defines the identification of the sensor quantities.

Table 6.36 Ultrasonic RSI sensor (Some of the values that define the virtual model may differ from real values, they have been determined via experimental studies or mathematical calculus)

Ultrasonic RSI sensor	Parameter name	
Field of view (h/v) (°)	120.0	60.00
Range (min/max) (m)	0.070	3.000
Rays (h/v) (–)	130.0	80.00
Echoes maxim (–)	200.0	
Frequency (kHz)	50.00	
Transmission time (ms)	0.300	
Separability (ms)	0.050	
Sound lobe	USonicRSI_Default	
Sound lobe parameter		
Dimension (h/v) (°)	180.0	180.0
Sampling period (h/v) (°)	5.000	5.000
Roundness (y-z) (–)	1.000	
Ratio (x-y) (–)	0.500	
Reference range (m)	0.300	
Kind	SPA	
Sound pressure amplitude		

Angle (°)	Amplitude (Pa)	
0.000	1.410	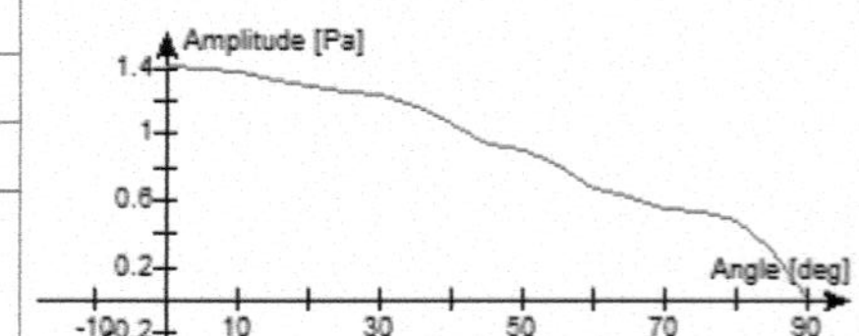
5.000	1.400	
10.00	1.370	

(continued)

Table 6.36 (continued)

20.00	1.280	
30.00	1.230	
40.00	1.050	
50.00	0.900	
55.00	0.810	
60.00	0.670	
65.00	0.620	
70.00	0.540	
75.00	0.550	
80.00	0.500	
85.00	0.300	
90.00	0.010	
Display data		

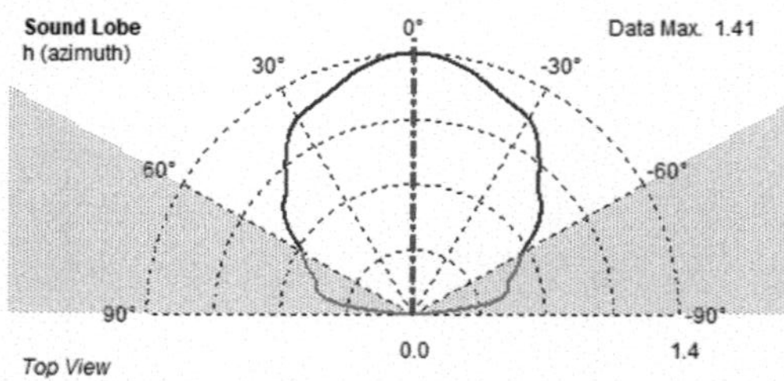

(continued)

Table 6.36 (continued)

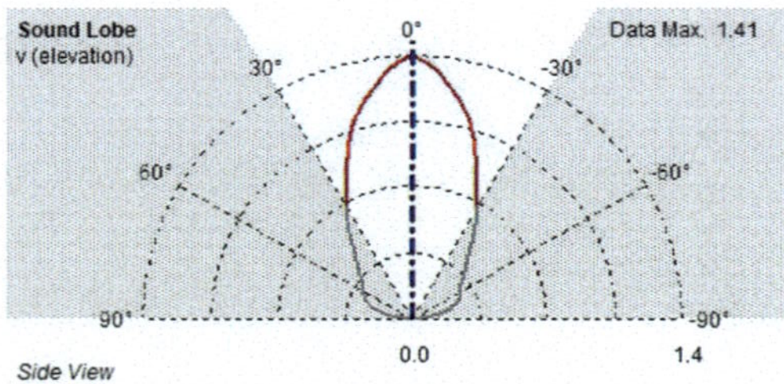

Scale (min/max)		0.000	1.410
Detection	SPA		
Detection threshold			

Time (ms)	Amplitude (Pa)
0.000	2.0e−2
1.000	2.0e−2
2.000	2.0e−2
4.000	1.5e−2
6.000	1.2e−2
8.000	1.0e−2
10.00	0.8e−2
15.00	0.4e−2
20.00	0.2e−2
25.00	0.1e−2

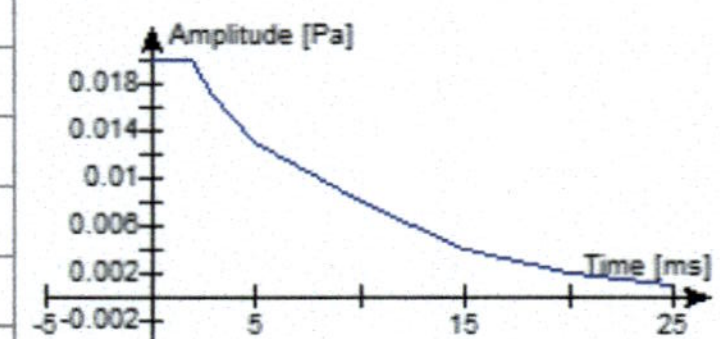

(i2) Field of view (h/v) defines the horizontal/vertical span of the sensor.
(i3) Range (min/max) defines the minimum and maximum range of the sensor beam.
(i4) Rays (h/v) represents the discretization of the vertical number of rays within the field of view. The horizontal number of rays will be derived by a combination of the horizontal field of view.
(i5) Echoes (max) defines the maximum value of possible echoes.
(i6) Frequency calculates the coefficient of atmospheric absorption, speed of sound and the relative phase coherent addition.
(i7) Transmission time is considered for coherent addition to determine the wave package overlap of echoes from an identical transmitter.
(i8) Separability represents the specification of the required difference in time of flight whether two echoes from an identical transmitter will be detected as separate or not.
(i9) Sound lobe is used to define the sensor lobe characteristics.

Sound lobe data set:

(i10) Dimension (h/v) defines the sensor lobes horizontal and vertical field of view. The elevation angle (v) is always half the azimuth angle (h).
(i11) Sampling period (h/v) defines the sensor's horizontal and vertical sampling period. Like the dimensions, the sampling period for vertical direction is always half the horizontal one.
(i12) Roundness (y-z) represents the grade of roundness, influences the shape of the sound lobe generated from the corresponding lookup table in (y) and (z) direction.
(i13) Ratio (x-y) represents the ratio of the vertical and horizontal dimensions of the sound lobe generated from the corresponding lookup table, in a similar way like the roundness, but in the (x) and (y) direction.
(i14) Reference range represents the sound pressure amplitude.
(i15) Display data generates the sound lobe in two sectional views: a horizontal section (x, y), azimuth, and a vertical section (x, z), elevation.
(i16) Detection specifies if SPA (Sound Pressure Amplitude) detection based on the sound pressure amplitude, or 3DFoV detection based on a 3D field of view, is used.

Lidar RSI Sensor

Lidar RSI sensor is a real sensor that simulates the propagation of a laser pulse to detect targets in the virtual environment using the tracking of the path of these rays. The sensor provides information for each received echo, namely the duration of the signal transmission-reception to the detected objects, the distance to these objects, the direction of the laser beam (azimuth, elevation), beam origin, beam intensity, pulse wavelength and number of reflections. The following options are available to configure the Lidar RSI sensor (Table 6.37) [3, 4]:

Table 6.37 Lidar RSI sensor (Some of the values that define the virtual model may differ from real values, they have been determined via experimental studies or mathematical calculus)

Lidar RSI sensor	Parameter name	
Range (min/max) (m)	0.300	300.0
Wavelength (nm)	900.0	
Separability distance (m)	0.100	
Transmit power (W)	60.00	
Reception area (cm^2)	0.400	
Detection threshold (nW)	1.000	
Angular aperture (°)	1.000	
Beams	LidarRSI_Default	
Lidar RSI beams data set		
Beams		
Type	Regular	
Field of view h (min/max) (°)	– 70.00	70.00
Field of view v (min/max) (°)	– 1.500	1.500
Beams (h/v) (–)	600.0	5.000
Single beams		
Beams width (h/v) (°)	0.100	0.800
Field of view coverage (h/v) (%)	40.00	100.0
Rays (h/v) (–)	1.000	10.00
Rays/beam		

(continued)

Table 6.37 (continued)

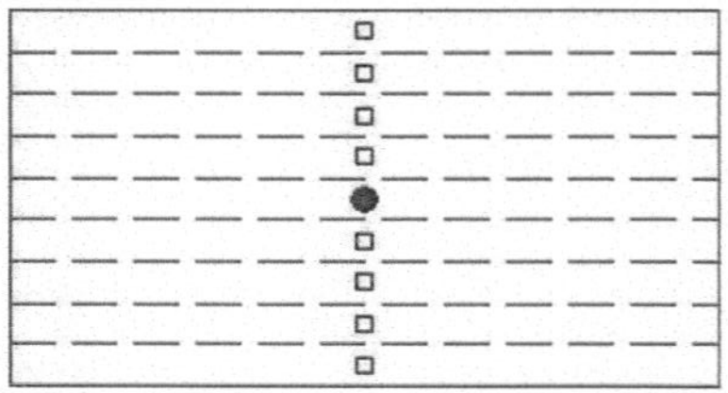

Beams data

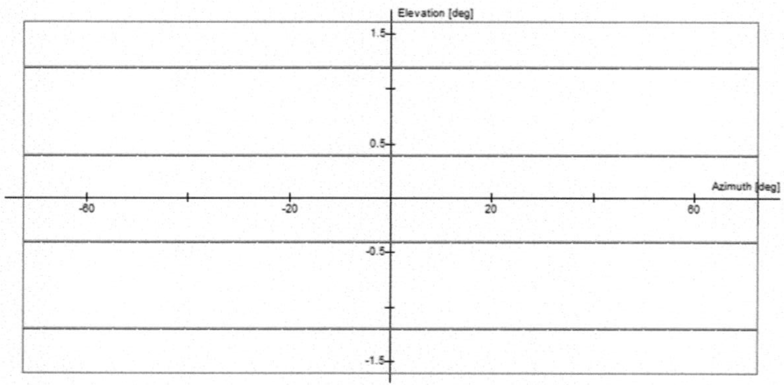

Intensity gain map

(continued)

Table 6.37 (continued)

Azimuth (°)	Intensity gain (–)
– 80.00	0.000
– 70.00	0.150
– 60.00	0.360
– 50.00	0.560
– 40.00	0.720
– 30.00	0.840
– 20.00	0.920
– 10.00	0.980
0.000	1.000
10.00	0.980
20.00	0.920
30.00	0.840
40.00	0.720
50.00	0.560
60.00	0.360
70.00	0.150
80.00	0.000

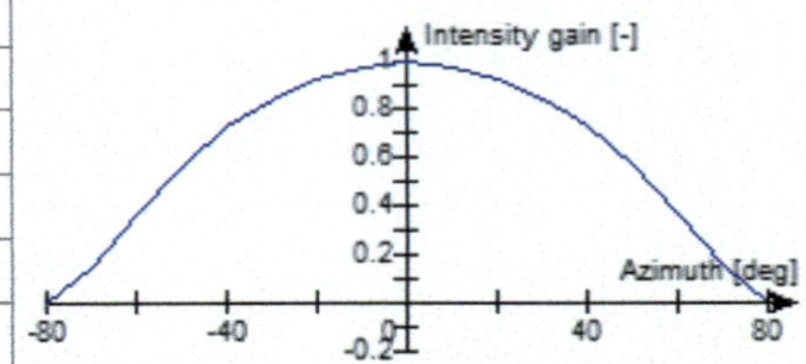

(i1) Lidar RSI sensor defines the identification of the sensor quantities.

Basic parameters:

(i2) Range (min/max) defines the longitudinal range of the sensor beam.
(i3) Wavelength sets the wavelength of the laser light pulse propagation.
(i4) Separability distance is the difference in length of flight whether two beams of the discretized laser beam echo from an identical transmitter will be detected as separate or not.
(i5) Transmit power is the maximum or the peak power of the transmitted laser beam.
(i6) Reception area represents the cross-section area of the receiver which is used for the calculation of the intensity of the received beam.
(i7) Detection threshold sets the intensity threshold which is only considered if the separability is greater than zero.
(i8) Intensity Gain Map specifies the intensity gain factor based on the azimuth angle.

Beams the Lidar RSI beams data set is used to define the characteristic of the laser beam grid:

(i9) Field of view h (min/max) defines the minimum and maximum range of the laser beam distribution in horizontal plane.
(i10) Field of view v (min/max) defines the minimum and maximum range of the laser beam distribution in vertical plane.
(i11) Beams (h/v) represents the discretization of the vertical number of beams within the field of view. The horizontal number of beams will be derived by a combination of the horizontal field of view and the condition of equidistant angles.
(i12) Beam width (h/v) defines the area covered by the laser beam during ray tracing by providing the width of the laser beam irradiation in azimuth and elevation plane.
(i13) Rays (h/v) defines discretization of each laser beam specified in the beam file.

6.2.4.9 Vehicle Control Model

Vehicle control assists the virtual driving system through predefined controller models as shown in Table 6.38 [3, 4].

The ACC controller controls the longitudinal acceleration of the vehicle by changing the position of the brake and gas pedal. If ACC is deactivated there is no manipulation of the pedal position by the controller. The controller distinguishes two cases (Fig. 6.35) [15]:

(1) No target detected—the velocity will be controlled.
(2) Relevant target detected—the distance will be controlled.

Table 6.38 Predefined vehicle control models (Some of the values that define the virtual model may differ from real values, they have been determined via experimental studies or mathematical calculus)

Model	Description
Acceleration control + ACC	Includes an advanced driver assistance system with acceleration control
Generic longitudinal control	Generic longitudinal controller with simplified AEB and FCW functions
Generic lateral control	Generic lateral controller with simplified LKAS (Lane Keeping Assist System) and LDW (Lane Departure Warning) functions

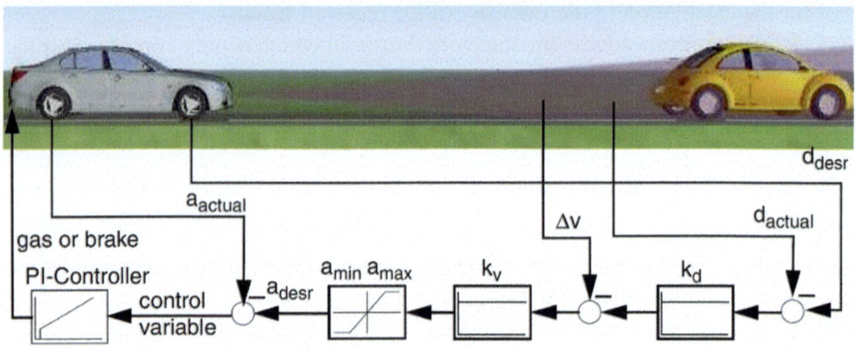

Fig. 6.35 Closed loop of the ACC controller

To modify the gas or brake pedal position depending on a desired longitudinal acceleration, a vehicle control model for acceleration control was implemented in CarMaker. This controller uses a PI controller. The desired longitudinal acceleration can be calculated using the CarMaker built-in ACC controller, via DVA (Direct Variable Access).

The predefined model for acceleration control and cruise control (ACC) ensures the assistance of the driving system by controlling the acceleration of the vehicle so that it maintains a minimum safety distance from the vehicle in front (Table 6.39).

The generic longitudinal control contains two functionalities AEB and FCW.

The AEB has the task to safely decelerate the vehicle to the velocity of the target object ahead. The system compares the time to collision with a time-threshold-brake to decide if a braking intervention is required.

The FCW system has the task to warn the driver by different degrees of warning level if time-to-collision falls below the defined time threshold. The simple generic model supports two different warning levels, which are activated before the AEB reaction. Model parametrization is described in Table 6.40 [3, 4, 15].

The generic lateral control model contains two functionalities: LKAS and LDW, which are explained below.

Table 6.39 Acceleration control + ACC (Some of the values that define the virtual model may differ from real values, they have been determined via experimental studies or mathematical calculus)

Model	Acceleration control + ACC
Acceleration function	ACC
Acceleration controller factor P (–)	0.001
Acceleration controller factor I (–)	1.000
Referenced object sensor	Object
Brake threshold (–)	0.200
Initial time distance (s)	1.800
Minimal distance (m)	20.00
Minimal acceleration (m/s^2)	– 4.000
Maximal acceleration (m/s^2)	2.500
Distance controller factor kd (–)	36.00
Distance controller factor kv (–)	2.000
Velocity controller factor kv (–)	13.00

Table 6.40 Generic longitudinal control (Some of the values that define the virtual model may differ from real values, they have been determined via experimental studies or mathematical calculus)

Model	Generic longitudinal control
Autonomous emergency braking	
Reference object sensor	Object
Maximal deceleration (m/s^2)	6.000
Acceleration controller factor P (–)	0.001
Acceleration controller factor I (–)	3.000
Minimal distance (m)	5.000
Time braking after standstill (s)	5.000
Time brake reacts (s)	0.200
Forward collision warning	
Time first warning level (s)	2.000
Time second warning level (s)	1.000

The LKAS system has the task to support the driver to keep the vehicle in the middle of the lane for security and comfort aspects. For this, the system needs information like deviation angle and deviation distance to the ideal path (middle). For the detection of the lane a camera-unit is used.

The LDW system has the task to warn the driver via a signal (visual, acoustic, or haptic) if the vehicle leaves the lane unintentionally.

The simple generic model uses the DLC-approach checking the distance to the line crossing. In this approach the system generates a warning if the lateral distance of

Table 6.41 Generic lateral control (Some of the values that define the virtual model may differ from real values, they have been determined via experimental studies or mathematical calculus)

Model	Generic lateral control
Initial line detection mode	Line sensor
Reference line sensor	LN00
Reference road sensor	RD00
Lane keeping assist system	
Minimal velocity (km/h)	55.00
Maximal assist torque (N m)	2.000
Time constant PT1 filter (s)	0.003
Maximal lane width (m)	7.000
Minimal lane width (m)	1.800
Curvature controller factor P (–)	2.000
Curvature controller factor I (–)	0.200
Maximal deviation distance (m)	10.00
Assist torque coefficient (N s^2)	2.000
Lane departure warning	
Minimal velocity (km/h)	55.00
Distance departure warning (m)	0.200

the tire to the detected line falls below a predefined threshold. Model parametrization is described in Table 6.41 [3, 4, 15].

6.2.5 Results of Simulation Process

6.2.5.1 Model Parameter Check

Model parameter check is a tool of CarMaker. It represents the parameterization done in CarMaker in a clear and well-arranged form. Diagrams of all vehicle modules can be generated automatically to check the data input visually and to give an overview. They can easily be compared, show false inputs, and can be used for presentations (Table 6.42) [4].

6.2.5.2 Results of Parameters Check

All subsystems of the vehicle modeling can be found in the dialog box. The submodels in detail are [4]:

(1) Aerodynamics specify the range of the minimum and maximum flow angle (Fig. 6.36).

Table 6.42 Model parameter check (Some of the values that define the virtual model may differ from real values, they have been determined via experimental studies or mathematical calculus)

Aerodynamics				
Flow angle (min/max) (°)	– 180.0		180.0	
Powertrain				
Engine speed (min/max) (rpm)	700.0		8000	
Gas pedal (min/max) (0…1)	0.000		1.000	
Drive unit mount				
Amplitude (min/max) (mm)	0.100		3.000	
Frequency (min/max) (Hz)	2.500		30.00	
Force element	Force 1			
Components (–)	x	y	z	
Brake				
Brake pedal (0…1)			1.000	
Brake increase (dt) (s)			0.200	
Brake decrease (dt) (s)			0.200	
Tire				
Vehicle speed (min/max) (m/s)	27.00		–	
Wheel speed (min/max) (m/s)	23.00		31.00	
Load (min/max) (N)	2000		6000	
Friction (min/max) (–)	0.100		1.000	
Slip angle (min/max) (°)	– 30.00		30.00	
Inclination angle (min/max) (°)	– 3.000		3.000	
Turn slip (min/max) (1/m)	– 1.000		1.000	
Selected tire	Front left			
IPGDriver				
Course range (min/max) (m)	0.000		–	
Suspension force elements	Front min/max		Rear min/max	
Compression parallel (m)	– 0.085	0.075	– 0.085	0.075
Compression antiparallel (m)	– 0.075	0.075	– 0.080	0.080
Compression velocity (m/s)	– 1.200	1.200	– 1.200	1.200
Display wheel compression on the x-axis				
Include external suspension forces (with limitations)				
Consider suspension force time variance (with limitations)				
Suspension kinematics and compliance				
Procedure	Standard		SPMM	
Compression parallel (m)	– 0.085	0.075	– 0.085	0.075
Compression antiparallel (m)	– 0.075	0.075	– 0.080	0.080

(continued)

Table 6.42 (continued)

Aerodynamics		
Acceleration force (min/max) (N)	0.000	6000
Deceleration force (min/max) (N)	0.000	– 6000
Side force (min/max) (N)	0.000	– 6000
Display wheel compression on the x-axis		
Include external suspension forces (with limitations)		
Consider suspension force time variance (with limitations)		
Vehicle characteristics		
Design configuration (before preprocessing)		
Equilibrium configuration (after processing)		
Compare with reference data		
Show reference data		

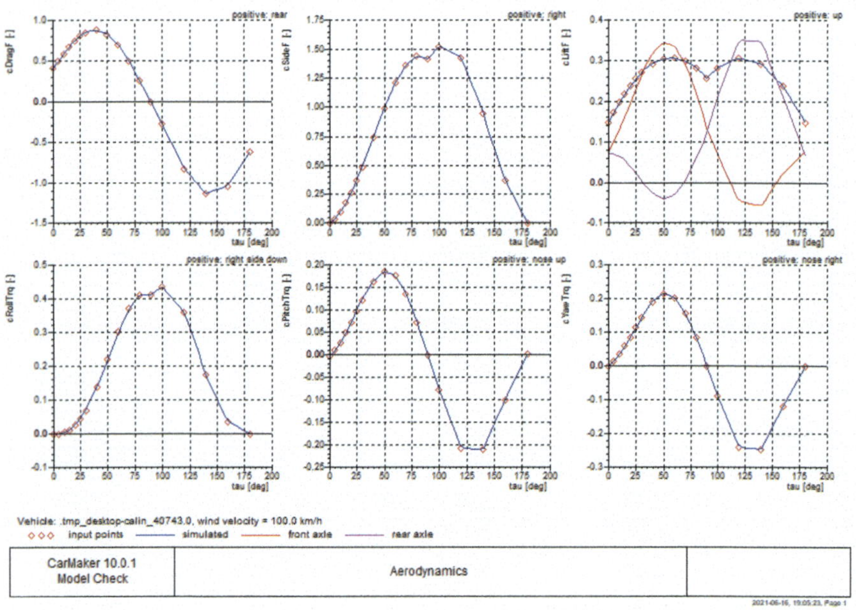

Fig. 6.36 Aerodynamics

(2) Powertrain the electric motor speed and the throttle position are varied (Fig. 6.37).

(3) Drive unit mount evaluates the influence of amplitude and frequency on the mounting characteristics in the case of an elastic mounted motor with hydromount.

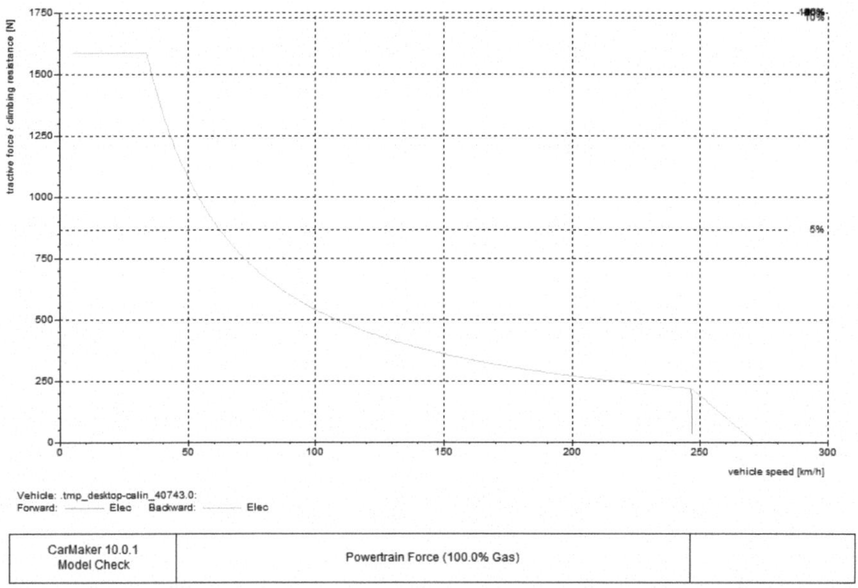

Fig. 6.37 Powertrain

(4) Brake reference quantities are the position of the brake pedal and the time required to reach this position, as well as the release time.
(5) Tire quantities are represented by vehicle speed, wheel speed, load, friction value, sliding angle for the selected tire (left front, right front, left rear, right rear) (Fig. 6.38).
(6) IPGDriver parameters valid for the section of the road (Fig. 6.39).
(7) Suspension force elements for parallel and reciprocal compression.
(8) Suspension kinematics and compliance at parallel and reciprocal compression, also acceleration, deceleration and side forces are applied.
(9) Vehicle characteristics calculates all parameters in the design and in the equilibrium configuration.
(10) Compare with reference data compares the current data with a reference TestRun.

Design configuration of the vehicle comprehends the whole parameter input done by the user in the vehicle parameterization. This includes the geometrical data as well as all other parameters such as masses, spring and damper stiffnesses etc. In most cases the design configuration is not the same as the static equilibrium configuration of the vehicle, it is a portrayal of the drawing sheet of the designer.

In the model check output, the part of the design configuration gives the possibility to check all input data collected. The sections for basic parameters present the most important vehicle parameters in equilibrium configuration (Table 6.43) [4].

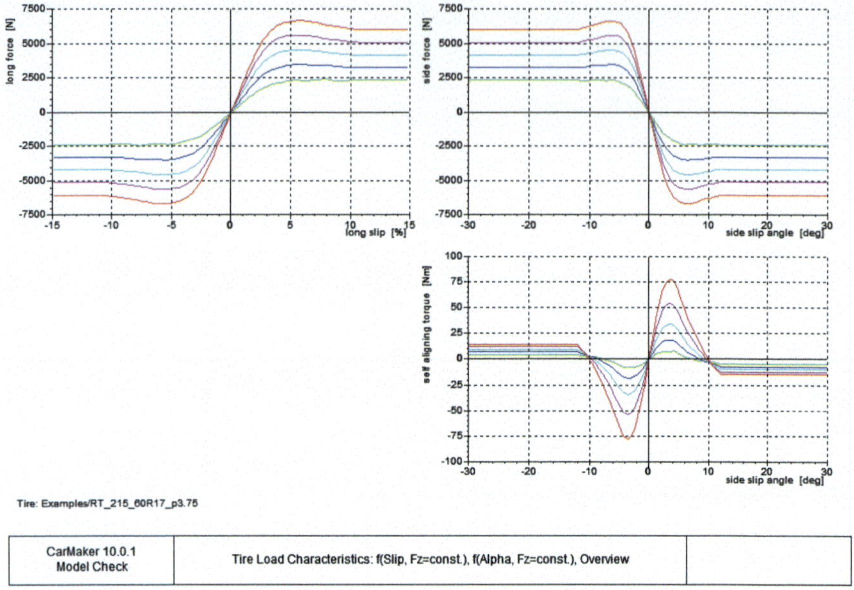

Fig. 6.38 Tire load characteristics

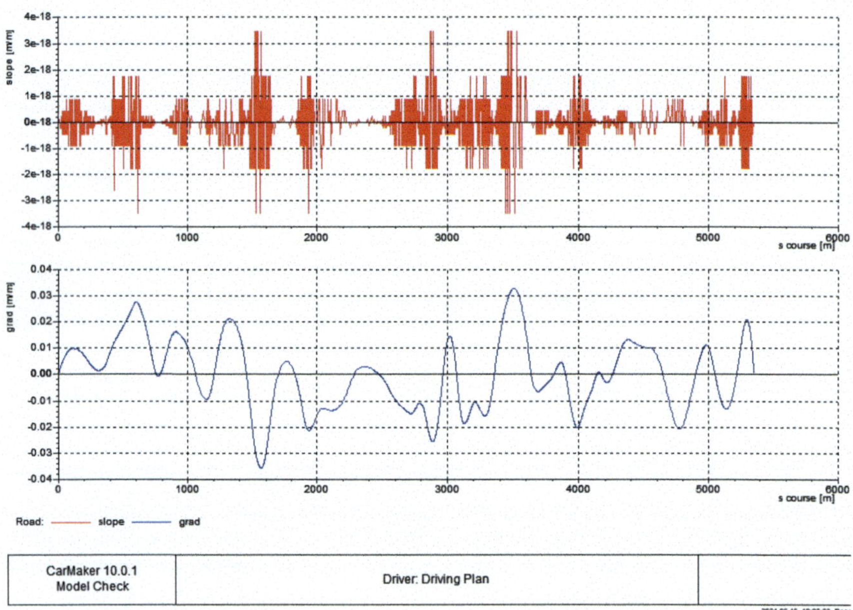

Fig. 6.39 IPGDriver

Table 6.43 Vehicle characteristics in equilibrium configuration (Some of the values that define the virtual model may differ from real values, they have been determined via experimental studies or mathematical calculus)

Basic parameters	Unit	Total	Left	Right
Total mass	kg	1770	–	–
Total weight	N	17,362	–	–
Total roll inertia	kg m^2	956.4	–	–
Total pitch inertia	kg m^2	3283	–	–
Total yaw inertia	kg m^2	3471	–	–
Total CoG height (FrD)	m	0.690	–	–
Sprung CoG height (Fr0)	m	0.653	–	–
Sprung mass	kg	1503	–	–
Sprung weight	N	14,742	–	–
Sprung roll inertia	kg m^2	726.5	–	–
Sprung pitch inertia	kg m^2	2677	–	–
Sprung yaw inertia	kg m^2	2708	–	–
Sprung CoG height (FrD)	m	0.750	–	–
Sprung CoG height (Fr0)	m	0.713	–	–
Unsprung mass (total)	kg	267.2	–	–
Wheelbase	m	2.854	–	–
Front axle				
Track width	m	1.713	–	–
Toe angle	°	– 0.187	– 0.187	– 0.188
Camber angle	°	– 2.722	– 2.718	– 2.726
Caster angle	°	10.05	10.06	10.05
Caster offset	mm	35.32	35.32	35.32
Kingpin angle	°	14.69	14.70	14.69
Kingpin offset	mm	– 3.542	– 3.542	– 3.542
Normal force	N	10,832	5406	5427
Loaded tire radius	m	0.339	0.339	0.339
Tire rate	N/mm	937.7	937.7	937.7
Rear axle				
Track width	m	1.692	–	–
Toe angle	°	0.000	– 0.000	0.000
Camber angle	°	0.000	0.007	– 0.007
Normal force	N	6510	3250	3261
Loaded tire radius	m	0.342	0.342	0.342
Tire rate	N/mm	937.6	937.6	937.6
Geometry equilibrium configuration		x	y	z

(continued)

Table 6.43 (continued)

Basic parameters	Unit	Total	Left	Right
Vehicle point of interest (FrD)	m	2.400	0.000	0.624
Vehicle point of interest (Fr1)	m	2.400	0.000	0.624
Vehicle point of interest (Fr0)	m	10.82	424.1	0.585
Origin (Fr1)	m	9.617	426.2	– 0.052
Roll (Fr1)	°	0.008	–	–
Pitch (Fr1)	°	– 0.303	–	–
Yaw (Fr1)	°	– 60.02	–	–
Generalized body (FrD)	m	2.755	0.000	0.690
Generalized body (Fr1)	m	2.755	0.000	0.690
Generalized body (Fr0)	m	10.99	423.8	0.653
Connected body (FrD)	m	2.810	0.000	0.750
Connected body (Fr1)	m	2.810	0.000	0.750
Connected body (Fr0)	m	11.02	423.8	0.713
Carrier wheel center	FL	FR	RL	RR
Carrier wheel center (tx) (FrD)	3.822	3.822	0.968	0.968
Carrier wheel center (ty) (FrD)	0.840	– 0.840	0.846	– 0.846
Carrier wheel center (tz) (FrD)	0.372	0.372	0.388	0.388
Carrier wheel center (tx) (Fr1)	3.822	3.822	0.968	0.968
Carrier wheel center (ty) (Fr1)	0.840	– 0.840	0.846	– 0.846
Carrier wheel center (tz) (Fr1)	0.372	0.372	0.388	0.388
Carrier wheel center (tx) (Fr0)	12.25	10.80	10.83	9.367
Carrier wheel center (ty) (Fr0)	423.3	422.5	425.8	425.0
Carrier wheel center (tz) (Fr0)	0.340	0.340	0.342	0.342
Carrier mounting (tx) (Fr1)	0.002	0.002	0.000	0.000
Carrier mounting (ty) (Fr1)	– 0.005	0.005	0.000	0.000
Carrier mounting (tz) (Fr1)	0.060	0.060	0.025	0.025
Compression coordinate (q0)	0.060	0.060	0.025	0.025
Compression coordinate (q1)	0.000	0.000	0.025	0.025
Kinematical spring deflection	– 0.060	– 0.060	– 0.025	– 0.025
Relaxed spring length	0.110	0.110	0.062	0.062
Kinematical damper deflection	– 0.060	– 0.060	– 0.025	– 0.025
Kinematical buffer deflection	– 0.060	– 0.060	– 0.025	– 0.025
Kinematic stabilizer deflection	0.060	0.060	0.025	0.025
Rotation angle of the wheel (x)	– 2.718	– 2.726	– 0.000	0.000
Rotation angle of the wheel (z)	– 0.187	– 0.188	0.000	0.000

(continued)

Table 6.43 (continued)

Basic parameters	Unit	Total	Left	Right
Rotation angle of the wheel (y)	− 2.118	− 2.216	0.000	0.000
Wheel center (Fx)	224.8	226.4	− 61.03	− 61.23
Wheel center (Fy)	619.7	− 623.5	0.212	0.213
Wheel center (Fz)	5377	5398	3249	3260
Wheel road (Fx)	1.408	1.430	0.077	0.078
Wheel road (Fy)	363.1	− 366.7	0.190	0.191
Wheel road (Fz)	5405	5426	3249	3260

References

1. Iclodean, C.: Interconectarea sistemelor virtuale de comandă și control pentru autovehicule. In: Risoprint (ed.) Software-in-the-Loop, vol. I. Cluj-Napoca, Romania (2018). ISBN: 978-973-53-2252-6
2. https://ipg-automotive.com/. 28 Aug 2020
3. IPG Automotive Group: Users Guide Version 9.1 Solution for Virtual Test Driving (2020)
4. IPG Automotive Group: Users Guide Version 10.2 Solution for Virtual Test Driving (2021)
5. IPG Automotive Group: Reference Manual Version 10.2 Solution for Virtual Test Driving (2021)
6. AVL Cruise M User Manual Version 2020
7. Iclodean, C., Cordoș, N., Todoruț, A.: Analysis of the electric bus autonomy depending on the atmospheric conditions. Energies **12**, 4535 (2019)
8. IPG Automotive: IPGDriver User Manual Version 10 (2020)
9. Winner, H., Hakuli, S., Lotz, F., Singer, C. (Eds.): Handbook of Driver Assistance Systems Basic Information, Components and Systems for Active Safety and Comfort. Springer International Publishing, Switzerland (2016). ISBN: 978-3-319-12351-6
10. Li, W., Stanula, P., Egede, P., Kara, S., Herrmann, C.: Determining the main factors influencing the energy consumption of electric vehicles in the usage phase. Procedia CIRP **48** (2016)
11. IPG Automotive: Quick Start Guide Version 9.1 (2020)
12. IPG Automotive: IPGControl User's Manual Version 3.0.9 Solution for Virtual Test Driving (2021)
13. Varga, B.O., Mariașiu, F., Moldovanu, D., Iclodean, C.: Electric and Plug-In Hybrid Vehicles Advanced Simulation Methodologies. Springer International Publishing, Berlin (2015). ISBN: 978-3-319-18638-2
14. Varga, B.O., Iclodean, C., Mariașiu, F.: Electric and Hybrid Buses for Urban Transport Energy Efficiency Strategies. Springer International Publishing, Berlin (2016). ISBN: 978-3-319-41248-1
15. IPG Automotive Group: Reference Manual CarMaker Version 10.2 Solution for Virtual Test Driving (2021)

Chapter 7
Legal Framework

7.1 European Legislation

Since there are means of transport using external energy sources, relevant rules and regulations have been established, but some of them have sometimes been inadequate and disproportionate to the harmful potential of these means of transport. For example, Locomotive Acts was passed by the British Parliament in 1865 and provided that the maximum speed at which any motor vehicle could travel was no more than 2 mph in the locality and a maximum of 4 mph outside the locality, but only if there was a person walking in front of the vehicle while blowing a whistle and carrying a red flag [1].

If we look at the content of this legislation in the present, it may seem like a joke, but it was nevertheless adopted by the British Parliament. The desire to change and update traffic laws and to improve them based on current technology is being felt more and more strongly today due to the fact that technology is changing every 3–5 years.

In the field of autonomous driving the technology is now mature enough to require dedicated legislation, but in order to generate and harmonize such legislation, research and development is needed to generate concrete results and to cover all the answers that the use of this technology raises. A real problem in the harmonization of international law on autonomous vehicles is due to the fact that the three major players in this market (the United States, Europe, and China) have not yet reached a joint decision. At the moment, autonomous vehicles, although active and operable on many public roads, are not yet protected by specific legislation, and their commissioning is often in test mode.

In retrospect, the British Parliament of 1865 sought to protect horses and carriages from the "Iron Beasts" as steam vehicles were called. However, in less than 100 years, there have been no horses and carriages on public roads in the UK (there is now clear legislation against access for horses and carriages on certain public roads). The same is true in the case of autonomous vehicles: there are currently some international regulations that prevent vehicles from operating without a driver, but it is possible that over

C. Iclodean et al., *Autonomous Vehicles for Public Transportation*, Green Energy and Technology, https://doi.org/10.1007/978-3-031-14678-7_7

the years, human-driven vehicles will no longer be accepted on certain roads due to possible disruptions to the pace and continuity of the traffic generated by autonomous vehicles [1]. Implementing these goals on a large scale requires fundamental changes to the road legislation, and derogations from the existing rules, as there are now clear requirements that stipulate that every vehicle must be accompanied by a driver.

For the time being, the legal aspects of public passenger transportation focus specially on safety, public accessibility, rhythmicity, and the reduction of pollutant emissions. These criteria are sustainable if the modernization of passenger transportation fleets is done with electric and autonomous vehicles.

Another issue based on the lack of drivers in autonomous vehicles is related to the situations in which traffic accidents occur, a situation which in the case of the involvement of at least one autonomous vehicle makes it difficult to establish the party that is at fault. Harmonization of legislation in the case of autonomous driving involves liability and legal coverage of the autonomous driving system, cyber security protection, and insurance issues. These legal issues, which are not currently covered, must be covered by regulations and normative acts that are flexible enough for them to be in force for at least a few decades (Fig. 7.1).

According to Allied Market Research estimations, the global autonomous vehicle market will be worth more than $550 billion by 2026. In terms of accident risk insurance, the main issue would be to determine liability for human safety in the event of an accident in traffic. For fully autonomous vehicles the responsibility in the event of an accident will lie entirely with the manufacturers. Thus, manufacturers will have to test the capabilities of artificial intelligence to cover all the variables used to develop the autonomous driving system and to develop a well-defined protocol

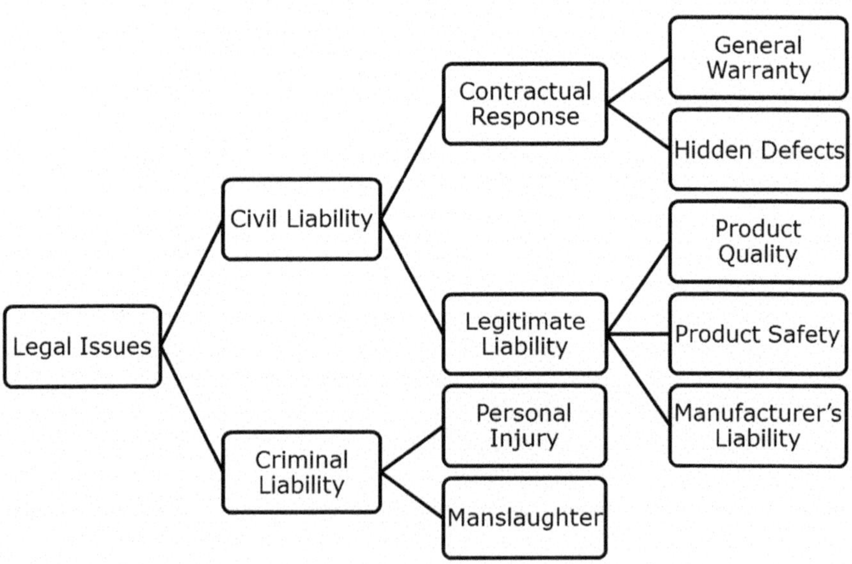

Fig. 7.1 Legal issues related to autonomous vehicles

that can unequivocally determine whether an accident was caused by a technical component problem or a software error.

The legislative framework for regulating the movement of autonomous vehicles on public roads is based on two essential components, namely: traffic rules, respectively the regulation of the technical equipment of these vehicles. According to [1, 2], in order to harmonize the specifications of UN Regulation no. 79 [1, 3] under the conditions of movement of autonomous vehicles, it is necessary that the functions of automation and monitoring of the driver's attention and the emergency maneuvers to ensure road safety in proportion to the level of automation. According to these regulations, the autonomous driving system can allow an autonomous vehicle to follow a defined trajectory or change its trajectory at any time in response to signals initiated and transmitted from outside the system. Consequently, compared to the autonomous driving system, the driver will not in all situations exercise primary control of the vehicle.

The regulatory framework for autonomous vehicles at international level (within the UN) was established on the basis of the 1968 Vienna International Convention on Road Traffic, which stipulates in Article 8 that "every driver must have constant control over his vehicle and be able to guide it" [1, 4]. Through the UNECE amendment ECE/TRANS/WP.1/145 on Article 8 of the Vienna International Convention the possibility that a driver can be assisted in the performance of his driving duties and in the control of the vehicle with driver assistance systems is ensured. Thus, in the Annex to the Amendment, paragraph 5b states that "from March 23, 2016, automated control systems shall be permitted on public roads, provided that they comply with the provisions of the United Nations Motor Vehicle Regulations or that they can be controlled or disabled by the driver" [1, 5].

In the particular case of autonomous driving there are several regulations of the United Nations (UN).

There are no clear regulations in Europe regarding the type-approval, registration, and entry into service on public roads intended for autonomous vehicles. Manufacturers of autonomous vehicles (especially those classified in level 3 and level 4) are based on Regulation (EU) 2018/858 of the European Parliament and of the Council of May 30, 2018 on the type-approval and market surveillance of motor vehicles and their trailers, as well as systems, components and separate technical units intended for the vehicles concerned, amending Regulation (EC) no. 715/2007 and (EC) no. 595/2009 and repealing Directive 2007/46/EC [6, 7]. An excerpt from the regulation is given below:

Chapter VII New technologies or new concepts.

Article 39 Exemptions for new technologies or new concepts:

> (1) "The manufacturer may apply for an EU type-approval in respect of a type of vehicle, system, component or separate technical unit that incorporates new technologies or new concepts that are incompatible with one or more regulatory acts listed in Annex II."
>
> (2) "The approval authority shall grant the EU type-approval referred to in paragraph 1 where all of the following conditions are met:

(a) the application for the EU type-approval states the reasons why the new technologies or new concepts make the vehicles, systems, components, or separate technical units incompatible with one or more regulatory acts listed in Annex II.
(b) the application for the EU type-approval describes the safety and environmental implications of the new technology or new concept and the measures taken in order to ensure at least an equivalent level of safety and environmental protection to that provided by the requirements in respect of which an exemption is sought.
(c) test descriptions and results are presented proving that the condition in point (b) is met."
(3) "The granting of EU type-approvals exempting new technologies or new concepts shall be subject to the authorization by the Commission. The Commission shall adopt implementing acts to decide whether to grant the authorization referred to in the first subparagraph of this paragraph. Those implementing acts shall be adopted in accordance with the examination procedure referred to in Article 83(2)."
(4) "Pending adoption of implementing acts referred to in paragraph 3, the approval authority may grant a provisional EU type-approval, valid only in the territory of the Member State of that approval authority, in respect of a type of vehicle covered by the exemption sought. The approval authority shall inform the Commission and the other Member States thereof without delay by means of a file containing the information referred to in paragraph 2. The provisional nature and the limited territorial validity of the EU type-approval shall be apparent from the heading of the EU type-approval certificate and the heading of the certificate of conformity."
(5) "Approval authorities of other Member States may accept the provisional EU type-approval referred to in paragraph 4 within their territory, provided they inform the approval authority that granted the provisional EU type-approval of their acceptance in writing."
(6) "Where appropriate, the implementing acts referred to in paragraph 3 shall specify whether authorizations are subject to any restrictions, in particular with regard to the maximum number of vehicles covered. In all cases, the EU type-approval shall be valid for at least 36 months."
(7) "Where the Commission adopts implementing acts referred to in paragraph 3 to refuse to grant the authorizations, the approval authority shall immediately inform the holder of the provisional EU type-approval referred to in paragraph 4 that the provisional EU type-approval approval shall be revoked six months after the date of the implementing act. However, vehicles that have been manufactured in conformity with the provisional EU type-approval before it ceased to be valid may be placed on the market, be registered, or enter into service in any Member State that has accepted the provisional EU type-approval in accordance with paragraph 5."

The type-approval process differs from country to country due to the fact that autonomous vehicles are considered special vehicles for which some states have issued specific regulations for entry into service, while others have granted exemptions or temporary authorizations for experimental purposes. The European Union Member States which have adopted legislation on autonomous vehicles, and which allow them to operate in test mode are Austria, Belgium, Finland, France, Germany, Greece, Italy, Luxembourg, the Netherlands, the United Kingdom, Spain, and Sweden [1].

The Technical Committee on Motor Vehicles (TCMV) in the EU on February 12, 2019 published a guide (version 4.1) with recommendations for the EU type-approval procedure for autonomous vehicles (Guidelines on the exemption procedure for the EU approval of automated vehicles). An excerpt with the main specifications of the guide is presented below [8]:

(C) Safety requirements.

(I) System performance in the automated driving mode:

(1) "When in the automated driving mode OD (Operational Domain), the automated vehicle drives and shall replace the driver for all the driving tasks under the situations which can be reasonably expected in the OD."

(2) "When in the automated driving mode, the vehicle shall not cause any traffic accidents that are rationally foreseeable and preventable."

(3) "When in the automated driving mode, the vehicle shall have a predictable and careful behavior and shall allow an appropriate interaction with other road users (e.g., obey to orders by authorities or communication with other road users when needed)."

(4) "When in the automated driving mode (OD), the automated vehicle shall drive in accordance with the traffic rules."

(5) "The manufacturer shall declare to the type-approval authority the scope of the automated driving mode (OD) where and when the automated driving system is designed to operate. This shall include at a minimum: road conditions (roads dedicated to automated driving vehicles); geographical area, environmental conditions, speed range, other conditions that must be fulfilled for the safe operation in the driving mode."

(6) "An automated driving system shall recognize whether or not the situation is within the set OD and operate only in that OD."

(7) "The system shall be safe by design to cope with any situation within the OD without continuous supervision by the driver."

(8) "In particular, the vehicle shall be able to keep a safe distance with other vehicles in front, exhibit caution in occluded areas, leave time and space for others in lateral maneuvers, be cautious with rights-of-way and if an accident can be safely avoided without causing another it shall be avoided."

(9) "The OD shall be set in a way that it allows the driver to take over safely from the automated system and in compliance with the relevant traffic rules."

(10) "The system shall detect when it is difficult to continue in the automated driving mode."

(II) Driver/operator/passenger interaction:

(11) “The activation of the automated driving mode shall only be possible when the conditions of the OD are met. Means shall be provided to humans (driver or no driver, passenger, operation control center) to deactivate or override immediately the automated mode in an easy manner.”
(12) “The vehicle shall always inform the driver or passengers about the operational status of the system in an unambiguous manner.”
(15) “For vehicles designed to operate only with no driver, a communication function shall be provided to send an emergency notification to an operation control center. A camera and voice communication device shall be provided in the vehicle so that an operation control center can monitor the situation inside the vehicle.”
(III) Transition of the driving tasks:
(16) “The system may request the driver to take over with a sufficient lead time in particular when the system determines that it is difficult to continue automated driving mode.”
(17) “The system shall remain in the automated driving mode as long as the driver has not taken over, and/or will otherwise transfer to a minimum risk maneuver.”
(19) “The system shall be able to determine whether or not the driver has taken over.”
(IV) Minimum risk maneuvers:
(20) “When the system detects that it is difficult to continue in the automated driving mode, it shall be able to transfer to a minimal risk condition (with or without take over request) through a minimal risk maneuver.”
(21) “Other road users shall be informed that the vehicle is performing a minimum risk maneuver in accordance with applicable traffic rules.”
(22) “The MRM (Minimum Risk Maneuver) shall comply with traffic rules.”
(V) Installation of event data recorders:
(23) “Automated vehicles should be equipped with an on-board device that records the operational status of the automated driving system and the status of the driver.”
(24) “This data collected shall allow to assign liability in case of accident and shall include information on the operation status of the automated driving system, state of the driver, information on surrounding, control information of the vehicle.”
(26) “The on-board device shall be able to store data in a secured manner, comply with EU data protection legislation and be protected against manipulation.”
(VI) Cybersecurity:
(28) “The Vehicle shall be designed to protect the vehicle against automated vehicle hacking using state of the art techniques against cyberattacks.”
(29) “Vehicle manufacturers shall take measures such as those related to updating of software etc., installed in automated vehicles necessary to ensure in-use cybersecurity over its lifetime.”
(VII) Safety assessment and tests:
(30) “Automated vehicles shall comply to the largest extent with the existing EU Safety Regulations listed in Annex IV to Directive 2007/46/EC.”

(33) "Systems shall in particular be designed to cope with risks that could impact safety critical functionality due to cyber-attacks and functional safety according to ISO 26262 or any other equivalent methods."
(34) "All design decisions shall be tested, validated and verified by the manufacturer as individual subsystem and as part of the entire vehicle architecture."
(35) "The manufacturer of the autonomous vehicle must evaluate the functional safety of the autonomous driving system using a number of test scenarios that include false negative and false positive ones.
Simulation method may be used, subject to their validation the approval authorities/technical services in accordance with the procedure for virtual testing in Directive 2007/46/EC or Regulation 858/2018."
(VIII) Information provision to automated vehicle users:
(38) "Vehicle manufacturers shall inform automated vehicle users of the following:
(-) Operational conditions of the system, scope of OD, functional limitations.
(-) Means to deactivate the automated driving mode.
(-) Driver's tasks to take over the control of the vehicle.
(-) Information related to indications by HMI (Human Machine Interface).
(-) User behavior to be adopted in case of urgency.
(-) Behaviors of the vehicle when a problem has occurred to the autonomous driving system."

Annex I Information to be provided by the vehicle manufacturer:

(1) System performance in the automated driving mode:
(a) Automated system type definition.
(b) Automated driving functions.
(c) Operational Domain:
(i) Speed, road type, country.
(ii) Environment.
(iii) Road conditions.
(d) Basic performance.
(e) Tasks other than driving technically enabled by the system.
(A) Environment perception:
(a) With respect to operation domain.
(b) Lanes/objects.
(c) System redundancy.
(d) Sensor monitoring:
(i) Plausibility check with respect to misuse.
(ii) Implemented monitoring system.
(e) Connectivity of autonomous vehicles.
(f) Digital maps.
(B) Dynamic driving task and interaction with other road users:
(a) Have a predictable and careful behavior:
(i) Driving in accordance with the speed limits.
(ii) Obeying passing restrictions.

(iii) Adapting the speed of the vehicle to environmental conditions affecting:
(-) Adhesion of the road.
(-) Viewing distance of the system.
(iv) Keeping the required minimum distance to other road users.
(v) Rules regarding the preferred lane of travel.
(vi) Compliance with relevant country specific traffic rules.
(b) Autonomous vehicles will react to:
(i) Other vehicles within the ego lane or in the neighboring lane or cutting into the ego lane from the neighboring lanes.
(ii) Other road users that may be vulnerable.
(iii) Police and Emergency vehicles.
(iv) Law enforcement injunctions (police control).
(2) Driver interaction with the autonomous vehicle:
(a) Activation/deactivation/modes (on/off/standby).
(b) Human Machine Interface (HMI):
(i) Driver information.
(ii) Optical warning signal (type and operation mode).
(iii) Acoustic/haptic warning signals (type and operation mode).
(d) Driver presence and responsiveness recognition system.
(e) Extract of the relevant part of the owner's manual.
(f) Means to prevent misuse and manipulation.
(3) Transition of the driving task:
(a) Planned:
(i) Boundary conditions.
(ii) System behavior.
(iii) System performance.
(b) Unplanned (vehicle systems failure):
(i) Boundary conditions.
(ii) System behavior.
(iii) System degradation.
(iv) System performance.
(c) Emergency situations (imminent danger of collision):
(i) Boundary conditions.
(ii) System behavior.
(iii) System performance.
(4) Minimum risk maneuver:
(a) Different risk maneuvers for different scenarios (planned or unplanned).
(5) Data storage system:
(b) Type of data stored.
(c) Storage location.
(d) Storage duration.
(e) Means to ensure date security and data protection.
(f) Access to the data.
(6) Cyber security:

(a) Description of the different risks and measures put in place to mitigate these risks.
(b) Description of update procedures.
(7) Safety assessment and testing:
(a) Design and validation process to be validated by the technical service and confirmed by the approval authority:
(i) Assessment of the functional and operational safety.
(ii) Test of functionality of the autonomous driving system.
(iii) Tests in case of autonomous driving system failure:
(-) Measurement equipment used.
(-) Test conducted by the technical service/type-approval authority.
(-) Description of in-use tests.

7.1.1 French Republic

The French Republic is the home country of two companies that develop autonomous Shuttle Buses for public passenger transportation (Navya and EasyMile). The French Republic is currently the country with the largest number of autonomous vehicle projects for public transportation operating on public roads under test [1]. In accordance with European law, the adoption of a framework for the type-approval of autonomous vehicles for public passenger transportation in a Member State of the European Union will automatically lead to the recognition of such type-approval in all other Member States [1, 6].

During 2015–2018, the French Republic issued a series of laws and regulations that allow autonomous motor vehicles on public roads on an experimental basis.

Another important requirement that applies to all Member States of the European Union is the protection of individuals with regard to the processing of personal data and the free movement of such data (EU Regulation No 679/2016) [1, 9].

The specific aspects of the field of autonomous vehicles are regulated in the French Republic by two normative acts, respectively Ordinance 2016–1057 of August 03, 2016 (Ordonnance 2016–1057 du 3 août 2016 relative à l'expérimentation de véhicules à délégation de conduite sur les voies publiques) [10] and Ordinance 2018–211 of March 28, 2018 (Décret 2018–211 du 28 mars 2018 relatif à l'expérimentation de véhicules à délégation de conduite sur les voies publiques) [11].

The terminology used in French legislation refers to the autonomous vehicle as the vehicle with partial or total driving delegation or the DPTC (Véhicule à Délégation Partielle ou Totale de Conduite), a term that refers to the advanced vehicle automation technologies, or to the fundamental change of the act of driving.

Ordinance 2016–1057 of August 03, 2016 has the role of regulating the conditions for putting DPTC vehicles on open public roads [10]: "The experimental use of a DPTC vehicle on an open public road is subject to the issuance of an authorization designed to ensure the safety of the experiment".

On the date of the publication of Ordinance 2016–1057, a report was published to the President of the French Republic on this ordinance on the testing of DPTC vehicles on public roads. This report specifies the measures that fall within the scope of the law to allow DPTC vehicles to travel on public roads, whether they are cars, freight cars or public passenger transportation cars, for experimental purposes, under conditions that ensure the safety of all users and to provide, where appropriate, for an appropriate liability regime, stating that "DPTC vehicle use may not be authorized on lanes reserved for public transportation vehicles; with the exception of vehicles used for public passenger transportation".

DPTC vehicles are an essential step towards defensive driving, traffic safety and more efficient and environmentally friendly transport. DPTC vehicles must have superior responsiveness and adaptability to those driven by a driver and thus have fewer errors than human-driven vehicles. Consequently, Ordinance 2016–1057 introduces the concept of authorization for traffic on public roads specific to the experimentation of DPTC vehicles [11].

The methods of applying the Ordinance 2016–1057 are described in the Decree 2018–211 of March 28, 2018 regarding the experimentation of DPTC vehicles on public roads. Decree 2018–211 stipulates that the authorization of DPTC vehicles is possible only for the following purposes: tests for the development of innovative technologies in the field of autonomous driving (software, sensors, mapping etc.) and/or tests on connected road infrastructure, evaluation performance of DPTC vehicles in real situations for the preparation of future uses and public demonstrations for information [12].

An extract with the main specifications of the Decree 2018–211 of March 28, 2018 is presented below [13]:

> Title I (issuing of the experimental authorization) contains the following sections.
> Section 1 "The authorization provided in Ordinance 2016–1057 refers to the testing of DPTC vehicles in one or more of the following cases:
> (1) Vehicle development tests.
> (2) Vehicle performance evaluation.
> (3) Public demonstrations during special events.
> Experimental authorization refers to vehicles intended for the public transportation of persons or goods."
> Section 3 "The authorization shall specify the sections of road on which DPTC vehicles are authorized to operate with a driving delegation as well as the driving delegation functions that can be activated. The authorization shall establish the routes on which the experimentation of motor vehicles intended for the transportation of persons, or the transportation of goods takes place."
> Section 4 "The authorization may be subject to some conditions to ensure safety during the experiment."
> Title II (granting the WW DPTC certificate) includes the following sections:
> Section 8 "The movement of a DPTC vehicle is authorized on the basis of a provisional registration certificate, which allows the experimental circulation of the vehicle (WW DPTC certificate)."

Title III (conditions of the experimental procedure) includes the following sections:
Section 10 "The holder of the authorization must take all necessary measures to avoid events which could affect traffic safety."
Section 11 "DPTC vehicles will be equipped with recording devices that allow permanent determination of the state of delegation of driving."
Section 12 "When driving delegation functions are activated, a person is responsible for driving the DPTC as a driver. The driver, whether on board the DPTC or not, must be able to take control at any time, especially in the event of an emergency. The deadline for resuming control of the DPTC vehicle will be clearly specified."
Section 13 "DPTC vehicles may only be used for the transportation of persons and equipment authorized by the driver. The transported persons will be informed about their participation in the experiment and will give their consent for this participation. When a DPTC vehicle is engaged in the operation of a public passenger transportation service, it must display a visible and accessible announcement for all passengers, indicating that it is an experimental vehicle with driving delegation."
Section 14 "The experiment will be monitored."
Title III (special provisions for transportation vehicles or which provide for the absence of the driver on board) includes the following sections:
Section 15 "If the experiment concerns a DPTC public passenger transportation vehicle or when the driver is expected to be on board, he or she will run a non-passenger trial period."

The French Traffic Code (Code de la Route) on DPTC vehicles contains the following legislative provisions [14]:

Book 1 General dispositions.
Title 2 Liability.
Chapter 3 Criminal liability applicable to a DPTC vehicle:
Article L123-1 "The provisions of Article L121-1 (Driver of a vehicle is criminally liable for offenses committed by him while driving that vehicle) do not apply to the driver, for offenses resulting from the handling of a DPTC vehicle whose driving functions are delegated to an autonomous driving system, when this system exercises, at the time of events, the dynamic control of the vehicle. The driver must at all times be able to respond to a request for takeover of the autonomous driving system."
Article L123-2 "During periods when the autonomous driving system exercises dynamic control of the DPTC vehicle in accordance with its conditions of use, the vehicle manufacturer is criminally liable for the offenses caused by the accidents in which the DPTC vehicle is involved".
Book 3 The vehicle.
Title 1 Technical dispositions.
Chapter 9 Provisions applicable to DPTC vehicles:

Article L319-1 "The autonomous driving system is subject to the conditions of use defined by the manufacturer of the DPTC vehicle."
Article L319-3 "The decision to activate an autonomous driving system is taken by the driver, who is informed in advance by the system that the latter is able to exercise dynamic control of the DPTC vehicle in accordance with its conditions of use. When the operating condition no longer allows it to exercise dynamic control of the DPTC or when the conditions of use are no longer met, the autonomous driving system must:
(1) Alert the driver.
(2) Submit a request for takeover.
(3) Perform a minimum risk maneuver in the event of a serious malfunction."

The Order of April 17, 2018 (Arrêté du 17 avril 2018 relatif à l'expérimentation de véhicules à délégation de conduite sur les voies publiques) on the experimentation of DPTC vehicles on public roads, establishes the authorization conditions for an experimental vehicle with autonomous driving on public roads and the methodology for monitoring the experiments and contains the specifications provided in the legislation on DPTC vehicles in force until the date of issue of the order. The order contains the following sections [15]:

Section 2 "The DPTC is a vehicle from the categories M, N, L, T, C, equipped with one or more functionalities that allow full or partial delegation of driving tasks while the vehicle travels. The delegation is partial when the driver delegates part of the driving tasks to the autonomous driving system of the vehicle but retains at least one physical driving action. The delegation is complete when the driver fully delegates all driving tasks to the autonomous driving system."
Chapter I The authorization request.
Section 3 "The application for authorization of a DPTC vehicle for experimental purposes describes the conditions under which the experiment will be conducted and will contain:"
Annex 1 Presentation of the experimental vehicle.
The type of experiment and their objectives:
(-) Technical tests.
(-) Performance evaluation.
(-) Public demonstration.
(-) Testing of a service.
Location of the experiment.
Public passenger transportation in DPTC.
DPTC vehicle description:
(-) Identification number VIN (Vehicle Identification Number).
(-) Engine type (thermal, hybrid, electric, hydrogen).
(-) Maximum speed of the DPTC vehicle from the factory.
(-) Maximum speed of the DPTC vehicle during the experiments.
(-) Maximum authorized mass.
(-) Total number of seats in the DPTC vehicle (seating down, standing up, driver's position).

(-) Additional equipment (sensors, devices, interfaces).
(-) Device for recording and accessing data to establish the status of delegation of the driving.
Passenger transportation or freight transportation services:
(-) Purpose of the transportation service (persons/goods).
(-) Integration of the service into an existing transportation service.
(-) Transportation service procedure.
(-) Transportation service operator.
(-) Description of the passenger transportation service (route map, stops, timetables, frequencies, prices, integration of the multimodal IT (Informational Technology) system, alternative solution in case of unavailability of DPTC vehicles).
(-) Description of the service for the transportation of goods (route map, transported goods, prices).
(-) Tasks of the driver (including if the driver is outside the DPTC).
(-) Tasks of the monitoring staff serving the management platform.
(-) Tasks of traffic surveillance personnel.
The autonomous driving system:
(-) Architecture of the autonomous driving system (presentation, functional diagram, interfaces, algorithms etc.).
(-) Sensor systems.
(-) The position of the driver who is not in the DPTC vehicle.
(-) Information available to the driver who is not in the DPTC vehicle.
(-) Interactions between surveillance systems and the driver who is not inside the DPTC.
(-) Interactions between the driver who is not inside the DPTC vehicle and a person in the vehicle.
(-) Interaction between the DPTC vehicle and other nearby vehicles.
(-) The interaction between the DPTC vehicle and the law enforcement.
(-) DPTC vehicle maintenance plan.
Maneuvers performed by the DPTC vehicle:
(-) Drivers' maneuvers inside the DPTC.
(-) Maneuvers performed by the driver located outside the DPTC vehicle.
(-) Travel conditions: direction, visibility, velocity etc.
(-) The role of the driver while the DPTC is moving.
(-) Devices for signaling the current state of the autonomous driving system.
(-) Conditions for starting, performing, and ending emergency maneuvers of the DPTC vehicle.
(-) Conditions for taking orders from the DPTC by the driver.
(-) Conditions for triggering remote maneuvers of the DPTC vehicle.
Movement of the DPTC vehicle:
(-) Identify the road sections on which the experiments will take place.
(-) Segmentation of the route into homogeneous sections in terms of the risks of traffic events.
(-) Road sections closed to public traffic.

(-) Specific equipment on the route for experimentation (horizontal and vertical signaling).
(-) Identify on the road sections the critical points from the point of view of the risks of some traffic events.
Risk analyses:
(-) Traffic risk event analysis.
(-) Analysis of the risk of failure of the elements of the autonomous driving system.
(-) Accident response plan.
(-) Closed circuit tests for the DPTC vehicle.
(-) Identical experiments performed in other locations with the DPTC vehicle.
Cyber security:
(-) Cyber security management plan.
(-) Approval of computer systems used in the experiment according to the methodology defined by the National Agency for Security of Information Systems (l'Agence Nationale de Sécurité des Systèmes d'Information).
(-) Cybersecurity analysis methodology related to the experiment.
(-) The main cyber security risks associated with the experiment.
(-) Cyber security risks if the driver is not inside the DPTC vehicle. Necessary measures to limit these risks.
(-) Cyber security risks in the case of the DPTC vehicle monitored from the management platform. Necessary measures to limit these risks.
(-) Measures applied to limit cyber security risks.
(-) Continuous improvement plan on the implemented cyber security (design, development, implementation, operation etc.).
(-) Specialized audit for the assessment of cyber security risks.
(-) Procedures for operating the computer systems during the experiment.
(-) Procedures for detecting, analyzing, and handling cyber security incidents.
(-) Implementation of a cyber security incident detection system in the DPTC vehicle.
(-) Implementation of a cyber security incident logging system in the DPTC vehicle.
Passenger and traffic pedestrian safety:
(-) Communication devices implemented between the driver who is not present inside the DPTC vehicle and the passengers.
(-) Broadcast of pre-recorded messages to the passengers.
(-) Informing passengers about the DPTC's involvement in the experiment.
(-) Conditions and restrictions on the access of minors in the DPTC during the experiment.
(-) Passenger evacuation procedure if the driver is not present inside the DPTC.
(-) Procedure for activating the emergency exits if the driver is not present inside the DPTC.
(-) Measures to ensure the safety of road users, especially vulnerable ones, in the vicinity of the DPTC vehicle.
Monitoring and evaluation:
(-) Procedure for collecting and processing data in the event of an incident.

(-) Specification of data collected and processed in the event of an incident.
(-) Specification of passenger data collected in the event of an incident.
Annex 2 Identification of road sections used for travel.
(-) Identifying the route used for travel.
Annex 3 Critical traffic events.
(-) Roadway geometry (current section, intersections and roundabouts, tunnels, maximum permissible speed).
(-) Development of traffic space (sharing of the traffic space, separation of traffic flows).
(-) Adjacent collision hazard generators (presence of other modes of transportation in the current section, crossing of pedestrian sectors, areas of public transportation stopping, exiting the parking area).
Annex 4 List of standard maneuvers.
Nominal maneuvers:
(-) Maintaining the lane at a constant speed.
(-) Changing the lane on a multi-lane road.
(-) Avoiding a fixed obstacle on the route.
(-) Overtaking a vehicle.
(-) Transiting a toll station.
(-) Exiting on a split lane.
(-) Crossing a pedestrian crosswalk.
(-) Crossing an intersection (with or without traffic lights).
(-) Turning right at an intersection.
(-) Turning left at an intersection.
(-) Leaving the parking or a private road and entering a public road.
(-) Crossing a roundabout.
(-) Crossing a pre-mapped area.
(-) Crossing an area for managing possible incidents.
(-) Crossing a lane reserved for pedestrians and bicycles.
Accidental maneuvers:
(-) Stopping on the road.
(-) Docking on the railway.
(-) Stopping at the nearest stop point or at the next safe area on the runway.
(-) Returning to the lane on the edge or from a stop.
(-) Entering in the parking area outside the lane.
(-) Exiting the parking area outside the lane.
(-) Driving in a straight line at low speed.
(-) Driving in a straight line at high speed.
(-) Turning right at an intersection.
(-) Turning left at an intersection.
(-) Avoiding an obstacle on the road.
(-) Opening the doors.
Annex 5 Monitoring criteria.
Technical elements during the experiment with the DPTC vehicle:

(-) Total traveled distance (in conventional driving mode and in autonomous driving mode).
(-) Incidents that may affect the safety of road users.
(-) Incidents that may affect cybersecurity.
(-) Accidents resulting in property and personal injuries.
(-) Assessment of risk control measures.
(-) Scenarios not identified in the initial risk analysis, and which proved to be critical during the experiment.
(-) Energy consumption of DPTC vehicles.
Evaluation elements on the behavior of participants and third parties, as well as on the perception and acceptability of the experiment:
(-) Assessing the driver's interaction with the DPTC vehicle.
(-) Assessing the behavior of other participants interacting with the experiment.
(-) Evaluation of the perception of the experiment by the users of the transportation services.
(-) Appreciation of acceptance of the experiment by participants and third parties.
(-) Future changes which are required to be performed to DPTC vehicles and/or to the test conditions.
Section 4 "The holder of the WW DPTC certificate is present inside the DPTC vehicle. By way of derogation, in the case of experiments the conditions of which require the driver not to be physically present inside the DPTC vehicle, the holder of the WW DPTC certificate will be present where the experiment takes place."
Chapter II Monitoring the experiments.
Section 5 "Information on authorized experiments will be recorded in a national register of state services responsible for pilot training and experiment evaluation. The registered information may only be made public with the consent of the holder of the WW DPTC authorization/certificate."

7.1.2 Federal Republic of Germany

In the Federal Republic of Germany, due to the high degree of development and technologization of the automotive industry, there is a real interest from both car manufacturers and the federal government in the development of the autonomous driving sector. Many of the big companies in the automotive field have developed autonomous vehicle models (Audi, BMW, Daimler, VW etc.). Thus, German car manufacturers have taken a step-by-step approach, focusing primarily on the introduction of automated driving functions on mass-produced classic vehicles, reserving for the category of autonomous vehicles (levels 3 and 4) a secondary testing role on public roads, university campuses, pedestrian areas etc. [1].

According to the Federal Ministry of Transport on July 27, 2022, the amendments to the Law on autonomous driving in 2017 came into force, amendments that also provide for the amendment of the Law on traffic on public roads, the Traffic Code, and the Law on car insurance.

In the Federal Republic of Germany, the testing phase of autonomous vehicles (level 3) on public roads has already been completed. Currently, autonomous vehicles (level 3) are only allowed on motorways, as there is no traffic in the opposite direction. Autonomous vehicles (level 3) can drive at a maximum velocity of 60 km/h and according to the new regulations, these vehicles must also be assisted by a driver in case of traffic congestion. For cases of heavy traffic where driving at a speed of up to 60 km/h, the driver can dedicate his time to other activities, he does not have to monitor the traffic.

The legal regulations that establish the legal liability in case of an incident/accident specify that the legal responsibility lies with the owner, and if an autonomous vehicle (level 3) was driven by the autonomous driving system, then the driver is not liable. He is only responsible for an accident if he ignored the system's request to take over the vehicle again.

The main legislations on autonomous driving in force in the Federal Republic of Germany are as follows [16]:

(-) Civil Code BGB (Bürgerliches Gesetzbuch).
(-) Federal Data Protection Act BDSG (Bundesdatenschutzgesetz).
(-) Directive 2007/46/EC of the European Parliament and of the Council of September 5, 2007 establishing a framework for the approval of motor vehicles and their trailers, and of systems, components and separate technical units intended for such vehicles.
(-) Regulation on the EC approval of motor vehicles and their trailers, as well as systems, components, and separate technical units for such vehicles EG-FGV (Verordnung über die EG-Genehmigung für Kraftfahrzeuge und ihre Anhänger sowie für Systeme, Bauteile und selbstständige technische Einheiten für diese Fahrzeuge (EG-Fahrzeuggenehmigungsverordnung)).
(-) Driving License Regulations FeV (Verordnung über die Zulassung von Personen zum Straßenverkehr, Fahrerlaubnis-Verordnung).
(-) Ordinance on the exemption of specific forms of transport from the norms of the Passenger Transportation Act FrStllgV (Verordnung über die Befreiung bestimmter Beförderungsfälle von den Vorschriften des Personenbeförderungsgesetzes).
(-) Vehicle Admission Ordinance FZV (Verordnung über die Zulassung von Fahrzeugen zum Straßenverkehr).
(-) Regulation (EU) 2016/679 of the European Parliament and of the Council of April 27, 2016 on the protection of natural persons with regard to the processing of personal data and on the free movement of such data (EU General Data Protection Regulation) GDPR (Datenschutz-Grundverordnung).
(-) Basic Law for the Federal Republic of Germany (German Constitution) GG (Grundgesetz).
(-) Passenger Transportation Act PBefG (Personenbeförderungsgesetz).
(-) Compulsory Insurance Law PflVG (Pflichtversicherungsgesetz).
(-) Product Liability Act ProdHaftG (Produkthaftungsgesetz).
(-) Criminal Code StGB (Strafgesetzbuch).

(-) Road Traffic Act StVG (Straßenverkehrsgesetz).
(-) Road Traffic Regulations StVO (Straßenverkehrsordnung).
(-) Road Traffic Licensing Regulation StVZO (Straßenverkehrs-Zulassungs-Ordnung).
(-) Agreement concerning the adoption of uniform technical prescriptions for wheeled vehicles, equipment and parts which can be fitted and/or be used on wheeled vehicles and the conditions for reciprocal recognition of approvals granted on the basis of these prescriptions (1958 Agreement).
(-) Vienna Convention on Road Traffic (Übereinkommen über den Straßenverkehr).
(-) Insurance Contract Act VVG (Versicherungsvertragsgesetz).

On May 20, 2021, the German federal legislature Der Bundestag passed a law on autonomous driving to amend the law on road traffic and the law on the compulsory insurances Gesetzentwurf "zur Änderung des Straßenverkehrsgesetzes und des Pflichtversicherungsgesetzes - Gesetz zum autonomen Fahren (19/27439)".

The purpose of the bill is to redefine the technical requirements for the construction and equipment of autonomous vehicles, as well as the examination and the procedure for granting the operating license for autonomous vehicles by the federal transport authority KBA (Kraftfahrt-Bundesamt). At the same time, the legislative amendments regulate the manipulation of the data necessary for the operation, respectively the conditions for the supervision of the technical operation. The draft law specifies the obligation of civil liability insurance.

The Federal Ministry of Transport and Digital Infrastructure (Das Bundesministerium für Digitales und Verkehr) has revised and amended the road traffic law StVG (Straßenverkehrsgesetz), introducing specific regulations for the circulation of autonomous vehicles on public roads ""Straßenverkehrsgesetz in der Fassung der Bekanntmachung vom 5 März 2003 das zuletzt durch Artikel 1 des Gesetzes vom Juli 12, 2021 geändert worden ist" [17, 18].

Section 1a Vehicles with highly or fully automated driving functions must meet the following technical requirements:

"The autonomous driving system must ensure the longitudinal and lateral guidance of the vehicle, recognize, and respect all traffic rules identical to a human driver. The autonomous driving system must be able to be deactivated at any time by the driver, respectively to acknowledge the driver's request to control the vehicle himself. The autonomous driving system must be able to indicate the need to hand over control of the vehicle to the driver. These conditions apply to vehicles which are type-approved in accordance with Article 20 of Directive 2007/46/EC."

Section 1b Obligations of the driver:

"The driver is required to take control of the vehicle if the autonomous driving system so requires or if he considers that due to obvious circumstances the preconditions for the use of the autonomous driving functions are no longer met."

Section 1d Operating areas defined:

"The defined operating area refers to the locally and spatially defined public street space in which an autonomous vehicle can be operated. The technical supervision of an autonomous vehicle is carried out by a person who can deactivate the autonomous driving function at any time. The minimum risk condition involves stopping the autonomous vehicle on its own initiative or at the request of the driver in a safe place, with the hazard warning lights activated, for the maximum safety of the vehicle occupants, respectively of the other traffic participants."

Section 1e Exploitation of autonomous vehicles:

"The autonomous vehicle is used in an area of operation approved by the responsible authority under federal law and the autonomous vehicle is type-approved for use on public roads. The autonomous vehicle must be equipped with technical equipment that can independently manage the task of autonomous driving, comply with traffic rules, and be equipped with an accident prevention system."

Section 1f Obligations to exploit autonomous vehicles:

"Owners of an autonomous vehicle have an obligation to ensure road safety, compatibility with the environment of the vehicle and to take the necessary precautions. The driver of the autonomous vehicle has the obligation to evaluate the maneuvers of the autonomous driving system and to take control of the vehicle at its request or if he deems it appropriate."

Section 1g Data processing:

"The manufacturer of an autonomous vehicle must equip the vehicle in such a way that it is possible to store the data recorded during operation. Software applications must provide options for storing and transmitting data processed by the autonomous driving system. The Federal Motor Vehicle Authority has the right to use the data collected for the purpose of scientific research as well as for the purpose of investigating road traffic accidents, the data being accessible to colleagues and universities, non-university research institutions and government agencies with research, development, transport planning or urban planning responsibilities."

Section 1h Subsequent activation of the autonomous driving functions:

"If an autonomous driving function is installed in a vehicle, which is not described in the international regulations applicable within the scope of this law, the license to operate the vehicle is granted only if by deactivating the autonomous driving function the operation of the autonomous driving system is not affected."

Section 1i Testing of autonomous driving functions:

"Vehicles used for testing the stages of development of autonomous driving functions may be put into circulation on public roads only if a test license has been issued for the vehicle, the vehicle is registered, the vehicle is used exclusively for testing and the vehicle is constantly monitored by a driver during the operation of the autonomous driving system."

With regards to the registration of autonomous vehicles, German legislation requires that vehicles subject to registration be type-approved in accordance with the specifications of the European legislation, namely Regulation (EU) 2018/858, establishing a framework for the approval of the vehicles, as well as the systems, components and technical units intended for those vehicles [1, 6, 7]. As autonomous vehicles do not comply with all the requirements of Regulation (EU) 2018/858 as laid down in their current form (with regard to the steering and braking systems), the German vehicle registration system allows an exemption to be applied for vehicles which are put into circulation for testing [1, 18, 19].

7.1.3 Italian Republic

In the Italian Republic, the vehicle traffic on public roads is regulated by Legislative Decree 285/1992 (Codice della Strada), which establishes specific rules on the movement of vehicles on public roads in the Italian regions [1, 20].

On February 28, 2018, the Minister of Infrastructure and Transport (Ministro delle Infrastrutture e dei Trasporti) issued Decree 18A02619 "Modalita' attuative e strumenti operativi della sperimentazione su strada delle soluzioni di Smart Road e di guida connessa e automatica" regulating the methods of implementing the operational tools needed to test intelligent, connected, and automated vehicles on public roads.

This decree establishes functional standards that, based on the new technologies introduced in the road infrastructure, can communicate with the users on board the vehicles in order to provide real-time information about traffic, accidents, weather conditions etc. At the same time, the decree establishes the conditions for testing autonomous vehicles on the infrastructure of public roads, in safe conditions for other road traffic participants. The main condition imposed by the regulation is that the autonomous vehicle must be permanently accompanied by a human operator who is present on board.

An extract with the main specifications of Decree 18A02619 of February 28, 2018 is presented below [21]:

> Article 1 Definitions of terms:
> (f) The autonomous driving vehicle (autonomous vehicle) is a vehicle equipped with technologies that allow it to adopt and implement driving behaviors without the active intervention of the driver on certain road sections.
> (g) Autonomous driving technologies are innovative technologies for autonomous guidance based on different types of sensors, software for processing the data provided by the sensors and interpreting traffic situations, software for autonomous driving decisions etc.
> (h) Autonomous mode is the mode of operation of the autonomous driving vehicle when the control belongs to the vehicle.
> Manual mode is the mode of operation of the autonomous driving vehicle when control is taken over by the driver.
> (j) The supervisor is the occupant of the vehicle (driver) who must always be able to take control of the vehicle at any time, regardless of its level of autonomy.
> (k) Tested system is an autonomous vehicle that performs an experiment, along with the monitoring, recording and communication systems.

(o) V2I cooperative systems are interaction systems between intelligent vehicles and intelligent infrastructure capable of transmitting information and services of interest for driving and safety.
(p) V2V cooperative systems are interaction systems between vehicles.
(q) Connected driving is the driving status of the vehicle in V2I or V2V cooperative systems.
(r) Protected experiments are experiments performed with autonomous vehicles on infrastructures that are not open to public traffic.
Article 9 Authorization to test autonomous vehicles:
"The experimentation of autonomous vehicles on public roads is done on the basis of an authorization. The authorization will be requested by the manufacturer of the autonomous vehicle and will be issued for the vehicles to be tested in partnership with specialized universities, respectively public or private research organizations for conducting experiments in order to validate new models of autonomous vehicles for mass production."
Article 10 Supervisor of the autonomous vehicle:
"The driving of the autonomous vehicle on public roads during the experiment is assisted by a supervisor with a driving license for the class to which the vehicle belongs, who has taken a course for autonomous vehicles in an accredited body in one of the countries of the European Union, and who has tested autonomous vehicles in a state where their traffic is regulated and has traveled at least 1000 km. The supervisor must be able to switch between the operation of the vehicle in autonomous mode and the operation in manual mode and vice versa, having the responsibility for the safety of the vehicle in both modes of operation."
Article 11 Application for authorization to experiment:
"The application for authorization must attest to the maturity of the autonomous driving technologies to be tested in the road sections for which the authorization is sought. Before testing in real conditions, it is necessary to perform simulations on a virtual model covering the actual application scenarios in a virtual environment similar to the real environment, highlighting any deviations from the real scenarios. Autonomous vehicles to be authorized must prove a distance of at least 3000 km before applying for a permit in real traffic conditions or a journey of at least 3000 km in simulated traffic conditions before applying for a permit.
The documentation for issuing the permit must show the operation of the vehicle in autonomous mode, the management of traffic situations, the behavior of the vehicle in various extreme weather conditions etc.
A description of the operation of the autonomous driving system and of the protection mechanisms to prevent unauthorized access to the guidance systems and an analysis of the risks associated with the use of the vehicle in the autonomous driving mode in road traffic are required."
Article 12 Characteristics of the autonomous driving system:
"No matter the situation, the autonomous driving system must comply with the traffic rules in accordance with the specifications of the Highway Code to interact safely with all road users and to allow at any time at the request of the supervisor the transition from autonomous driving mode to manual driving mode.

The autonomous vehicle must be equipped with protection systems to ensure the integrity of the data and the security of the communications, so as to prevent unauthorized access. At least the following data must be recorded during the experiment:
(-) Duration of the experiment.
(-) The operating mode during the experiment: automated or manual.
(-) Date, time, and position in WGS84 coordinates.
(-) Instant velocity and acceleration.
(-) The distance traveled since the beginning of the experiment.
(-) Activation of the controls for the lateral dynamics of the vehicle.
(-) Activation of the controls for the longitudinal dynamics of the vehicle.
(-) Electric motor speed and the transmission ratio.
(-) The current value of the pitch, roll and yaw angle.
(-) Use of visual and acoustic lighting and signaling devices.
(-) Data generated by sensors that equip the autonomous vehicle.
(-) V2V and V2I communications messages (received and transmitted).
Article 13 Characteristics of the road sections:
"Highways and main urban roads that have the following characteristics:
(-) Possibility of braking and maneuvers at high velocities.
(-) Controlled presence of pedestrians on the road.
(-) Controlled and limited access to ramps.
(-) Rainwater runoff conditions.
Secondary urban roads that have the following characteristics:
(-) Presence of bicycles, motorcycles, pedestrians, animals.
(-) Traffic control systems (roundabouts or traffic lights).
(-) Congested traffic conditions, especially at intersections.
(-) Uncontrolled access to ramps.
Urban roads that have the following characteristics:
(-) Possible work in progress, objects, or obstacles on the road.
(-) Presence of bicycles, motorcycles, pedestrians, animals.
(-) Traffic control systems (roundabouts or traffic lights).
(-) Congested traffic conditions, traffic jams at intersections.
(-) Parking areas accessible in two rows.
(-) Road sections with speed limits.
(-) Areas with limited traffic and pedestrian areas.
(-) Unpaved intersections, unpaved roads."
Article 14 Applicant for the authorization to experiment:
"If the applicant has not documented the tests for a total of 10,000 km traveled for the autonomous vehicle, then the applicant must present similar results for tests performed in the laboratory, in a controlled test environment or with a virtual model using software platforms, which describe situations similar to real situations."
Article 19 Liability insurance:

"The applicant for the authorization to experiment must take out a liability insurance of the autonomous vehicle with a minimum limit four times higher than the one established for the vehicle of a similar class but which is not equipped with autonomous driving technologies."

7.1.4 Netherlands

Testing of autonomous vehicles is possible on all public roads in the Netherlands, on the basis of approval from the Dutch Department of Road Traffic RDW (Rijksdienst voor het Wegverkeer) and the road sector administrator RWS (Rijkswaterstaat) [22].

The Dutch vehicle authority RDW is responsible for registering vehicles for entry onto public roads. Vehicles eligible for registration must be type-approved in an EU Member State in accordance with European regulations and be admitted by a national authority of an EU Member State for usage on public roads [23]. These vehicles do not require a separate approval from RDW for usage on the roads in the Netherlands. Vehicles without EU type-approval may be allowed on public roads in the Netherlands but require individual approval from RDW.

The operational test procedure on public roads can be performed with a driver on board the autonomous vehicle, respectively without a driver on board the autonomous vehicle.

According to WVW, vehicles with autonomous functions can travel on public roads provided that a driver is present to control the movement of these vehicles [24]. The autonomous functions are: Automatic Following, Connected Adaptive Cruise Control, Lane-Keeping Assist, Vehicle Following, Lane Change, Traffic Jam Assist, Overtaking, Valet Parking, Collision Avoidance, Emergency Stop, Self-Driving Vehicles.

The Law IENW/BSK-2019/134685 (Regeling vergunningverlening experimenten zelfrijdende auto) on the issuance of a license for experiments on autonomous driving vehicles entered into force on July 1, 2019 [25]. This law provides for the possibility for autonomous vehicles to be tested without a driver on board. The driver can be at a distance, for example in a control room. An excerpt with the main specifications of this law is presented below:

Chapter 1 General dispositions.
Article 1 Definitions:
"The transport system is the whole vehicle and its behavior in relation to the infrastructure and the connection between them, including the monitoring of the task of driving the vehicle by the driver.
An emergency scenario is an action due to an interruption of the driving process that endangers the movement of the transport system and/or the safety of the passengers, respectively the integrity of the vehicle.
ADS is the autonomous driving system in which the hardware and software elements of a vehicle in a specific operational area are able to control the vehicle and the perform dynamic driving tasks."

Chapter 2 Application for authorization.
The license authorization request for experiments on autonomous driving vehicles shall be submitted to RDW and the following steps shall be required to obtain the test approval:
(1) Intake is the pre-assessment and admission stage in the approval procedure which will include an action plan with the following information:
(-) Type of vehicles to be tested.
(-) Purpose of the experimental tests.
(-) Current time and the duration of the tests.
(-) Proposed route on which the tests are to be carried out.
(-) Driver's location.
(-) If public passenger transportation is provided.
(-) Traffic safety and environmental sustainability measures.
(-) Functional description of the transportation system.
(-) Risk analysis containing:
(i) EMC (ElectroMagnetic Compatibility).
(ii) ISO 26262 Functional Safety Management Method.
(-) Risk analysis of environmental factors.
(-) Civil liability insurance of the autonomous vehicle.
(-) Declaration of cooperation on knowledge exchange, monitoring, and evaluation of tests.
(-) Detailed description of the presence and operation of a device for recording data from the ADS sensors and control systems.
(-) Detailed description of the means of monitoring the operation of the transportation system.
(2) Desk research is the theoretical evaluation stage of the testing application that will go through the following stages:
(-) Evaluation of the autonomous vehicle. Its performance must be comparable to that of a vehicle without an autonomous driving system.
(-) Evaluation of the infrastructure on which the tests will be performed.
(-) Evaluation of the behavior of the vehicle in autonomous driving mode. Evaluation of the behavior of the vehicle in the road traffic.
(3) Testing on a closed proven ground is the stage of evaluating the real vehicle model, in the field, which will involve the physical inspection of the autonomous vehicle and the performance of tests on the autonomous driving system, including a risk control analysis.
(4) Admittance based on an exemption is the stage of granting approval for traffic on public roads for a maximum of one year.
(5) Evaluation is the stage of evaluation of tests on public roads which involves the monitoring by RDW together with the parties involved in the tests performed with the autonomous vehicle throughout the validity of the approval.

Traffic on public roads in the Netherlands is governed by the 1994 WVW (Wegenverkeerswet) Road Traffic Act, which contains the basic legislative regulations on road traffic.

7.1.5 United Kingdom

In the United Kingdom, the law on autonomous and electric vehicles ("The Act") came into force in July 2018. This law adapts the existing framework for liability insurance by extending the compulsory insurance for both the autonomous vehicle and the driver. The main specifications of the law are presented below [26]:

Section 1 "Autonomous vehicles will be identified and listed by type, according to the information recorded in the registration document, or in another way."
Section 2 "The insurer is liable in the event of an accident caused by an autonomous vehicle (when the autonomous driving system is active) and is liable both for damage to the autonomous vehicle and for injuries to passengers and/or the driver. If the autonomous vehicle is not insured (under the conditions of exemption according to the law), in the event of an accident when the autonomous driving system is active, the owner of the vehicle is liable both for damage to the autonomous vehicle and for damage to passengers and/or driver."
Section 4 "The insurer's liability in the event of an accident is excluded if the software of the autonomous driving system is not updated or has been modified from the original version supplied by the manufacturer."

United Kingdom has established a code of rules and procedures for testing autonomous vehicles Code of Practice: automated vehicle trialing amended on January 28, 2022 which provides details and recommendations to ensure safety and reduce potential risks especially for testing autonomous vehicles used in public passenger transportation.

The United Kingdom is not a signatory to the 1968 Vienna Convention, and therefore does not apply the provisions relating to the need for a human driver on board a vehicle. The Code of Practice sets out requirements for the assistance of the autonomous vehicle by a test driver or an authorized operator who must supervise the vehicle permanently and be prepared to take control of the vehicle if necessary. The main specifications of the Code of Practice are presented below [27]:

Legal requirements: "It is possible to carry out tests with autonomous vehicles in the United Kingdom at any level, under the following conditions:
(-) A driver is present on board or outside of the vehicle and is prepared, able and willing to take control of the vehicle.
(-) The vehicle is compatible with the concept of autonomous driving.
(-) A liability insurance has been taken out for the vehicle.
It is the responsibility of the entities carrying out tests on autonomous vehicles to ensure that they comply with all applicable legal requirements.
Insurance: "Testing of autonomous vehicles on public roads is conditioned by a liability insurance being taken out."
Advanced tests: "Testing of autonomous vehicles can be carried out without a driver aboard the vehicle, but it is necessary to have a safety operator who can use a remote connection to the vehicle, through which he can take control if necessary."

Safety plan: "Test organizers should develop a detailed safety plan that demonstrates that the test activity can be performed safely without involving unreasonable risks."
Passenger transportation services: "Organizations wishing to test autonomous vehicles for passenger transportation services must comply with the legal regulations in force."
Software versions: "The versions of the software running on the autonomous vehicles to be tested need to be updated with the latest versions and revisions available from the manufacturer."
Data protection: "Test organizers must ensure that data protection legislation is complied with, including the requirements that personal data be used correctly and legally, kept safe and no longer than necessary (in accordance with the 2018 Data Protection Act)."
Cyber security: "Manufacturers of autonomous vehicles must ensure that all vehicle systems have adequate security measures in place to manage data security and the risk of unauthorized access to the data."
Publication of safety measures: "Test organizers should develop detailed safety precautions before conducting the tests and publish an abbreviated version of them for the education and information of road users, as well as for the safety of the test."
Report: "Test organizers shall report on the performance of the autonomous vehicle containing the results of the tests performed. The reports will be available for public consultation."
General requirements for the vehicles: "The autonomous vehicles used for testing on roads open to public traffic must meet the legal requirements for construction and use. The autonomous vehicle through the sensor systems or through the control of the driver or the safety operator, will have to respond appropriately to all interactions with all factors and with all the events in the environment."
Tests on virtual models: "Before performing tests on roads open to public traffic, tests shall be performed on virtual models under conditions similar to the real ones. These tests will be performed by accredited organizations and must provide a wide range of scenarios to which the autonomous driving system will have to respond safely. Test organizers will need to ensure that vehicles have successfully completed virtual and real tests on closed roads or test tracks before being tested on roads open to public traffic."
Remote operation: "Remote operation of an autonomous vehicle is possible if the remote operating system provides a two-way real-time communication link, is redundant and is designed to deal with any malfunctions and is able to provide the same level of security as a driver located inside the vehicle."
Data recording: "The tested autonomous vehicles must be equipped with data recording devices or devices that capture data from the sensors and the autonomous driving system of the vehicle, as well as other information regarding the movement of the vehicle. This data must be stored and contain at least the following information:

(-) Details of the autonomous system, the software version, hardware specifications.
(-) Vehicle mode: autonomous or manual.
(-) Longitudinal acceleration in the direction of travel of the vehicle.
(-) Lateral acceleration when the vehicle is moving sideways.
(-) Vertical acceleration when the vehicle passes over an object that causes it to lift.
(-) Vehicle velocity.
(-) Steering system control.
(-) Braking system control.
(-) Operation of vehicle lights and indicators.
(-) Vehicle geolocation.
(-) Connectivity, network access and latency.
(-) Usage of the vehicle's audible warning system.
(-) Sensor data regarding the presence of other road users or objects in the vicinity of the vehicle.
(-) Remote controls that affect the movement of the vehicle.
(-) Any intervention carried out by the driver or safety operator, including the time of such intervention."

The transition system between autonomous and manual modes: "For the safety of tests on autonomous vehicles, it is important to easily manage the transitions from the autonomous control mode to the manual control mode and vice versa. Alerts issued by the autonomous vehicle must be clear and easy to understand by the driver or the safety operator."

7.1.6 Kingdom of Spain

In the Kingdom of Spain, tests have been carried out for the usage of autonomous vehicles on public roads with both closed-circuit and normal traffic. From a legislative point of view, a series of normative acts have been issued with the aim of the progressive introduction of new technologies, based on existing regulations and procedures.

Instruction 15/V-113 din November 13, 2015 (Autorización de pruebas o ensayos de investigación realizados con vehículos de conducción automatizada en vías abiertas al tráfico en general) aims to regulate the granting of special authorizations for tests and research performed with autonomous vehicles of level 3 or higher according to SAE 3016™ on roads open to public traffic. The instruction sets out a number of requirements that must be met for the vehicle, the driver, and the applicant for the special authorization. An excerpt with the main specifications of instruction 15/V-113 is given below [28]:

Paragraph 1 Authorization to test autonomous vehicles.
"Requirements to be met by an autonomous vehicle in order to obtain the necessary authorization to carry out the tests:

An autonomous vehicle is a vehicle equipped with technology that allows its operation or driving without the need for active control or supervision of a driver, whether such automated technology is activated or deactivated, permanently, or temporarily. Active safety or driver assistance systems and equipment included in vehicles that require human control or supervision to operate are not considered autonomous technology.

Only vehicles incorporating functions and technology associated with levels 3, 4 and 5 (levels of autonomy 0–5 specified in the instruction are similar to those specified in the SAE J3016™ standard) are subject to this instruction.

Autonomous mode is the driving mode which consists in operating or driving the autonomous vehicle without the active control of a driver when its autonomous technology is activated.

Manual mode is the driving mode of an autonomous vehicle in which autonomous technology is deactivated and its operation or driving must be performed by the active control of a driver.

The motor vehicle must have a civil liability insurance for events which may result in injury or damage to persons or property during tests on roads open to public traffic.

The autonomous vehicle must carry out the approval procedures established by the national accreditation body ENAC (Entidad Nacional de Acreditación) or have passed the equivalent approval procedures established by the competent authorities of a Member State of the European Union.

The driver must be prepared to take control of the vehicle in the event of an incident that poses a risk to the passengers of the vehicle or to other road users."

Paragraph 3 Documents required to submit the application for authorization to test autonomous vehicles.

"Descriptive report containing a description of the technology fitted to the autonomous vehicle, of the principles used to ensure its functional safety and of the activation system of the autonomous driving mode, the degree of autonomy according to the equivalents of the SAE J3016™ standard. Identification and detailed description of the road sector selected for testing and general description of the test plan on the road sector open to public traffic.

Certificate of autonomous driving tests issued by a technical service accredited by the competent Spanish bodies or an equivalent body of a Member State of the European Union.

Technical specifications of the autonomous vehicle: mass, construction dimensions, powertrain, transmission system, suspension system, steering system, braking system, lighting and signaling system, bodywork etc. according to the technical specifications necessary to obtain the type-approval of vehicles.

ISO 26262 functional safety assessment documentation according to HARA (Hazard Analysis and Risk Assessment). The risk analysis must cover all situations that present a potential risk based on their probability of occurrence, driver controllability, and severity to ensure that the autonomous vehicle is tested in maximum safety conditions."

Evaluation documentation of the software version and software applications that are integrated in the autonomous vehicle and that have been previously tested, validated, and certified.
Documentation certifying the emergency stop procedure (control of powertrain, steering and braking systems) accessible to both the driver and each occupant of the vehicle. When operating any of these three systems, the autonomous driving system shall switch off automatically.
Description of the functions of the autonomous driving system that must cover all possible scenarios that the autonomous vehicle may face while driving on public roads. Traffic situations which the autonomous vehicle cannot deal with must be clearly specified in order to define the appropriate measures to be taken when planning the tests in order to avoid such situations.
Test scenarios on roads open to public traffic in real traffic conditions (urban, interurban and on the highway), on different types of infrastructure (urban roads, secondary and main roads, highways etc.) for variable traffic densities, for different moments of time (day, night), respectively for different extreme climatic conditions.
Cyber security assessment and validation certificate for the autonomous vehicle and all its systems.
EMC electromagnetic interference compatibility validation certificate."

7.1.7 Hellenic Republic

In the Hellenic Republic, the competent authorities for testing and regulating the use of automated, autonomous, and connected vehicles are the Ministry of Infrastructure and Transport (Υπουργείο Υποδομών και Μεταφορών) and the Ministry of Development and Investment (Υπουργείο Ανάπτυξης & Επενδύσεων), through the General Secretariat for Research and Innovation (Γενική Γραμματεία Έρευνας και Τεχνολογίας)—Hellenic Institute of Transport (Ινστιτούτο Βιώσιμης Κινητικότητας & Δικτύων Μεταφορών).

The testing of autonomous vehicles on roads open to public traffic was not carried out under normal traffic conditions, the activities so far targeting only the segment of public transportation. These tests were performed under controlled conditions, in pilot evaluation projects, means of passenger transportation without driver.

In view of the topicality of autonomous driving, the Hellenic authorities have adopted the necessary framework legislation to regulate the field, especially in the public transportation segment, through the Law 4784 of March 16, 2021 on supporting urban mobility, micro-mobility, modernization and digitization of the Ministry of Transport, Infrastructure and other provisions (Βιώσιμη Αστική Κινητικότητα—Μικροκινητικότητα—Ρυθμίσεις για τον εκσυγχρονισμό, την απλούστευση και την ψηφιοποίηση διαδικασιών του Υπουργείου Υποδομών και Μεταφορών και άλλες διατάξεις).

An extract with the main specifications of Law 4784 of March 16, 2021 is presented below [29]:

> Article 42 Regulations on the movement of autonomous vehicles (amendment of Article 13 of Law 2696/1999):
> "Every vehicle in circulation must have a driver.
> By way of exception to paragraph 1, traffic on public roads is permitted without the presence of a driver on board an urban passenger bus in the context of a pilot application or research tests. Its circulation is allowed only for public transportation of persons, by a decision of the municipal council after the approval of the traffic police, for a determined period of time, on routes established as a result of a traffic study carried out according to the legislation in force.
> Urban passenger buses must be equipped with autonomous mechanisms and systems (autonomous buses) to ensure movement, braking, and stopping, identical to the vehicles operated by a driver.
> In order to ensure the smooth operation without a driver of these autonomous buses and to ensure the safety of passengers and all road users, it is necessary to monitor the vehicles and the route they travel through a control center, using electronic means and audio/video recording devices. These devices must be located outside the autonomous buses, respectively on the route at predetermined points and in addition to the monitoring function they must facilitate the takeover of the vehicles by the surveillance staff from the control center according to the traffic conditions and/or in case of emergency situations. Remote control functions of autonomous buses involve stopping them or making maneuvers to avoid obstacles.
> For reasons of passenger safety and public safety, it is possible to video monitor inside the autonomous buses throughout the journey through audio/video recording devices under the conditions specified by the legislation on the protection of personal data.
> The supervising person from the control center who is responsible for monitoring the autonomous buses must have a driving license corresponding to the category of the vehicles monitored and has the responsibility to ensure their smooth movement.
> In the event of an emergency, the supervising person is responsible, as a driver, for stopping the autonomous bus or carrying out maneuvers to avoid obstacles.
> A supervising person from the control center is responsible for the simultaneous monitoring of up to four autonomous buses unless, for reasons of high speed or heavy traffic, it is necessary to reduce the number of monitored vehicles.
> The autonomous buses defined in this article operate on established routes as a result of a traffic study carried out in accordance with the legislation in force. These buses must comply with all traffic rules, road signs and signals, respectively traffic police signals.
> The operation of autonomous buses without a driver in the context of a pilot application or a test operation will be carried out in successive stages, after evaluating the results of each stage, which will determine the nature and conditions of implementation of the next stage.

With the exception of the autonomous buses described above, cars with a maximum mass of 3500 kg may be used on public roads without the presence of a driver on board them only for test operations for research purposes, after obtaining the necessary approvals."

7.1.8 Romania

According to Article 288 of the Treaty on the Functioning of the European Union (TFEU), the institutions shall adopt regulations, directives, decisions, recommendations, and notices for the exercise of Union powers, all of which shall constitute legal acts. Mandatory European legal acts (regulations, decisions, and directives) are adopted at Union level through legislative procedures (the European decision-making process).

Given their directly applicable nature provided by TFEU, it is not permitted to duplicate the provisions of regulations or decisions by copying them into the law of the Member States (legislative parallelism). Thus, the competent national institutions have the obligation to repeal or amend the existing national normative acts that "duplicate", violate or add impermissibly to the provisions of some regulations, so that there are no parallels, violations, or other obstacles in their direct application [30].

In order to be put into service on public roads, autonomous vehicles must be made in accordance with the regulatory framework and the national and European rules laid down for the fulfillment of the operating conditions, with all subsequent amendments and completions for the following (but not limited to excluding those that do not apply to autonomous vehicles, respectively those whose autonomous driving technology does not ensure compatibility).

A bill was submitted in 2022 to the Romanian Parliament for amending and supplementing the Government Emergency Ordinance no. 195/2002 regarding traffic on public roads. This bill facilitates the introduction of autonomous vehicles equipped with automated driving systems for usage on public roads. An excerpt with the main specifications of the bill is presented below [31]:

"Article 6 point 13^2 operational domain is defined by the environment, time of day, traffic, existing infrastructure, geographical and meteorological conditions, as well as other conditions under which an automated driving system is designed to operate."

"Article 6 point 29^2 automated driving system is defined by the function of a vehicle through which its lateral and longitudinal movements, road monitoring, response to road traffic events, planning and signaling maneuvers are performed in real time."

"Article 13^1 (1) In the case of testing vehicles with an automated driving system, the authorization and trial plates shall be issued on the basis of proof of compliance with the technical requirements for the operation of automated driving systems

established by order of the Minister of Transport and Infrastructure, proof issued by the Romanian Automotive Register (Registrul Auto Român).
(2) The routes on which motor vehicles equipped with an automated driving system may be tested, as well as the time interval, shall be established by the traffic police, at the request of the legal person holding the test authorization."
(3) It is forbidden to test the emergency brake function of the vehicle provided with an automated driving system on the public road.
(4) The legal person referred to in paragraph 2 shall undertake in writing to comply with the established routes and time limits, the proper functioning of the test and the ability of the person designated to supervise the driving of the vehicle to take control of it according to the operational domain."
"Article 23 (1^1) During the road test of a vehicle equipped with an automated driving system, the condition that a motor vehicle is driven by a person shall be deemed to be fulfilled if the provisions of Article 13^1 paragraphs (1) and (2) are met.
(1^2) The provisions of paragraph (1) shall apply to the person designated to supervise the driving of the vehicle equipped with an automated driving system.
(1^3) During the test of a vehicle equipped with an automated driving system on a public road, the legal person holding the authorization referred to in Article 131 paragraph (1) shall be liable for a breach of the traffic rules, unless control of the vehicle is taken over by the person referred to in paragraph (1^2) according to the operational domain.
(1^4) In the event of a road accident, as well as in the event of non-compliance with traffic rules, at the request of the traffic police the legal person referred to in paragraph (1^3) is obliged to communicate to it the data recorded by the technical device installed onboard the automated driving vehicle, during its testing on the public road, regarding the operational conditions of the system, the operational status of the system, as well as the condition of the driver."
"Article 105 point (31) non-compliance with the undertaking referred to in Article 13^1 paragraph (4) and point (32) failure to provide the data referred to in Article 23 paragraph (1^4)."

The following legislative amendments are proposed in the bill presented above [32]:

"(1) Creation of the legal premises for the authorization for tests of the vehicles equipped with an automated driving system, in order to test these vehicles on the routes previously established by the traffic police.
(2) Definition of the automated driving system as a function of a vehicle through which the lateral and longitudinal movements of this vehicle, road monitoring, response to road traffic events, as well as planning and signaling maneuvers are performed in real time.
(3) Obligation of the legal person holding the authorization issued for the testing of vehicles equipped with an automated driving system to designate a person to supervise the driving of the vehicle.

(4) Regulation of liability for violation of traffic rules of the legal person holding the authorization issued in the case of testing of vehicles equipped with automated driving system, unless the control of the vehicle is taken over by the person designated to supervise the driving of the vehicle."

In Romania, the regulations regarding the type-approval and issuance of the identity card of road vehicles (RNTR2) are established by Order of the Ministry of Transport, Infrastructure and Communications with the last modification on December 23, 2020, which is published in the Official Gazette (Monitorul Oficial) no. 43 of January 14, 2021.

According to these regulations, autonomous vehicles fall under Chapter I, Section 3 (Definition of categories of use), paragraph 5 "special vehicle (special purpose vehicle) - vehicle of category M, N or O having specific technical characteristics for the purpose to perform a function that requires special adaptations and/or equipment".

Chapter II, Section 3 provides for the derogations set out in the Annex to this Chapter and in Part III of Annex II to Regulation (EU) 2018/858 which applies cumulatively to the type of special purpose vehicle subject to type-approval.

Technical requirements for the national type-approval of vehicles produced in small series in category M2, a category which also includes autonomous vehicles of the Shuttle Bus type (technical requirements applicable in view of the explanatory notes in the table in Part I to Annex II of Regulation (EU) 2018/858) are as follows [6]:

(-) EU Regulation 540/2014 The sound level of silencers - establishes the technical requirements for the type-approval of all new vehicles with regard to their sound level and silencers [33].
(-) UNECE Regulation 58/2019 Ground protection devices - specifies that all vehicles must be constructed and/or be equipped in such a way as to provide effective protection over their entire width against a rear-end collision [34].
(-) EU Regulation 1003/2010 Installation of rear registration plates - designates the dimensions of the space for the location and installation of the rear registration plate, the location of the space for the location and installation of the rear license plate and the shape of the surface for the location and installation of the rear license plate for all types of vehicles specified in the regulation [35].
(-) UNECE Regulation 79/2008 The steering equipment - establishes the uniform provisions for the configuration and operation of steering systems fitted to vehicles admitted to service. The main requirement is that the main steering system contains a direct mechanical connection between the command, normally the steering wheel, and the wheels, to determine the trajectory of the vehicle. This regulation has been amended by UNECE Amendment ECE/TRANS/WP.1/145, which provides for the possibility for a driver to be assisted in the performance of his driving duties and in the control of the vehicle with driver assistance systems [5, 36].
(-) EU Regulation 130/2012 Vehicle accessibility and maneuverability - establishes the technical requirements for the type-approval of all new vehicles with

regard to vehicle accessibility and maneuverability in terms of the characteristics of steps, access ladders and handles, the characteristics of reversing devices. Respectively their minimum height for opening the door through which a person must step to enter the passenger compartment [37].
(-) UNECE Regulation 28/2010 Audible warning devices - establishes the technical requirements for the type-approval of all new vehicles with regard to audible warning devices, supplied with direct current, alternating current or air, intended to be fitted to vehicles [38].
(-) UNECE Regulation 46/2010 Indirect visibility and the visibility of vehicles - establishes the technical requirements for the type-approval of all new vehicles with regard to mandatory and optional devices for indirect visibility for motor vehicles which are devices for observing the traffic area adjacent to the vehicle, which cannot be observed by direct visibility. These may be conventional mirrors, monitor cameras, or other devices capable of displaying information about the driver's indirect field of vision [39].
(-) UNECE Regulation 13/2016 Braking of vehicles - establishes the technical requirements for the type-approval of all new vehicles with regard to the braking system. The braking system is the set of components whose function is to gradually reduce the velocity of a moving vehicle or to stop it or to keep it stationary if the vehicle is already stopped [40].
(-) UNECE Regulation 10/2012 Electromagnetic compatibility - establishes the technical requirements for the type-approval of all new vehicles with regard to immunity to radiated and conducted disturbances for functions related to direct control of the vehicle, related to driver, passenger and other road users' protection and related to disturbances, which would cause confusion to the driver or other road users, as well as related to the functionality of the transmission channel for data from the vehicle systems [41].
(-) UNECE Regulation 18/2010 Protection against unauthorized use - establishes the technical requirements for the type-approval of all new vehicles with regard to protection devices intended to prevent the unauthorized normal operation of the engine or other power source of the main engine of the vehicle in combination with at least one system that locks the steering, or the transmission, or the control of the gearbox, or the brakes [42].
(-) UNECE Regulation 17/2010 Seats, their anchorages, and any headrests - establishes the technical requirements for the type-approval of all new motor vehicles with regard to the strength of their seats and anchorages and their headrests [43].
(-) UNECE Regulation 39/2004 The speedometer - establishes the technical requirements for the type-approval of all new vehicles with regard to the speedometer and its installation. The speedometer is a device designed to indicate to the driver the velocity of the vehicle at any time [44].
(-) EU Regulation 19/2011 The manufacturer's plate - establishes the technical requirements for the type-approval of all new vehicles with regard to a plate or label, affixed by the manufacturer to a vehicle containing the main technical characteristics necessary to identify the vehicle and provide relevant information on

the maximum permissible total masses, the identification number (VIN) assigned to the vehicle by the manufacturer in order to ensure its identification [45].
(-) UNECE Regulation 14/2015 Seat belt anchorages - establishes the technical requirements for the type-approval of all new vehicles with regard to seat belt anchorages for adult occupants of front-facing or rear-facing or side-facing seats [46].
(-) UNECE Regulation 48/2007 Lighting and signaling devices - establishes the technical requirements for the type-approval of all new vehicles with regard to the number and installation of lighting and light-signaling devices, their type, location, and geometric visibility of lamps and angle the inclination of the passing beam provided for the concerned type of vehicle [47].
(-) UNECE Regulation 3/2011 Retro-reflecting devices - establishes the technical requirements for the type-approval of all new vehicles with regard to retro-reflecting devices for vehicles [48].
(-) UNECE Regulation 7/2014 Position lamps and stop lamps - establishes the technical requirements for the type-approval of all new vehicles with regard to the front and rear position lamps and the stop lamps for vehicles. The front/rear position lamps are the lamps indicating the presence and the width of the vehicle seen from the front/rear, and the stop lamps are the lamps indicating other road users who are behind the vehicle, highlighting the activation of the service brake [49].
(-) UNECE Regulation 87/2010 Daytime running lights - establishes the technical requirements for the type-approval of all new vehicles with regard to daytime running lamps, which are front-facing lamps used to make the vehicle more easily visible during use, during the day [50].
(-) UNECE Regulation 91/2010 The side-marker lamps - establishes the technical requirements for the type-approval of all new vehicles with regard to the side-marker lamps for vehicles used to indicate the presence of the vehicle when viewed from the side [51].
(-) UNECE Regulation 6/2010 The direction indicator lamps - establishes the technical requirements for the type-approval of all new vehicles with regard to the direction indicator lamps for vehicles, which are devices mounted on a vehicle and which, when actuated by the driver, signal the intention to change the direction of the vehicle [52].
(-) UNECE Regulation 4/2009 Rear registration plate lamp - establishes the technical requirements for the type-approval of all new vehicles with regard to the rear registration plate lamps for vehicles, which are devices for illuminating rear registration plates and which illuminate the rear license plate by reflection [53].
(-) UNECE Regulation 112/2011 Headlamps for LED vehicles - establishes the technical requirements for the type-approval of all new vehicles with regard to headlamps for vehicles equipped with LED modules [54].
(-) UNECE Regulation 123/2019 Adaptive front lighting systems - establishes the technical requirements for the type-approval of all new vehicles with regard to adaptive front lighting systems, which are beam-emitting lighting devices with different characteristics for automatic adaptation to various conditions of use of the

asymmetrical beam (passing beam) and, if applicable, of the main beam (driving beam) [55].
(-) UNECE Regulation 19/2010 Front fog lamps - establishes the technical requirements for the type-approval of all new vehicles with regard to front fog lamps, which may incorporate glass or plastic lenses which allow light to pass through the illuminating surface [56].
(-) EU Regulation 1005/2010 Towing devices - establishes the technical requirements for the type-approval of all new vehicles with regard to vehicle towing devices for attaching a connecting piece, such as a towing bar or cable [57].
(-) UNECE Regulation 38/2010 Rear fog lamps - establishes the technical requirements for the type-approval of all new vehicles with regard to rear fog lamps for vehicles, which increase rear-view visibility by producing a red signal of a higher intensity than that of the rear position lamps [58].
(-) UNECE Regulation 23/2010 Reversing lamps - establishes the technical requirements for the type-approval of all new vehicles with regard to reversing lamps for vehicles, which illuminate the road behind the vehicle and warn other road users that the vehicle is reversing or is about to go back [59].
(-) UNECE Regulation 77/2010 Parking lamps - establishes the technical requirements for the type-approval of all new motor vehicles with regard to parking lamps for vehicles which are used to indicate the presence of a stationary vehicle [60].
(-) UNECE Regulation 16/2011 Seat belts - establishes the technical requirements for the type-approval of all new vehicles with regard to the installation of seat belts and restraint systems intended to be used separately as individual accessories by adults occupying front-facing or rear-facing seats [61].
(-) UNECE Regulation 121/2016 Location of manual controls - establishes the technical requirements for the type-approval of all new vehicles with regard to the requirements for the location, identification, color and lighting of the vehicle manual controls, tell-tales, and indicators [62].
(-) UNECE Regulation 122/2010 Heating systems - lays down the technical requirements for the type-approval of all new vehicles with regard to the passenger compartment heating system, which is any type of device intended to increase the temperature in the passenger compartment [63].
(-) UNECE Regulation 43/2014 Materials for safety glass panes - establishes the technical requirements for the type-approval of all new vehicles with regard to the materials for safety glass panes to be installed as windscreens or other exterior windows or as partitions on vehicles [64].
(-) EU Regulation 458/2011 Tire Fitting - establishes the technical requirements for the type-approval of all new vehicles with regard to tire fitting, definition of tire types, maximum and minimum tire size indices, wheel dimensions, and speed and load characteristics for tire fitting [65].
(-) UNECE Regulation 54/2008 Tires for motor vehicles - establishes the technical requirements for the type-approval of all new vehicles with regard to pneumatic tires identified by the symbols of the speed categories intended for vehicles [66].

(-) UNECE Regulation 117/2011 Noise emissions - establishes the technical requirements for the type-approval of all new vehicles with regard to noise emissions and rolling resistance of new tires and their adhesion on wet surfaces [67].
(-) UNECE Regulation 89/2002 Maximum speed limitation - establishes the technical requirements for the type-approval of all new vehicles with regard to the speed limitation of vehicles by means of a system whose essential function is to regulate the fuel supply of the engine or by means of the engine's electronic control system [68].
(-) EU Regulation 1230/2012 Mass and dimensions of vehicles - establishes the technical requirements for the type-approval of all new vehicles with regard to the masses and dimensions determined by the manufacturer [69].
(-) UN/ECE Regulation 55/2010 Mechanical coupling devices-establishes the technical requirements for the type-approval of all new vehicles with regard to the requirements that mechanical coupling devices and components must meet in order to be internationally compatible [70].
(-) UNECE Regulation 107/2015 Vehicles of categories M2 and M3 - establishes the technical requirements for the type-approval of all new vehicles with respect to any motor vehicle of category M2, Class A, with regard to its general construction [71].
(-) EC Regulation 661/2009 Safety - establishes the technical requirements for the type-approval of all new vehicles with regard to the safety criteria for their components [72].
(-) EC Regulation 347/2012 AEBS automatic emergency braking - establishes the technical requirements for the type-approval of all new vehicles with regard to the equipment with the Automatic Emergency Braking System (AEBS) [73].
(-) UN/ECE Regulation 351/2012 Involuntary crossing of dividing lines - establishes the technical requirements for the type-approval of all new vehicles with regard to the warning system for inadvertent crossing of the lines separating the lanes [74].
(-) UNECE Regulation 100/2011 Electric powertrains - establishes the technical requirements for the type-approval of all new motor vehicles with regard to the electric powertrains of motor vehicles with one or more electrically driven traction motors which are not permanently connected to the grid, as well as their high voltage components and systems which are galvanically connected to the high voltage bus of the electric powertrain [75].

7.2 United States Legislation

In the United States (US), the regulations regarding the safety of autonomous vehicles are subject to state governments, and their registration and entry into public roads are subject to the laws of the federal government. As early as 2017, the Department of Transportation and the National Highway Transportation Safety Association (NHTSA) launched the "Automated Driving Systems 2.0: A Vision for Safety" program, which was designed to promote and improve the safety, mobility, and efficiency of autonomous driving systems. The NHTSA criteria for setting the safety criteria for autonomous vehicles refer to the following [1, 18, 76]:

(1) System Safety compliance with and implementation of ISO (International Standards Organization) safety standards and standards imposed by SAE International.
(2) Operational Design Domain (ODD) documentation regarding at least the following information: road types, geographical area, velocity, environmental conditions, and other constraints.
(3) Object and Event Detection and Response - the detection and reaction of autonomous vehicles in the event of static or dynamic objects, or events.
(4) Backup plan (Minimal Risk Condition) ensuring the minimum risk conditions of the autonomous management system.
(5) Validation Methods - methods of validating the control algorithm of the autonomous driving system.
(6) Human Machine Interface (HMI) - the interaction between the autonomous vehicle and the driver.
(7) Vehicle Cybersecurity - ensuring the security protocols implemented in the control of the autonomous driving system.
(8) Active and passive safety systems (Crashworthiness) - ensuring the protection of passengers in the event of an accident.
(9) Post-Crash Autonomous Driving System Behavior - protection of passengers after the accident (disconnection of electricity and main systems).
(10) Data Recording - recording data for monitoring events and malfunctions.
(11) Consumer Training - training programs for staff involved in the operation of the autonomous vehicles.
(12) Federal, State, and Local Laws—compliance with applicable federal, state, and local laws, rules, and regulations in the operation of autonomous vehicles.

Because each state has an independent set of laws in the United States Constitution that seek to promote its own interests, the law on autonomous vehicles differs significantly from state to state. Thus, while some states (Arkansas, Colorado, Connecticut, District of Columbia, Illinois, Massachusetts, New York, Pennsylvania, Vermont) clearly specify the requirement for a human operator to be on board the autonomous vehicles throughout the test, other states (Alabama, California, Louisiana, Michigan, Nevada, North Carolina, Ohio, Tennessee, Texas, Utah, Washington) allow autonomous vehicles on public roads without being accompanied by a human operator.

There are currently 38 states in the United States that have adopted a legislation or issued executive orders regarding autonomous vehicles.

According to data collected from the National Conference of State Legislatures (NCSL) and the Insurance Institute for Highway Safety (IIHS), among these states: [77–79]:

(-) Six states authorize studies on autonomous vehicles (Delaware, Idaho, Minnesota, New Jersey, Oregon, Wisconsin).
(-) 15 states authorize testing of autonomous vehicles (Arkansas, Connecticut, Hawaii, Illinois, Maine, Maryland, Massachusetts, New Hampshire, New Mexico, New York, Ohio, Pennsylvania, Vermont, Virginia, Washington).
(-) 18 states authorize the full implementation of autonomous vehicles on roads open to public traffic (Alabama, Arizona, California, Colorado, District of Columbia, Florida, Georgia, Iowa, Louisiana, Michigan, Nebraska, Nevada, North Carolina, North Dakota, Tennessee, Texas, Utah).
(-) Four states regulate truck grouping through truck platooning (Indiana, Kentucky, Mississippi, South Carolina).

An extract with specifications of the legislation in force for the states that have adopted normative acts for autonomous motor vehicles is presented below [80].

7.2.1 Alabama (AL)

Alabama authorizes the full implementation of autonomous vehicles without a human operator on public roads.

AL S 47 (2019) Motor Vehicles [81].
Summary: "Relates to motor vehicles, authorizes autonomous commercial vehicles operated by an automated driving system and commercial motor vehicles with teleoperation systems. Automated commercial motor vehicles, operation of authorized under certain conditions."
AL S 125 (2018) Motor Vehicles [82].
Summary: "Relates to motor vehicles, relates to the prohibition against following a motor vehicle too closely, exempts truck platoons under certain conditions, provides that the intent of this exemption relates to research for truck platooning technology and to exempt the trailing trucks from receiving a citation for following too closely."

7.2.2 Arizona (AZ)

Arizona authorizes full implementation of autonomous vehicles with or without a human operator depending on the level of automation of the vehicle.

AZ H 2813 (2021) Autonomous Vehicles [83].

Summary: "Definitions, operation on public roads, operator requirements relates to autonomous vehicles."

7.2.3 Arkansas (AR)

Arkansas authorizes testing of autonomous vehicles with a human operator, depending on the level of automation of the vehicle.

AR H 1562 (2021) Autonomous Vehicles Regulations [84].
Summary: "Amends the law concerning autonomous vehicles, provides that the program shall include proof that the autonomous vehicle program complies with the minimum liability insurance coverage requirements for a motor carrier of property."
AR H 1561 (2019) Operation of Autonomous Vehicles [85].
Summary: "Authorizes the operation of autonomous vehicles or fully autonomous vehicles on the streets and highways of this state under the Autonomous Vehicle Pilot Program, outlines operating requirements for autonomous vehicles."
AR H 1822 (2019) Operation of Autonomous Vehicles at Railroad Crossings [86].
Summary: "An act concerning the operation of autonomous vehicles at railroad crossings, to declare an emergency, and for other purposes."
AR H 1754 (2017) Testing of Vehicles With Autonomous Technology [87].
Summary: "Regulates the testing of vehicles with autonomous technology, relates to vehicles equipped with driver-assistive truck platooning systems."

7.2.4 California (CA)

California authorizes full implementation of autonomous vehicles without a human operator on public roads.

CA S 500 (2022) Autonomous Vehicles: Zero Emissions [88].
Summary: "This bill, commencing January 1, 2030, and to the extent authorized by federal law, would prohibit the operation of certain new autonomous vehicles that are not zero-emission vehicles, as defined. The bill would also prohibit the Department of Motor Vehicles from commencing rulemaking for the adoption of regulations implementing this provision until January 01, 2027."
CA S 570 (2022) Vehicles: Equipment [89].
Summary: "Exempts autonomous vehicles, as defined, that are not capable of operation with a human driver or occupant in the vehicle, as specified, from certain standards and requirements for the equipment of motor vehicles, provides specified alternative standards for those vehicles, provided that those exemptions or alternative standards are consistent with applicable federal laws or regulations."
CA A 87 (2018) Vehicles: Removal: Autonomous Vehicles [90].

Summary: "This bill would additionally authorize a specified public employee, as specified, to remove a vehicle that uses autonomous technology without a valid permit that is required to operate the vehicle on public roads. The bill would authorize the release of the vehicle after the registered owner of, or person in control of, the autonomous vehicle furnishes the storing law enforcement agency with proof of current registration and a valid driver's license, and either a valid permit that is required to operate the autonomous vehicle using autonomous technology on public roads or a declaration or sworn statement to the Department of Motor Vehicles that states that the autonomous vehicle will not be operated using autonomous technology, as specified. The bill would make additional technical, no substantive changes."

CA S 145 (2018) Autonomous Vehicles: Testing on Public Roads [91].

Summary: "This bill would repeal the requirement that the department notify the Legislature of receipt of an application seeking approval to operate an autonomous vehicle capable of operating without the presence of a driver inside the vehicle. The bill would also repeal the requirement that the approval of such an application not be effective any sooner that 180 days after the date the application is submitted. The bill would require the department to provide public notice when it adopts the autonomous vehicle regulations, would repeal the 120-day prohibition against limiting or expanding the authority to operate autonomous vehicles, and instead would prohibit the department from approving an application submitted pursuant to the regulations until 30 days after public notice of the adopted regulations is provided."

CA A 669 (2018) Department of Transportation: Vehicle Testing [92].

Summary: "Extends the repeal date of provisions authorizing the Department of Transportation to conduct testing of technologies that enable drivers to safely operate motor vehicles with less than 100 feet between each vehicle or combination of vehicles. Prohibits a person from operating a motor vehicle participating in this testing unless the person holds a valid driver's license of the appropriate class for the participating vehicle."

7.2.5 Colorado (CO)

Colorado authorizes full implementation of autonomous vehicles with a human operator on public roads.

CO S 260 (2021) Sustainability of the Transportation System [93].

Summary: "Concerns the sustainability of the transportation system in State, creates new sources of dedicated funding and new state enterprises to preserve, improve, and expand existing transportation infrastructure, develop the modernized infrastructure needed to support the widespread adoption of electric motor vehicles, and mitigate environmental and health impacts of transportation system use, expands authority for regional transportation improvements."

CO S 213 (2017) Automated Driving Motor Vehicles [94].
Summary: "Concerns authorization for automated driving systems to control motor vehicles throughout Colorado, provides that a person may use an automated driving system to drive or control a function of a motor vehicle if the system us capable of complying with every state and federal law that applies to the function that the system is operating, requires the department of transportation to submit a report on the testing of automated driving systems."

7.2.6 Connecticut (CT)

Connecticut authorizes testing of autonomous vehicles with a human operator.

CT S 924 (2019) Department of Motor Vehicles Recommendations [95].
Summary: "Implements recommendations by the Department of Motor Vehicles regarding motor vehicle registration notice, the International Registration Plan, carriers, the Medical Advisory Board, other motor vehicle statutes."
CT S 260 (2017) Autonomous Vehicles [96].
Summary: "Establishes a task force to study fully autonomous vehicles, includes an evaluation of the standards established by the National Highway Traffic Safety Administration regarding state responsibilities for regulating fully autonomous vehicles, relates to evaluation of laws, legislation and regulations proposed or enacted by other states, provides for recommendations on how the state should regulate."

7.2.7 Florida (FL)

Florida authorizes full implementation of autonomous vehicles with or without a human operator depending on the automation level of the vehicle.

FL H 1289 (2021) Autonomous Vehicles [97].
Summary: "An act relating to autonomous vehicles, revising the definition of the term "autocycle", defining the term "low-speed autonomous delivery vehicle", revising the definition of the term "personal delivery device", authorizing the Department of Transportation to adopt rules, authorizing the operation of a low-speed autonomous delivery vehicle on certain streets and roads, providing construction, authorizing the operation of a low-speed autonomous delivery vehicle on streets or roads with a posted speed limit of up to 45 miles per hour under specified conditions, providing requirements for low-speed autonomous delivery vehicles, providing that certain fully autonomous vehicles are not subject to certain provisions of law or regulations, providing that statutory provisions regarding the

authorized use of golf carts, low-speed vehicles, and utility vehicles are not applicable to low-speed autonomous delivery vehicles, conforming cross-references, providing an date."

FL H 107 (2019) Use of Wireless Communications Devices While Driving [98]. Summary: "An act relating to wireless communications while driving; revising legislative intent; requiring a law enforcement officer to inform a motor vehicle operator of certain rights; prohibiting certain actions by such officer; requiring such officer to record the race and ethnicity of a violator when issuing a citation; requiring law enforcement agencies to report such information to the Department of Highway Safety and Motor Vehicles; requiring the department to annually report certain data to the Governor and Legislature; removing the requirement that enforcement be accomplished as a secondary action; defining the term "wireless communications device"; providing legislative intent; prohibiting a person from operating a motor vehicle while using a wireless communications device in a handheld manner in a designated school crossing, school zone, or work zone; providing applicability; providing construction; authorizing a law enforcement officer during a specified period to stop motor vehicles to issue warnings to persons who are driving while using a wireless communications device in a handheld manner in a designated school crossing, school zone, or work zone; providing for repeal of that authorization; authorizing a law enforcement officer, on and after a specified date, to stop motor vehicles and issue citations to persons who are driving while using a wireless communications device in a handheld manner in a designated school crossing, school zone, or work zone; providing exceptions to such prohibition; requiring a law enforcement officer who stops a motor vehicle for a violation of driving while using a wireless communications device in a handheld manner in a designated school crossing, school zone, or work zone to inform the motor vehicle operator of his or her right to decline a search of his or her wireless communications device; prohibiting the law enforcement officer from taking specified actions; requiring certain consent to search a motor vehicle operator's wireless communications device; providing that a user's billing records for a wireless communications device or the testimony of or written statements from certain authorities are admissible as evidence in crashes resulting in death or serious bodily injury for certain purposes; requiring that law enforcement officers indicate specified information in the uniform traffic citation; providing penalties for driving while using a wireless communications device in a handheld manner in a designated school crossing, school zone, or work zone; authorizing first-time offenders to participate in a wireless communications device driving safety program, in lieu of the imposition of penalties; authorizing a clerk of the court to dismiss a case and assess court costs under certain circumstances; requiring the deposit of fines into the Emergency Medical Services Trust Fund of the Department of Health; requiring law enforcement officers to record the race and ethnicity of violators when issuing a citation for a violation of this section; requiring all law enforcement agencies to maintain such information and report it to the Department of Highway Safety and Motor Vehicles in a form and manner determined by the department; beginning on a specified date, requiring the department to annually report the data to the Governor and Legislature; providing requirements for the report; authorizing the department, in consultation with the Department of Transportation, to implement a statewide campaign to raise awareness of and encourage compliance with the prohibitions on operating a motor vehicle while using a wireless communications device; authorizing the department to use certain messaging to implement the campaign; authorizing the department to contract with certain entities for certain purposes; providing effective dates."

FL H 311 (2019) Autonomous Vehicles [99].

Summary: "An act relating to autonomous vehicles; revising and providing definitions; providing applicability; exempting a vehicle being operated with the automated driving system engaged from a prohibition on the active display of television or video; exempting a motor vehicle operator who is operating an autonomous vehicle from a prohibition on the use of wireless communications devices; providing that a licensed human operator is not required to operate a fully autonomous vehicle; authorizing a fully autonomous vehicle to operate in this state regardless of whether a human operator is physically present in the vehicle; requiring the automated driving system to be deemed to be the operator of an autonomous vehicle operating with the automated driving system engaged; providing construction; providing requirements for operation of on-demand autonomous vehicle networks; providing insurance requirements; authorizing an autonomous or fully autonomous vehicle equipped with a teleoperation system to operate without a human operator physically present in the vehicle when the system is engaged; providing application to certain statutory provisions; providing for uniformity of laws governing autonomous vehicles; providing construction with respect to certain fees charged and staging or pickup locations designated by an airport or seaport; revising requirements for autonomous vehicles registered in this state; providing applicability; authorizing the Florida Turnpike Enterprise to enter into agreements to fund, construct, and operate certain facilities; conforming provisions to changes made by the act; providing definitions; providing insurance requirements for autonomous vehicles; conforming a cross-reference; providing an effective date."

FL H 725 (2019) Commercial Motor Vehicles [100].

Summary: "An act relating to commercial motor vehicles; revising regulations applicable to owners and drivers of commercial motor vehicles; exempting persons who operate a commercial motor vehicle solely in intrastate commerce which does not transport hazardous materials in amounts that require placarding from certain requirements related to electronic logging devices and hours of service supporting documents until a specified date; deleting a limitation on a civil penalty for falsification of certain time records; deleting a requirement that a motor carrier maintain certain documentation of driving times; extending an exemption from specified commercial motor vehicle requirements for a commercial vehicle having a certain gross vehicle weight rating and gross combined weight rating, under certain circumstances; deleting such exemption for a person transporting petroleum products; deleting an exemption from specified regulations relating to diabetes for certain drivers of commercial motor vehicles; revising length and load extension limitations for stinger-steered automobile transporters; authorizing automobile transporters to backhaul certain cargo or freight under certain circumstances; authorizing an unladen power unit to tow a certain combination of trailers or semitrailers under certain circumstances; providing for the calculation of specified fines for vehicles fueled by electric batteries; requiring the Department of Transportation in conjunction with the Department of Highway Safety and Motor Vehicles to develop, by a specified date, a permitting program that authorizes the operation of any combination of truck tractor, semitrailer, and trailer combination coupled together so as to operate as a single unit, subject to certain requirements; providing that the permitting program expires in five years unless reauthorized by the legislature; prohibiting a permitting program from being implemented or continued under certain circumstances."

7.2.8 Georgia (GA)

Georgia authorizes full implementation of autonomous vehicles based on the vehicle's automation level.

GA S 165 (2022) Motor Vehicles [101].
Summary: "Relating to the annual license fees for operation of vehicles, so as to reduce the registration fee charged to low-speed alternative fueled vehicles; relating to equipment and inspection of motor vehicles, so as to exempt certain autonomous vehicles from certain equipment standards; to require the display of a reflective emblem or the use of strobe light while operating a slow-moving or low-speed vehicle; to provide for issuance of a permit relating to operation of an amber strobe light upon a low-speed vehicle; to provide for related matters; to provide for an effective date and applicability; to repeal conflicting laws; and for other purposes."
GA S 219 (2018) Automated Driving Systems [102].
Summary: "Relating to motor vehicles, so as to provide for definitions; to exempt persons operating a fully autonomous motor vehicle with the automated driving system engaged from the requirement to hold a driver's license; to provide for satisfaction of requirement to notify law enforcement in certain instances of accidents by fully autonomous motor vehicles; to provide for certain equipment and insurance requirements for fully autonomous motor vehicles; to provide for registration requirements for such vehicles; to provide for exemptions from seat belt requirements; to provide for applicability; to provide for limitations on adoption of certain rules and regulations; to provide for related matters; to repeal conflicting laws; other purposes."
GA H 472 (2018) Following Requirements for Motor Vehicles [103].
Summary: "Relates to motor vehicles and traffic, provides for an exception for following requirements for vehicles following in a procession when speed of the non-leading, participates vehicles are coordinated automatically, provides for related matters, repeals conflicting laws. To amend Title 40 of the Official Code

of Georgia Annotated, relating to motor vehicles and traffic, so as to provide for an exception for following requirements for non-leading vehicles following in a coordinated platoon; to provide for a definition; to provide for related matters; to repeal conflicting laws; other purposes."

7.2.9 Hawaii (HI)

Hawaii authorizes testing of autonomous vehicles.

HI HCR 220 (2019) Autonomous Vehicle Legal Preparation Task Force [104].
Summary: "Requests the Attorney General to convene an Autonomous Vehicle Legal Preparation Task Force."

7.2.10 Illinois (IL)

Illinois authorizes testing of autonomous vehicles with a human operator.

IL H 791 (2018) Vehicle Code [105].
Summary: "Amends the Vehicle Code, provides that a unit of local government, including a home rule unit, may not enact an ordinance prohibiting the use of autonomous vehicles on its roadways, limits the concurrent exercise of home rule Powers, defines autonomous vehicle."

7.2.11 Iowa (IA)

Iowa authorizes full implementation of autonomous vehicles without a human operator on public roads.

IA S 302 (2020) Automated Driving System Motor Vehicles [106].
Summary: "Establishes regulations for motor vehicles operated by an automated driving system. Relating to motor vehicles operated by an automated driving system and making penalties applicable."

7.2.12 Kentucky (KY)

Kentucky regulates truck grouping through truck platooning.

KY S 116 (2018) Platoon Operation of Commercial Motor Vehicles [107].
Summary: "Sets forth requirements for commercial motor vehicles to operate as a platoon, defines the term platoon as a group of commercial vehicles traveling in a unified manner at electronically coordinated speeds at following distances

that are closer than would ordinarily be allowed, defines a prearranged ride with a transportation network."

7.2.13 Louisiana (LA)

Louisiana authorizes full implementation of autonomous vehicles without a human operator on public roads.

LA H 455 (2019) Autonomous Commercial Motor Vehicles [108].
Summary: "Relative to autonomous commercial motor vehicles; to provide for definitions; to establish the controlling authority for autonomous commercial motor vehicles; to provide relative to applications to operate an autonomous commercial motor vehicle; to provide relative to requirements of operators; to establish the reporting requirements following an accident; to provide relative to remote drivers and teleoperation systems; to provide relative to liability and jurisdiction; and to provide for related matters."
LA H 308 (2018) Non-Lead Motor Vehicles In a Platoon [109].
Summary: "Authorizes non-lead motor vehicles in a platoon to follow other motor vehicles in a platoon closely; to exempt non-lead motor vehicles in a platoon from operating such vehicle in a manner that allows sufficient space to enable any other vehicle to enter and occupy the space between any motor vehicle in a platoon; to authorize platoon operation upon approval of an operational plan by the Department of Public Safety and Corrections and the Department of Transportation and Development; to provide for rulemaking authority; to provide for a prohibition against platoon operation; to provide for definitions; and to provide for related matters."

7.2.14 Maine (ME)

Maine authorizes the testing of autonomous vehicles.

ME H 1204 (2018) Use Of Autonomous Vehicles For Public Transportation [110].
Summary: "Establishes the Commission on Autonomous Vehicles to coordinate efforts among state agencies and knowledgeable stakeholders, relates to the development of a process to allow an autonomous vehicle tester to demonstrate an automated driving system on a public way."

7.2.15 Maryland (MD)

Maryland authorizes the testing of autonomous vehicles.

MD S 291 (2019) Motor Vehicle Offenses—Following Too Closely—Unified Truck Travel [111].
Summary: "For the purpose of establishing that certain motor vehicle offenses related to following too closely do not apply to a group of individual trucks that are traveling in a unified manner with electronically coordinated speed and braking systems and are being operated in a certain manner; requiring the Motor Vehicle Administration to adopt certain regulations governing truck travel in a unified manner; and generally relating to the application to trucks of motor vehicle offenses related to following too closely."

7.2.16 *Mississippi (MS)*

Mississippi regulates truck grouping through truck platooning.

MS H 1343 (2018) Motor Vehicle Laws [112].
Summary: "Defines the term platoon, exempts certain vehicle operators in a platoon from provisions of law requiring certain distances to be maintained between traveling vehicles, requires plans for the operation of a platoon to be filed with the Department of Transportation; and for related purposes."

7.2.17 *Nebraska (NE)*

Nebraska authorizes full implementation of autonomous vehicles with or without a human operator on public roads.

NE L 989 (2018) Autonomous Vehicle Testing [113].
Summary: "A bill for an act relating to transportation; to authorize automated-driving- system-equipped vehicles, automated driving systems, driverless-capable vehicles, and on-demand driverless-capable vehicle networks as prescribed; to provide requirements, restrictions, and limitations; to provide powers and duties for the Department of Motor Vehicles; and to define terms."

7.2.18 *Nevada (NV)*

Nevada authorizes full implementation of autonomous vehicles without a human operator on public roads.

NV S 288 (2021) Transportation Network Company Agreements [113].
Summary: "An act relating to transportation; authorizing a monitored autonomous vehicle provider to enter into an agreement with a transportation network company

to provide transportation services through the digital network or software application of the company; imposing certain requirements on a transportation network company and monitored autonomous vehicle provider relating to the provision of transportation services by a monitored autonomous vehicle provider; authorizing a transportation network company to charge a fare for such services on behalf of a monitored autonomous vehicle provider; prohibiting a local governmental entity from imposing certain taxes or fees relating to such services; revising provisions relating to transportation network company insurance; and providing other matters properly relating thereto."

NV A 412 (2021) Fully Autonomous Vehicles [114].

Summary: "Relates to motor vehicles, revises provisions governing fully autonomous vehicles, provides other matters properly relating thereto."

NV A 69 (2017) Use of an Autonomous Vehicle [115].

Summary: "An act relating to transportation; revising requirements for the testing or operation of an autonomous vehicle on a highway within this State; authorizing the use of driver-assistive platooning technology; authorizing the use of a fully autonomous vehicle to provide transportation services in certain circumstances by persons licensed by the Department of Motor Vehicles, Nevada Transportation Authority or Taxicab Authority; providing for the regulation of autonomous vehicle network companies; providing penalties; and providing other matters properly relating thereto."

7.2.19 New Hampshire (NH)

New Hampshire authorizes testing of autonomous vehicles.

NH S 216 (2020) Automated Vehicle Testing and Deployment Commission [116].

Summary: "Establishes an Automated Vehicle Testing and Deployment Commission and an automated vehicle testing pilot program, requires the State transportation council to conduct a review of automated vehicle testing and deployment, establishing an automated vehicle testing pilot program, and providing requirements for automated vehicle deployment."

7.2.20 New Jersey (NJ)

New Jersey authorizes a study on autonomous vehicles.

NJ AJR 164 (2020) Advanced Autonomous Vehicle Task Force [117].

Summary: "This joint resolution establishes the "New Jersey Advanced Autonomous Vehicle Task Force" (task force). The task force is to conduct a study of advanced autonomous vehicles and to make recommendations on laws, rules, and regulations that this State may enact to safely integrate advanced autonomous vehicles on the State's highways, streets, and roads. In this bill, an advanced autonomous vehicle means a motor vehicle equipped with autonomous technology that has a driving automation level of three, four, or five, as defined in the SAE J3016, which is an information report related to automated driving systems issued by the Society of Automotive Engineers International and is used by the United States Department of Transportation for autonomous vehicle policy guidance. The bill requires the task force to issue a report to the Governor and Legislature, which is to include, but not be limited to, an evaluation of the National Highway Transportation Safety Administration's safety standards for advanced autonomous vehicles and whether this State may enact stricter safety standards, an evaluation of existing state and federal law concerning advanced autonomous vehicles with a focus on safety standards, an evaluation of existing legislation and regulations in other states concerning advanced autonomous vehicles with a focus on safety standards, recommendations on how New Jersey could safely integrate advanced autonomous vehicles on the highways, streets, and roads of this State, any other information relevant to the subject of the report, and any draft legislation the task force deems appropriate to implement the purposes of this bill. The bill states that the task force is to expire upon the issuance of the report to the Governor and to the Legislature."

7.2.21 New Mexico (NM)

New Mexico authorizes testing of autonomous vehicles.

NM H 270 (2021) Autonomous Vehicles [118].
Summary: "Relates to motor vehicles, adds definitions to the motor vehicle code, provides for autonomous motor vehicles, requires permits, allows platooning of motor vehicles."
NM SJM 3 (2018) Autonomous Vehicles In New Mexico [119].
Summary: "Requests the department of transportation to create a committee that includes relevant state agencies and private entities to review the current and developing state of autonomous vehicle technology and develop a proposal to allow autonomous vehicle to use in New Mexico while ensuring public safety."

7.2.22 North Carolina (NC)

North Carolina authorizes full implementation of autonomous vehicles without a human operator on public roads.

NC H 814 (2022) Neighborhood Occupantless Vehicles [120].
Summary: "Authorizes the operation of neighborhood occupantless vehicles, provides for restrictions, regulations, and equipment requirements, defines neighborhood occupantless vehicle."

NC H 469 (2018) Fully Autonomous Motor Vehicles [121].
Summary: "Amends the General Statutes, establishes regulations for the operation of fully autonomous motor vehicles on public highways of this state, establishes the Fully Autonomous Vehicle Committee."
NC H 716 (2018) Use of Platoons [122].
Summary: "Makes an amendment, relates to any non-leading commercial motor vehicle. An act to provide that the laws regulating the distance between motor vehicles traveling on the road do not apply to any non-leading commercial motor vehicle traveling within a platoon where authorized by the Department of Transportation."
NH H 469 (2017) Fully Autonomous Motor Vehicles [123].
Summary: "Amends the General Statutes, establishes regulations for the operation of fully autonomous motor vehicles on public highways of this state, establishes the Fully Autonomous Vehicle Committee."

7.2.23 North Dakota (ND)

North Dakota authorizes full implementation of autonomous vehicles with or without a human operator on public roads.

ND H 1199 (2019) Motor Vehicle Road Rules [124].
Summary: "Relates to following a motor vehicle too closely, relating to platoons, and to provide a penalty."
ND H 1418 (2019) Automated Vehicle Network Companies [125].
Summary: "Relates to automated vehicle network companies and autonomous vehicle operations, relates to nonmoving violations, relates to unattended motor vehicles."
ND H 1202 (2017) Operation and Regulation of Autonomous Vehicles [126].
Summary: "Relates to the operation and regulation of autonomous vehicles, requires authorization to operate an automated driving system equipped vehicle on a highway in this state, regulates the integration and testing of automated driving system equipped vehicles on the public highway system."

7.2.24 Ohio (OH)

Ohio authorizes testing of autonomous vehicles without human operator.

OH H 26 (2017) Transportation and Public Safety Appropriations [127].
Summary: "Makes appropriations for programs related to transportation and public safety for the biennium for a specified time frame, provides authorization and conditions for the operation of those programs."

7.2.25 Oregon (OR)

Oregon authorizes a study on autonomous vehicles.

OR H 4063 (2018) Autonomous Vehicle Operation [128].
Summary: "Permits operation of autonomous vehicles on highways in this state under certain circumstances, directs owners of commercial autonomous vehicles to obtain additional motor vehicle liability insurance policies, directs Department of Transportation to adopt rules for operation of autonomous vehicles on highways in this state, takes effect on 91st day following adjournment sine die. Declares emergency, effective on passage."

7.2.26 Pennsylvania (PA)

Pennsylvania authorizes testing of autonomous vehicles with a human operator.

PA H 1958 (2018) Autonomous Vehicles [129].
Summary: "Amends statutes relating to vehicles, provides additional rules of the road in general, relates to platooning, provides for highly automated vehicles."

7.2.27 South Carolina (SC)

South Carolina regulated truck grouping through truck platooning.

SC H 3289 (2018) Distance Maintained Between Vehicles on Highway [130].
Summary: "Relating to the distance that must be maintained between vehicles traveling along a highway, so as to revise the term "driver" to "operator" in regard to these vehicles, and to provide that this section does not apply to the operator of any nonleading commercial motor vehicle subject to federal motor carrier safety regulations and traveling in a series of commercial vehicles using cooperative adaptive cruise control or any other automated driving technology."

7.2.28 Tennessee (TN)

Tennessee authorizes full implementation of autonomous vehicles without a human operator on public roads.

TN S 151 (2018) Motor Vehicles [131].
Summary: "Relates to motor vehicles, authorizes the operation of autonomous vehicles on the public roads of this state, provides for related requirements, including insurance requirements."
TN S 676 (2018) Operate a Platoon on the Streets and Highways [132].

Summary: "Relates to motor vehicles, authorizes a person to operate a platoon on the streets and highways of this state after the person provides notification to the Department of Transportation and the Department of Safety."

7.2.29 Texas (TX)

Texas authorizes full implementation of autonomous vehicles without a human operator on public roads.

TX S 1308 (2021) Study on Impacts of Certain Motor Vehicle Technologies [133].
Summary: "Relates to a study on the impacts of using certain motor vehicle technologies."
TX H 3026 (2021) Operation and Regulation of Automated Motor Vehicles [134].
Summary: "Relates to the operation and regulation of certain automated motor vehicles."
TX H 1791 (2017) Connected Braking Systems [135].
Summary: "Relates to the use of connected braking systems to maintain distance between vehicles."
TX S 2205 (2017) Automated Motor Vehicles [136].
Summary: "Relates to automated motor vehicles, Relates to the research and testing of motor vehicles with automated driving systems on highways in this state."

7.2.30 Utah (UT)

Utah authorizes full deployment of autonomous vehicles without a human operator on public roads.

UT H 101 (2019) Autonomous Vehicle Regulations [137].
Summary: "Amends provisions regarding traffic laws, licensing, and titling requirements, and adds provisions regarding the operation of autonomous vehicles, defines terms related to autonomous vehicles, allows the operation of a vehicle in the state by an automated driving system, exempts a vehicle with an engaged automated driving system from licensure, provides protocol in case of an accident involving an autonomous vehicle."
UT S 56 (2018) Vehicle Platooning Amendments [138].
Summary: "Modifies provisions of the Traffic Code related to safe following distance."

7.2.31 *Vermont (VT)*

Vermont authorizes testing of autonomous vehicles with human operator.

VT S 149 (2020) Vehicles [139].
Summary: "Makes miscellaneous changes to laws related to vehicles, establishes an automated vehicle testing program, requires the Department of Motor Vehicles to provide written forms, applications, and tests translated into certain languages and to allow an applicant to have an individual of his or her choosing at the oral examination or road test required for licensing to serve as an interpreter, requires the Commissioner of Motor Vehicles to provide training on cultural differences to employees."

7.2.32 *Washington (WA)*

Washington authorizes testing of autonomous vehicles without a human operator.

WA S 5460 (2022) Recommendations of the Autonomous Vehicle Work Group [140].
Summary: "Implements recommendations of the autonomous vehicle work group, defines autonomous, removes a provision that prevents a person from driving a vehicle equipped with certain video screens that are visible to the driver."
WA H 2676 (2022) Autonomous Vehicles Testing [141].
Summary: "Establishes minimum requirements for the testing of autonomous vehicles."
WA H 2970 (2018) Autonomous Vehicle Work Group [142].
Summary: "Establishes an autonomous vehicle work group to develop policy recommendations to address the operation of autonomous vehicles on public roadways in the state."

7.2.33 *Wisconsin (WI)*

Wisconsin authorizes a study on autonomous vehicles without a human operator.

WI S 695 (2018) Motor Vehicle Distances [143].
Summary: "Revises provisions relating to distances between motor vehicles, makes technical corrections."

The number of states that have adopted the legislation on autonomous vehicles has gradually increased in recent years. Figure 7.2 shows the current situation in the United States [1, 78, 79, 144, 145]. Legend: blue—full implementation of autonomous vehicles, grey—testing with a human operator, yellow—testing without a human operator, green—study on autonomous vehicles, orange—truck platooning.

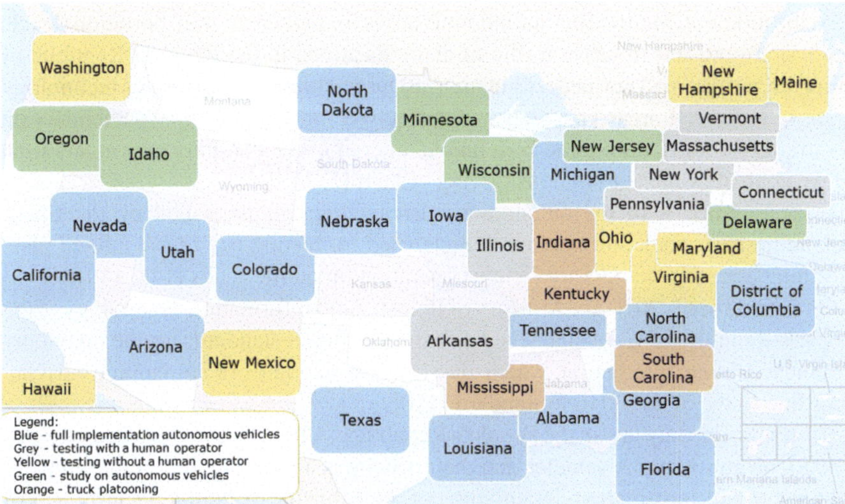

Fig. 7.2 United States and the legislation on autonomous driving

The concept of autonomous vehicle is perceived and defined differently for each of the American states. Most US states make the testing of autonomous vehicles conditional on the demonstration of the financial stability of the operator performing the tests, and the existence of insurance to cover any incidents that may occur during their entire operation period.

7.3 People's Republic of China Legislation

In the People's Republic of China, government authorities have been issuing local regulations for the testing of autonomous vehicles since 2017 and promulgated the "National Road Testing Guide" strategic plan to modernize the car industry by 2025 [1, 18]. The first autonomous Shuttle Bus (Baidu), which was launched for testing in Beijing in March 2018 had a 105-km-long public road in the suburbs, away from heavy traffic areas.

On February 10, 2020, 11 ministries of the People's Republic of China government jointly enacted the Innovative Development Strategy of Intelligent Vehicle. This strategy aims for innovative technologies to work together for the development of autonomous and intelligent vehicles by 2025, so that by 2035, a fully intelligent vehicle can be developed and operational. To this end, the Chinese state will issue policies to promote and develop autonomous driving and will support the research, development and industrialization of intelligent technologies and infrastructure for vehicle, which will be integrated as major projects in the smart city [146].

At the moment, Shuttle Bus autonomous buses for passenger transportation (levels 3 and 4) are present in urban traffic in many locations in the People's Republic of China: on public roads or pedestrian areas in large cities, in software or technology parks, in industrial areas [1]. They travel freely or on request, with passengers on pre-arranged routes, monitored by a human operator who can intervene at any time to take control of the autonomous driving system [1, 147].

In accordance with the autonomous driving regulations issued by the Shenzhen authorities, each autonomous vehicle used for testing must have a fixed plate informing about the temporary right of operation on public roads [148, 149].

On September 3, 2020, the Ministry of Industry and Information Technology of the People's Republic of China launched a draft of national standards on the taxonomy of vehicle driving automation, which classifies the standards of autonomous driving according to the SAE J3016™ standard [150].

Currently, the only law in the People's Republic of China that was enacted on April 3, 2020, on autonomous vehicles and their driving is a joint circular of the Ministry of Industry and Information Technology, of the Ministry of Public Security and of the Ministry of Transportation of People's Republic of China on Issuing the Norms on the Administration of Road Testing of Intelligent Connected Vehicles (ICV Road Test Norms for Trial Implementation). An excerpt with the main specifications of the standards contained in the ICV Road Test Norms is presented below [151]:

Chapter II Test Subject, Test Driver, and Test Vehicle.

Article 5 "The subject of the test refers to the organizations that organize the test, assume the relevant responsibilities, and comply with the following conditions:

(I) Legal entity registered in the People's Republic of China.

(II) Possesses business capabilities: manufacturing, research, technical development, and testing of the autonomous vehicle.

(III) Financial capacity to materially bear civil liability in the event of any possible personal and/or material damage that may occur during the testing of the autonomous vehicle.

(IV) Develops the methodology for conducting and evaluating tests for the autonomous driving function of the autonomous vehicle.

(V) Ability to perform remote and real-time monitoring of the tested autonomous vehicle.

(VI) Ability to record, analyze and reproduce test events."

Article 6 "The driver who operates the autonomous vehicle throughout the tests has responsibilities for the implementation of emergency measures in case of critical situations and must meet the following conditions established by the legislation in force (employment contract, driving license, driving experience, no criminal record etc.)."

Article 7 "The vehicle used for the test is an intelligent connected vehicle that must meet the following conditions:

(I) Proof of registration/exemption of the autonomous vehicle.

(II) Mandatory technical inspection.

(III) Manual (operator) and autonomous (system) operating capability.

(IV) Ability to monitor, record and save and operate throughout the real-time testing. The following data will be recorded and saved:
(1) Vehicle control mode (manual or autonomous).
(2) Travel route coordinates.
(3) Vehicle velocity and acceleration.
(4) Environmental conditions.
(5) Real-time signals from the vehicle.
(6) 360° video monitoring outside the vehicle.
(7) Audio–video monitoring inside the vehicle.
(8) Remote control of the vehicle.
(9) Situations of malfunction of the vehicle.
(V) The testing of the autonomous vehicle in real conditions must be carried out in a protected area (road closed to public traffic) in accordance with the relevant national and industry standards.
(VI) The autonomous driving function of the vehicle used for the performance of the tests must comply with the provisions of Annex 1."
Chapter III Test Application and Review
Article 13 "The vehicle used for testing must be temporarily registered by the competent authorities.
Article 14 The test area must be restricted and properly arranged."
Chapter IV Test Management
Article 18 "The driver who operates the autonomous vehicle throughout the tests must be permanently in the autonomous vehicle, to monitor the operating condition of the vehicle and the environment and be prepared to take control of the vehicle at all times. When the driver finds that the autonomous vehicle is no longer safe to remain in the state of autonomous driving or if the system requires manual operation, the driver must take control of the vehicle.
Article 23 When government authorities consider that the tests performed on the autonomous vehicle pose a risk to the safety of the test participants, the driving license of the autonomous vehicle must be suspended or revoked."
Chapter V Traffic Violations and Accident Handling
Article 25 "In the event of traffic accidents during the tests, the liability of the test organizers will be established in accordance with the laws and regulations on road safety and compensation for damages will be decided in accordance with applicable law."
Chapter VI Supplementary Provisions
Article 28 "The autonomous vehicle of the Intelligent Connected Vehicle type referred to in this circular is a new generation of vehicle, equipped with advanced generation sensors, with intelligent communication technology that allows the sharing of information between the vehicle and everything (X) (where X can be it other vehicles, infrastructure, people etc.), with environmental perception systems and is able to make intelligent decisions about its operation and movement. Autonomous driving (conditionally autonomous and fully autonomous) refers to the system that can partially or completely replace the driver and is able

to complete all driving operations on the road instead of the driver and without his intervention."

Annex 1 Testing Projects of Autonomous Driving Function of ICV:

(1) Identification of traffic signs and route marking lines.
(2) Identification of traffic lights.
(3) Identification of the driving condition of the vehicle in front (including the one in the opposite direction).
(4) Identification of obstacles.
(5) Identification of pedestrians and non-motorized vehicles.
(6) Tracking of the vehicle moving in front (stop/start).
(7) Parking on the side of the road.
(8) Overtaking.
(9) Driving on mixed lanes.
(10) Passing through an intersection.
(11) Passing through a roundabout.
(12) Automatic emergency braking.
(13) Taking control manually.
(14) Network communication.

In recent years, the local governments of some provinces of the People's Republic of China have issued regulations allowing the development of autonomous vehicle test routes in pilot areas (Fig. 7.3) over distances of up to 100 km on routes equipped with 5G communication infrastructure, allowing shuttle bus operation with passengers on board [152].

(1) The Beijing-Xiong'an express way is a 100-km expressway that connects Beijing with the neighboring technology city of Xiong'an and it has two inner lanes for testing autonomous vehicles. The express way has been operational since May 29, 2021 and is equipped with charging stations and induction lanes for electric vehicles [153].
(2) Cangzhou, Hebei province is an urban road network of over 400 km long open for testing with Apollo Go Robotaxi autonomous vehicles starting from November 2020. Apollo Go Robotaxi service currently covers 55 boarding stations in Cangzhou, including railway stations, schools, hotels, museums, business areas, industrial areas, and other public spaces [154].
(3) Wuhan economic and technological development zone has developed an autonomous vehicle testing area of more than 100 km, which by the end of 2022 will extend the tests to the city's central urban areas, implementing autonomous driving buses, sanitation services and Robotaxi [155].
(4) Changsha, Hunan became the first province of the People's Republic of China to launch an autonomous passenger shuttle bus service in April 2020 with Apollo Go Robotaxi autonomous vehicles on a demonstration route covering an area of 130 km^2 and having 22 passenger boarding stations [156].

The legal provisions on the protection of personal data are governed by the law on the protection of personal data of the People's Republic of China. The principle

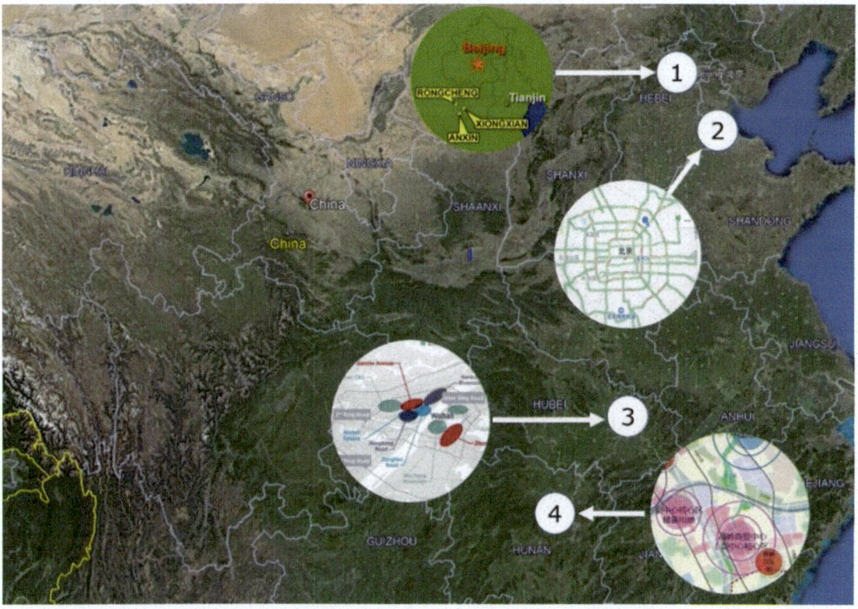

Fig. 7.3 Test routes for autonomous vehicles in pilot areas

of personal data protection is included in the People's Republic of China General Rules of Civil Law which entered into force on October 01, 2017 and which provides for the obligation of any organization or person who obtains personal information of other persons to ensure the security of such information lawfully and not to collect, use, process or transmit the personal information of others, nor to purchase, sell, supply, or publish the personal information of others.

Personal data and critical data collected and generated by infrastructure operators in the People's Republic of China may be stored only in the People's Republic of China and may not be transferred abroad unless the data export has passed a security assessment by the competent Chinese authority [157].

The movement of autonomous vehicles generates digital maps based on the location data provided by the sensors. In the People's Republic of China, data mapping, digital mapping can only be done by a Chinese organization authorized by the Ministry of Natural Resources [158].

7.4 Commonwealth of Australia Legislation

Autonomous vehicles offer the opportunity to fundamentally change the concept of transportation by improving road safety, increasing the mobility and productivity of freight transportation and by reducing traffic congestion. Autonomous vehicles (see Table 1.2 Autonomous shuttle buses for public transportation) are currently operating in Australia for experimental purposes, without any legislation to allow these vehicles to be put on public roads [159].

The National Transport Commission of Australia published on September 1, 2020 the document Automated Vehicle Program Approach which presents the current program for the implementation of autonomous vehicles, respectively the planned legislative reforms to develop a set of unitary normative acts to regulate the safe operation of autonomous vehicles throughout Australia. The key terms that the National Transport Commission considers important to be implemented in future legislation are as follows [160]:

(i) Automated Driving System (ADS) is the autonomous driving system made of hardware and software that has the ability to collectively perform the dynamic driving task in accordance with SAE J3016™ rules (level 3, level 4 and level 5).
(ii) Automated Driving System Entity (ADSE) is the entity owning ADS and seeking the approval of the Commonwealth of Australia Government to certify the use of this system.
(iii) Automated Vehicle is the autonomous vehicle with the level of automation between level 3 and level 5, without the intervention of a human operator.
(iv) Conditional Automation (SAE Level 3) is the autonomous driving function for which ADS assumes the entire dynamic driving task for sustained periods under defined circumstances.
(v) Dynamic Driving Task consists of the operational and tactical functions necessary to drive the autonomous vehicle in road traffic.
(vi) Fallback-Ready User represents the human operator, who is in the vehicle and who is receptive to all ADS requests to intervene to take control of the vehicle in critical situations.
(vii) Full Automation (SAE level 5) is the autonomous driving function for which ADS assumes the entire dynamic driving task for any situation and at any time.
(viii) High Automation (SAE level 4) is the autonomous driving function for which ADS assumes the entire dynamic driving task for sustained periods in certain situations in defined places.
(ix) In Service Safety is the degree of safety of the autonomous vehicle in operation on the public roads.

The most important studies and publications developed by the Australian authorities to facilitate the adoption of autonomous vehicle legislation are the following:

(-) Minimum Physical Infrastructure Standard for the Operation of Automated Driving, published on January 25, 2022 which includes guidance on the necessary investments in infrastructure to create an environment suitable for the safe operation of autonomous vehicles [161].
(-) Agency Business Capability Model to Support Connected Vehicles, published on January 12, 2022 which presents a model for the classification and management

of priority data sets held by government agencies that can support the implementation and operation of autonomous and connected vehicles: management and security of databases, monitoring of the compliance with the principles of ethics of the data collected, the architecture, models and design of the data collected, the storage, integration and interoperability of the data collected, respectively the rules of use of the data collected [162].

(-) Future Vehicles Forecasts Update 2031 (update for Future Vehicles 2030), published on September 06, 2021 which presents an analysis of new and emerging technologies in the field of autonomous driving, provides a forecast of the evolution of these technologies in the next 10 years. The main attributes pursued in this analysis refer to the active safety systems, the autonomous driving ability in traffic, cooperative connectivity C-ITS (Cooperative Intelligent Transport Systems) etc. [163].

(-) Guidelines for tests of automated vehicles in Australia, published on December 2, 2020, presents an analysis of the criteria to be met in authorizing the use of autonomous vehicles for experimental purposes on public roads in Australia based on regulations published by the National Transport Commission in 2017.

As evidenced by the studies and analyzes conducted by Australian authorities in recent years, autonomous driving is largely based on C-ITS co-operative driving. C-ITS systems will allow both intelligent vehicles to communicate with each other in real time and to communicate with the smart and interconnected infrastructure, constantly and consistently generating data with maximum implications for reducing the risks to which the vehicles are exposed in road traffic [164].

Big Data Control presents complex legal and regulatory issues for contributors in Australia's transportation infrastructure space. These issues are related to the confidentiality of vehicle data, the elimination of cyber security threats and the creation of a legislative framework for the civil insurance of autonomous vehicles.

Data generated by the autonomous vehicles, processed, and shared by C-ITS will be classified as personal information containing data on the current location of a vehicle, respectively on its history of travel and personal identification data of the human operator accompanying the autonomous vehicle (fingerprint and/or facial print etc.).

According to the Privacy Act of 1988 [165] this data is confidential and must be protected. The Privacy Act prohibits the use and disclosure of personal information generated by autonomous vehicles.

Australian Privacy Principles guidelines in Chapter 3 APP 3 Collection of solicited personal information [166] specifies that personal information may be collected from a person only with the consent of that person and only if it is necessary to enforce the applicable law.

According to Chapter 5 APP 5 Notification of the collection of personal information [167] any person must be notified if an entity collects personal information about him/her.

Future legislation on the use of data collected from autonomous vehicles will need to be harmonized to comply with these regulations.

Cybercrime is regulated in Australia by the Criminal Code Act 1995 which criminalizes activities such as unauthorized access to computer systems and DoS attacks [168]:

Division 478 Other computer offences
Section 478.1 "Unauthorized access to, or modification of, restricted data, restricted data means data (a) held in a computer; and (b) to which access is restricted by an access control system associated with a function of the computer."
Section 477.3 "Unauthorized impairment of electronic communication and unauthorized modification of data to cause impairment."

7.5 Japan Legislation

In Japan, autonomous driving at level 3 on public roads is regulated according to SAE J3016™ through the revised versions of two normative acts:

The Road Vehicles Act, 1951-Law No. 185 [169] which establishes safety and pollution control requirements for motor vehicles, and also establishes a motor vehicle registration and inspection system.
Road Traffic Act, 1960-Law No. 105 [170] which provides for traffic control by signals, road signs and traffic rules and provides requirements for the motor vehicle drivers.

The amendments to the two laws came into force on April 1, 2020, with Japan practically being the first country in the world to approve the use of level 3 autonomous vehicles (according to SAE J3016™) on public roads. The Ministry of Land, Infrastructure, Transport and Tourism (Japan) has issued an authorization for operation for the Legend model produced by Honda in 100 copies starting with November 2020 [171].

The Road Vehicles Act, 1951-Law No. 185 (revised) defines the ADS autonomous driving system as consisting of a set of sensors managed by AI artificial intelligence that replaces the driver with all his abilities and records videos while functioning.

The same normative act mentions the mandatory technical specifications for a vehicle equipped with ADS technology, as they were adopted as international standards at the 181st session of The UNECE World Forum for Harmonization of Vehicle Regulations (WP.29) from June 2020 [172].

With regard to the updating of software applications used by an autonomous driving system, this can only be done with the approval of relevant Japanese entities in the field. While for level 0/1/2 vehicles (according to SAE J3016™) the legislation recommends vehicle manufacturers and importers to provide users with the technical specifications necessary for technical inspections and maintenance, in the case of autonomous vehicles level 3 vehicles this is mandatory.

Level 3 autonomous driving supposes the existence of conditions of use established by the Ministry of Land, Infrastructure, Transport and Tourism (Japan) in the Guideline regarding Safety Technology for Automated Vehicles in Japan [173].

These conditions of use are those related to the route used by the autonomous vehicles (highway or other type of road, number of lanes per direction, road intended exclusively for autonomous vehicles etc.), geographical conditions (urban, mountainous area etc.), environmental conditions (weather conditions, night driving restrictions etc.) and others (speed limits, infrastructure adjustment, driving license category limitations etc.). Without meeting these conditions of use, the autonomous driving system must not be activated, the user can only operate up to level 2 of automation in which he has to make all decisions on maneuvers performed in road traffic, being solely responsible in the event of an accident. The level 3 autonomous driving characteristic supposes the existence of a human operator in the vehicle, the liability for any road event with material damage or victims falling on both the autonomous driving system and the human operator depending on the cause of this incident. Thus, in case of a system error the manufacturer will be liable, and in case of a human error, the operator.

If the autonomous driving system is activated, and during its movement it finds, by means of sensors and Artificial Intelligence, that one of the conditions of use is no longer met, the system will warn the operator that he/she must take control of the vehicle.

It is essential that the human operator be able to take control of the vehicle at all times, even if Japanese law has removed restrictions on the use of mobile phones or other electronic devices during the movement. From the moment the operator takes control of the autonomous vehicle, he/she becomes fully responsible for its driving. In order to determine the party responsible in the event of an accident, Japanese law provided for the obligation to permanently record the movement of the vehicle. If the trip was not recorded or deleted, it is assumed that the operator is responsible in the event of an accident and not the manufacturer, even if there was a system error.

Automobile Liability Security Act, 1955-Law No. 97 [174] obliges vehicle owners to take out liability insurance and provides government assistance to victims of uninsured drivers.

The Road Act, 1952-Law No. 180 [175] establishes construction and maintenance standards for the road infrastructure. The law establishes the government authorities in charge of the road infrastructure management.

Level 4 autonomous driving is permitted under the Road Traffic Act, but only for testing experimental autonomous vehicle models with the prior approval of the National Police Agency. The Agency published the rules for approval in September 2019 (for example, the velocity of the autonomous vehicle must not exceed 20 km/h, operators will be required to have capabilities to monitor the autonomous vehicles remotely, to assign the monitoring staff and to implement a system that can move the monitoring personnel to the vehicle in the event of an accident or other emergencies [176]).

7.6 Insuring Autonomous Vehicles

According to a study by Accenture in conjunction with the Stevens Institute of Technology in [177], autonomous vehicle insurance is a $81 billion annual opportunity given the rapid development of the Autonomous Driving field. Thus, according to current estimates, up to 2335 million autonomous vehicles will run on US highways alone by 2035 (the total number of autonomous vehicles that will be in operation worldwide may exceed 10,000 million units [178]). This huge number of vehicles equipped with a totally new technology, which is completely different from the current technology, from everything that is known and regulated in the field of vehicles on public roads, presents major challenges in the car insurance industry, but also significant short-term opportunities.

Increasing the number of autonomous vehicles on public roads will lead to major changes in the way insurance policies and premiums are currently generated. Because most autonomous vehicles will be owned by public or private service providers (public transportation companies, Ride Sharing companies that will use these units for public passenger transportation, respectively companies, universities, hospitals that will own and operate for the benefit of these units), the number of individual insurance policies will decrease, while the number of insurance packages will increase. As autonomous vehicles are considered much safer than vehicles operated by human drivers, the number of road accidents will also be reduced, which will lead to lower prices for insurance policies. Studies conducted by Accenture [177] estimate that the frequency of claims in the event of accidents involving autonomous vehicles will decrease significantly compared to claims for accidents involving vehicles operated by human drivers. In conclusion, it is estimated that if the insurers of autonomous vehicles will conclude a smaller number of insurance policies, the payments for damages resulting from traffic accidents will be reduced at the same time.

Accident insurance must cover both the autonomous vehicles and the transported passengers, respectively third-party vehicles, persons, movable and immovable property etc. with which the autonomous vehicles may possibly interact with, against any possible incidents/accidents that may occur during transportation, in accordance with the provisions of Directive 2009/103/EC, with all subsequent amendments and completions [179].

In addition to the standard risks (damage resulting from incidents/accidents on vehicles and passengers), the insurance of autonomous vehicles must cover the risks due to the following error factors that may occur during operation as well [180]:

(-) Errors of the autonomous driving system decision-making algorithms.

(-) Errors due to telecommunications services outage (telematics, GPS etc.), due to communication networks outage, respectively due to unfavorable weather conditions.

(-) Errors due to unscheduled defects of any assemblies or subassemblies, systems or subsystems, aggregates, or other components.

(-) Errors due to the interventions of the monitoring/operation staff serving the management platform.

(-) Errors due to possible cyberattacks, respectively unauthorized intrusions into the autonomous driving system.

According to studies conducted, the Stevens Institute of Technology predicted that following the implementation of autonomous vehicles in operation on public roads, in terms of classifying the risk categories to which insurers will have to adapt their insurance instruments, these categories will be: risks due to cyberattacks (Cyber Risk), malfunctions due to malfunctioning of Software (Algorithms) and Hardware (Sensors) elements that are part of the autonomous vehicles and problems due to infrastructure elements (Infrastructure) [181].

(1) Risks due to cyberattacks (Cyber Risk) generate opportunities that include protection against theft of control of the vehicle, unauthorized entry into the software architecture of the vehicle and the use of Ransomware attacks to block the control of the vehicle etc. This category of risks also includes the hijacking of the vehicle's drive systems with the possibility of using them for criminal or terrorist purposes. Other risks that will be insured in this category are protection against identity theft, confidentiality of data and information insurance, respectively theft or misuse of personal information.
(2) Risks due to software malfunctions (Algorithm) cover the lack (outage) of internet connection, programming errors, memory overflow, algorithm defects etc. Risks due to Hardware failures (Sensors) that are part of autonomous vehicles cover errors or malfunctions due to sensory systems (Lidar, radar, camera, GPS/GNSS, IMU, Ultrasound).
(3) Risks due to infrastructure elements (Infrastructure) cover potential problems with the smart infrastructure that has the role to guide the connected vehicles and the traffic flow (infrastructure sensor systems, Edge, Fog and Cloud servers), other communication problems between the autonomous vehicles and intelligent infrastructure systems.

The cumulative impact of the autonomous vehicles on insurance premiums will generate huge potential revenue for insurers (up to $81 billion by 2025) provided they change and adapt their business models and meet these new categories of risks in the shortest possible time and with a policy that is flexible and adaptable to the new requirements.

In Germany, the damage resulting from vehicle collisions has legal consequences. Liability for damage is imposed on the party responsible for causing it. The liable party is generally the driver, the holder, or the vehicle manufacturer. If the driver was at fault in the accident, he or she may be held liable for the damage under §18 StVG. Under §18 StVG, the driver is liable if it would have been possible for him or her to intervene in the collision. A party is considered capable of intervention if it has a minimum level of control over the vehicle. If the vehicle operator of an automated vehicle can stop and steer the vehicle in case of an emergency, he or she satisfies the criteria for minimum control. In this case, the driver must prove that he or she is not at fault because the damage was caused by system failure. Under §63a StVG, he or she may access the automated vehicle's mandatory data storage system for this purpose. Without a vehicle operator, however, there is no driver to hold accountable. As a result, §18 StVG is not applicable.

The injured party can also claim damages from the holder of the vehicle under §7 StVG. The holder is the person who pays the bills and has the power of disposal over the vehicle but is not necessarily the owner or the driver. Under §7 StVG, the holder is held liable without personal fault. The only exception is §7(2) StVG, which exempts the holder from liability if the damage was caused by force majeure. Technical failures of automated driving systems do not qualify as force majeure [178].

Thus, the liability standards for automatic driving systems are not necessarily different from those for vehicles with manual operation.

The manufacturer is liable for damages if the damage results from a defective product. Manufacturer liability is regulated under the Product Liability Act (ProdHaftG). These regulations are based on European Directive 85/374/EWG, which all EU Member States are legally bound to implement. According to 4 ProdHaftG, the manufacturer is the party that produces the final product, pieces of the final product, or basic parts of the final product. The product is defective if it cannot be used correctly for its intended purpose. Liability is limited to a maximum of 85 million Euro under 10(1) ProdHaftG. Claims based on the ProdHaftG do not require personal fault. As a result, the manufacturer is always liable for damage caused by driverless vehicles. The injured party is not required to prove any misconduct of, or production error by the manufacturer. This means that the introduction of automated, driverless cars will shift liability to the manufacturer [16, 182].

According to EU Common Approach on the liability rules and insurance related to the Connected and Autonomous Vehicles, an analysis of how product liability could be applied to casualties with autonomous vehicles is described in the theory of the six scenarios [8]:

> Scenario 1: hardware failure. This scenario concerns the operator of an autonomous vehicle who is confronted with a sensor that has not properly detected a sudden lane closure for repair work with conflicting road lines. This raises the question who would or should be liable if the failing sensor caused the vehicle to slip. The first scenario involved the driver/operator of an autonomous vehicle, who suffered damage from one of its sensors that had not detected a sudden lane closure. Following the PLD (Product Liability Directive), he may sue the vehicle manufacturer or the hardware manufacturer (or another component manufacturer), assumed that these are two different parties. His right to compensation will depend on the reason for the sensor to miss the lane closure.
> Scenario 2: software failure. This scenario is different from the former in that instant malfunctioning of its software that caused the vehicle to slip. Undefinable symbols appear on the user's display. This scenario introduces the software producer as a possible cost bearer. As will be seen below, this case also stands for the intricate complexities of the question as to defectiveness.

The second scenario involves malfunctioning of the software (undefinable symbols appear on the display). It is yet uncertain if software can be qualified as a product. But the more intricate question is whether an unexpected interruption of the operating software will make it defective. If such interruptions happen on a large scale, affecting many vehicles with the same software or production series, this will point to defectiveness. But if the software interruption only affected the vehicle in this individual case, the reasonable expectation test calls for a careful balance to be made. Users may not expect software in general to be without interruptions. On the other hand, as concerns the software of their motor vehicle the reasonableness standard may be higher than it is for regular software in private computers and alike. In our estimation, it will be reasonable for consumers to at least expect a back-up system that in principle protects them against malfunctioning software. Nevertheless, inherent to the use of software is that it will have flaws, weaknesses, or vulnerabilities and that not all of these will exist or may be detected at the moment of production or during updates. The development risks-defense may entail that in the cases where there is defectiveness, the scientific technical knowledge was not such, as to prevent this disruption or to detect it prior to its market launch (and its updates). One might further argue that the defect was in the vehicle. The vehicle was defective for not responding in a safer way to the software interruption (with a safety back-up). The claimant would then have to direct his claim solely to the manufacturer of the vehicle, not to the software manufacturer.

Scenario 3: wireless network collapse. In this scenario, there was an unplanned construction of a roundabout during the night. The next morning the vehicle is, due to network problems, unable to obtain these data before and while it is used. It fails to download the information in time. As a result, it must make an emergency stop, causing it to slip and crash into the barrier. Next to the operator and the manufacturer of the vehicle (or its software), this scenario introduces the wireless network provider as a potentially liable party. The situation may be different in scenario 3, where the wireless telecom network is unavailable and, contractually, was not part of the package offered by the manufacturer. If the latter has committed to responsibility for the wireless connection of the vehicle, its unavailability will in principal fall within the manufacturer's sphere of risk. But if being connected was not an integral part of the package offered by the manufacturer, the question will be whether the vehicle can be regarded as 'defective'. If the vehicle is defective for reasons as those that were just given, then liability of its manufacturer will not be limited by the mere argument that the telecom provider is (also) liable for the network lapse.

Scenario 4: users' personal fault. Here, the driver/operator of the automated vehicle is at fault for not following the instructions properly, by not updating the software or not keeping his hands on the steering wheel. In scenario 4, the damage was caused by the fact that the driver/operator has violated the instructions in the user guide or manual. For product liability to arise here, again, defectiveness must be proven by the injured party. Dependent on more facts than were given, there could be an information defect: did the manufacturer provide instructions that were adequate, clear, unambiguous, repeated, etcetera? The autonomous vehicle may have an information defect and/or a design defect if it is not equipped with a system that gives adequate and repeated timely warnings and the technology to intervene at the right point if the driver/operator remains passive. If there is defectiveness (but this does not immediately appear to be the scenario as sketched), then the manufacturer can raise a contributory negligence defense, by arguing that the user was at fault. The fact that the driver/operator has ignored instructions may be qualified as contributory negligence (and limit or even exclude the manufacturer's liability). This can limit or, dependent on each national law system, even exclude the manufacturer's liability.

Scenario 5: third parties' hacking. This concerns the 'hacking' of the programming software of the autonomous vehicle by third parties, due to which the autonomous vehicle is no longer responding and crashes. The fifth scenario concerns hacking of the operating software of the vehicle. This case raises questions as to the programming of the software (or software updates), used in the vehicle. As motor vehicles are (and will be) made more autonomous and connected, it seems reasonable of users to expect adequate protection from the producer. The technology used (which for vehicles currently on the market includes internet-connected entertainment systems and radio's) must of course be protected against hacking attacks and malware, also in the form of periodic updates. Drivers/operators of the motor vehicle must further be adequately warned of the risks and other knowledge that the manufacturer may have and that can be shared, and recalls must be made, if need be. But such equipment could also be installed in the vehicle and functioning as should be, while the vehicle was nevertheless hacked. The user may, again, not expect 100% safety and the development-risks-defense may excuse the manufacturer from liability. This makes the outcome of these hypotheticals uncertain and heavily reliant on expert testimonies.

Scenario 6: programming choice. This scenario presents a case in which the autonomous vehicle tries to avoid hitting a house, but collides with a pedestrian, who is seriously injured. The sixth scenario asks whether the manufacturer must bear responsibility for the fact that the autonomous vehicle has severely injured a pedestrian, as it tried to avoid hitting a house. The central question is whether this is a design defect: should the vehicle have prioritized to avoid hitting the pedestrian instead of avoiding hitting the house. If this programming choice would qualify as defective, the manufacturer might invoke the development-risks defense.

An overview about fault liability rules and limited role for damage caused by autonomous vehicles are [8]:

(1) The PLD (Product Liability Directive):

"(i) Imposes liability on (a) automotive manufacturer, (b) developer/tester whose brand is on the vehicle, (c) importer, (d) supplier (if manufacturer is unidentified) and (e) component producer (for manufacturing defects only), including the software producer.

(ii) Autonomous vehicle or component part has manufacturing defect, design defect or information defect. Although PLD-liability is risk-based, the idea seems to be that the risk presented by the defect was or should have been avoidable by these parties.

(iii) The PLD only protects damage consequential to personal injuries, death, and property bigger than 500 Euro."
(2) Cases not (or partially) protected by the PLD:
"(i) The claim is directed against a party who is not liable under the PLD (software programmers behind the scenes, rental companies etc.).
(ii) The defectiveness cannot be proven.
(iii) It is probable that the autonomous vehicle or its technology became defective after being on the market.
(iv) The defect concerns scientifically unknown and unforeseen risks at the time of production or it was due to mandatory regulations.
(v) The case is for property damage less or equal to 500 Euro or purely financial harm.
(vi) The case is for damage to the vehicle itself or professional property.
(vii) The manufacturer lacks financial resources, and the vehicle has a design or information defect (component makers are not liable).
(viii) The victim was contributory negligent (partially loses protection)."
(3) National traffic liability rules:
"(i) Countries with risk-based traffic liability:
(-) Sweden, UK, France, and Belgium generally offer protection in the circumstances of case (a), (b), (c), (d), (g), and (h) that are mentioned in (2) above.
(-) The other strict liability systems may cover the cases (a), (b), (c), and (g) in (2). Case (d) in (2) is covered if the external cause defense cannot be invoked for any defects of the motor vehicle. As for case (h) in (2), in most rules on risk-based liability for motor vehicles, the victim partially loses his right to compensation if the liable party proves contributory negligence.
(ii) Countries with fault-based traffic liability: only if user (operator) knew or should have known about defects or was otherwise to blame."
(4) Not protected by special traffic liability rules:
"(i) Countries with risk-based traffic liability except in Sweden and the UK, France and Belgium, strict liability for motor vehicles generally excludes certain external cause situations (sudden lapse of connectivity), damage to driver, property damage more limited, purely financial harm excluded, and contributory fault/intent of victim (degree of protection varies).
(ii) Countries with fault-based traffic liability cases of (2) will not be protected, unless fault of either operator/driver, the manufacturer or others can be proven.
(iii) Single-vehicle accidents except in the Swedish and the UK insurance, the national traffic liability laws under review offer no protection for single-vehicle accidents."
(5) Rely on general (fault) liability:
"(a) Claim against the manufacturer. The manufacturer may incur fault liability if he:
(-) Knew or ought to have known of serious risks to health.
(-) Did not provide information.
(-) Did not properly update/protect the software.

(b) Claim against the developer or tester. The developer or tester may, inter alia, incur fault liability if he:
(-) Acted as under (a).
(-) Was negligent in the design/testing.
(c) Claim against driver/operator. The user may, incur fault liability if he:
(-) Violated instructions/legal duties.
(-) Took unacceptable risks (vehicle use in extreme weather, crowded places).
(-) For bad maintenance of the vehicle and/or its technology (ignoring messages to update software).
(d) Public authorities may, inter alia, incur fault liability if they:
(-) Grant permits for unacceptable risks.
(-) Do not protect road safety (connectivity unstable or road signs not fit)."
(6) No solution found:
"(a) Risk-based traffic liability. Some countries (excepted Sweden, UK, France, and Belgium cases (d) and (h) in (2) are covered) there is no protection for cases (e), (f) and (h) in (2) and excluded are:
(-) Damage of the driver.
(-) An external cause.
(-) Contributory negligence.
(b) Fault-based regime for traffic liability:
(-) Entire (2) not protected unless fault can be proven.
(-) None of the situations gives rise to compensation if the concrete facts of the individual case at hand remain unclear."

References

1. Iclodean, C., Cordos, N., Varga, B.O.: Autonomous shuttle bus for public transportation: a review. Energies **13**, 2917 (2020)
2. https://www.ecologique-solidaire.gouv.fr/sites/default/files/90p%20VDEF.pdf (April 20, 2020)
3. https://eur-lex.europa.eu/eli/reg/2018/1947/oj (April 20, 2020)
4. https://www.unece.org/fileadmin/DAM/trans/conventn/crt1968e.pdf (April 11, 2020)
5. https://www.unece.org/fileadmin/DAM/trans/doc/2014/wp1/ECE-TRANS-WP1-145e.pdf (April 11, 2020)
6. https://eur-lex.europa.eu/legal-content/EN/TXT/?uri=CELEX:32018R0858&from=ro (April 11, 2020)
7. https://eur-lex.europa.eu/legal-content/EN/TXT/?uri=CELEX%3A52018DC0283 (April 20, 2020)
8. https://www.europarl.europa.eu/RegData/etudes/STUD/2018/615635/EPRS_STU(2018)615635_EN.pdf (April 20, 2020)
9. https://eur-lex.europa.eu/eli/reg/2016/679/oj (April 20, 2020)
10. https://www.legifrance.gouv.fr/jorf/id/JORFTEXT000032966695/ (February 25, 2022)
11. https://www.gouvernement.fr/conseil-des-ministres/2016-08-03/experimentation-de-vehicules-a-delegation-de-conduite-sur-le (February 25, 2022)
12. https://www.insidetechlaw.com/autonomous-vehicles/06_france (March 03, 2022)
13. https://www.legifrance.gouv.fr/loda/id/JORFTEXT000036750342/ (February 25, 2022)

14. https://www.legifrance.gouv.fr/codes/id/LEGITEXT000006074228/ (February 25, 2022)
15. https://www.legifrance.gouv.fr/loda/id/JORFTEXT000036868691/ (February 26, 2022)
16. Ainsalu, J., Arffman, V., Bellone, M., Ellner, M., Haapamäki, T., Haavisto, N., Josefson, E., Ismailogullari, A., Lee, B., Madland, O., Madžulis, R., Müür, J., Mäkinen, S., Nousiainen, V., Pilli-, E., Rutanen, E., Sahala, S., Schønfeldt, B., Smolnicki, P.M., Soe, R.-M., Sääski, J., Szymańska, M., Vaskinn, I., Åman, M.: State of the art of automated buses. Sustainability **10**, 3118 (2018)
17. https://www.gesetze-im-internet.de/stvg/BJNR004370909.html (February 28, 2022)
18. https://www.automotive-iq.com/content-auto-download/5bcecf6ed1d92e3b5f3ec797 (April 18, 2020)
19. https://assets.kpmg/content/dam/kpmg/xx/pdf/2019/02/2019-autonomous-vehicles-readiness-index.pdf (April 20, 2020)
20. https://www.edscuola.it/archivio/norme/decreti/dlvo285_92.htm (April 20, 2020)
21. https://www.gazzettaufficiale.it/eli/gu/2018/04/18/90/sg/pdf (April 20, 2020)
22. https://www.connectedautomateddriving.eu/regulation-and-policies/national-level/eu/netherlands/ (March 01, 2022)
23. http://publications.europa.eu/resource/cellar/f851bb74-fa41-4b30-9d37-8adbffbf6e7e.0014.03/DOC_1 (March 01, 2022)
24. https://www.insidetechlaw.com/-/media/files/nrf/inside-tech-law/autonomous-vehicles-volume-3.pdf?revision=4c010c82-05ef-4c76-bfe3-312ebb49b248&revision=4c010c82-05ef-4c76-bfe3-312ebb49b248 (March 01, 2022)
25. https://wetten.overheid.nl/BWBR0042343/2019-07-01 (March 01, 2022)
26. https://www.gov.uk/government/publications/trialling-automated-vehicle-technologies-in-public/code-of-practice-automated-vehicle-trialling (March 03, 2022)
27. https://www.legislation.gov.uk/ukpga/2018/18/contents/enacted (March 03, 2022)
28. http://catalonialivinglab.com/wp-content/uploads/2019/06/15v-113-vehiculos-conduccion-automatizada-english.pdf (March 02, 2022)
29. https://www.taxheaven.gr/law/4784/2021 (March 02, 2022)
30. https://www.mae.ro/node/27929#null (March 03, 2022)
31. https://webapp.mai.gov.ro/frontend/documente_transparenta/459_1650348699_proiect%20Lege%20testare%20autovehiculelor%20automatizate_03.02.2022.pdf (April 21, 2022)
32. https://webapp.mai.gov.ro/frontend/documente_transparenta/459_1650348699_expunere%20de%20motive%20proiect%20Lege%20autovehicule%20automatizare_03.02.2022.pdf (April 21, 2022)
33. https://eur-lex.europa.eu/eli/reg/2014/540/oj (March 03, 2022)
34. https://eur-lex.europa.eu/legal-content/EN/TXT/PDF/?uri=CELEX:42019X0272&from=SL (March 03, 2022)
35. https://eur-lex.europa.eu/eli/reg/2010/1003/oj (March 03, 2022)
36. https://eur-lex.europa.eu/legal-content/EN/TXT/?uri=CELEX%3A42008X0527%2801%29 (March 03, 2022)
37. https://eur-lex.europa.eu/eli/reg/2012/130/oj (March 03, 2022)
38. https://eur-lex.europa.eu/legal-content/EN/TXT/?uri=CELEX:42010X0717(01) (March 03, 2022)
39. https://eur-lex.europa.eu/legal-content/EN/TXT/?uri=CELEX%3A42010X0710%2806%29 (March 03, 2022)
40. https://eur-lex.europa.eu/legal-content/EN/TXT/?uri=CELEX%3A42016X0218%2801%29 (March 03, 2022)
41. https://eur-lex.europa.eu/legal-content/EN/TXT/?uri=CELEX%3A42012X0920%2801%29 (March 03, 2022)
42. https://eur-lex.europa.eu/legal-content/EN/ALL/?uri=CELEX:42010X0513(02) (March 03, 2022)
43. https://eur-lex.europa.eu/legal-content/EN/ALL/?uri=CELEX:42010X0831(03) (March 03, 2022)

44. https://eur-lex.europa.eu/legal-content/EN/ALL/?uri=CELEX%3A42004X0331%2801%29 (March 03, 2022)
45. https://eur-lex.europa.eu/eli/reg/2011/19/oj (March 03, 2022)
46. https://eur-lex.europa.eu/eli/reg/2015/1406/oj (March 03, 2022)
47. https://eur-lex.europa.eu/legal-content/EN/TXT/?uri=CELEX%3A42007X0530%2801%29 (March 03, 2022)
48. https://eur-lex.europa.eu/legal-content/EN/TXT/?uri=CELEX:42011X1206(01) (March 03, 2022)
49. https://eur-lex.europa.eu/legal-content/EN/ALL/?uri=CELEX:42014X0930(01)&from=EN (March 03, 2022)
50. https://eur-lex.europa.eu/legal-content/EN/TXT/?uri=CELEX%3A42010X0630%2803%29 (March 03, 2022)
51. https://eur-lex.europa.eu/legal-content/EN/ALL/?uri=CELEX%3A42010X0630%2804%29 (March 03, 2022)
52. https://eur-lex.europa.eu/legal-content/en/ALL/?uri=CELEX:42010X0710(02)&from=NL (March 03, 2022)
53. https://eur-lex.europa.eu/legal-content/EN/TXT/?uri=CELEX%3A42009X0131%2802%29 (March 03, 2022)
54. https://eur-lex.europa.eu/legal-content/EN/ALL/?uri=CELEX:42010X0831(06)&from=IT (March 03, 2022)
55. https://eur-lex.europa.eu/legal-content/EN/ALL/?uri=CELEX:42019X0273&from=EN (March 03, 2022)
56. https://eur-lex.europa.eu/legal-content/EN/ALL/?uri=CELEX%3A42010X0710%2804%29 (March 03, 2022)
57. https://op.europa.eu/ro/publication-detail/-/publication/c47e68f4-23a8-4cf3-8a4a-389f24 6fb630/language-EN (March 03, 2022)
58. https://eur-lex.europa.eu/legal-content/EN/ALL/?uri=CELEX:42010X0612(03)&from=EN (March 03, 2022)
59. https://eur-lex.europa.eu/legal-content/EN/ALL/?uri=CELEX:42010X0612(02)&from=EN (March 03, 2022)
60. https://eur-lex.europa.eu/legal-content/EN/TXT/?uri=CELEX%3A42010X0528%2801%29 (March 03, 2022)
61. https://eur-lex.europa.eu/legal-content/EN/TXT/?uri=CELEX%3A42011X0909%2801%29 (March 03, 2022)
62. https://eur-lex.europa.eu/legal-content/EN/TXT/?uri=CELEX%3A42016X0108%2801%29 (March 03, 2022)
63. https://eur-lex.europa.eu/legal-content/EN/TXT/?uri=CELEX%3A42010X0630%2808%29 (March 03, 2022)
64. https://eur-lex.europa.eu/legal-content/en/TXT/?uri=CELEX%3A42014X0212%2801%29 (March 03, 2022)
65. https://op.europa.eu/en/publication-detail/-/publication/4dbaab80-a4c4-4b59-a7f5-22c385 3a3344/language-en (March 03, 2022)
66. https://eur-lex.europa.eu/legal-content/EN/ALL/?uri=CELEX%3A42008X0711%2801%29 (March 03, 2022)
67. https://eur-lex.europa.eu/legal-content/EN/ALL/?uri=CELEX%3A42011X1123%2803%29 (March 03, 2022)
68. https://eur-lex.europa.eu/legal-content/EN/TXT/?uri=CELEX%3A42007X0619%2801%29 (March 03, 2022)
69. https://op.europa.eu/en/publication-detail/-/publication/20d6793c-4b66-11e2-9294-01aa75 ed71a1/language-en (March 03, 2022)
70. https://eur-lex.europa.eu/legal-content/EN/TXT/?uri=CELEX%3A42010X0828%2801%29 (March 03, 2022)
71. https://eur-lex.europa.eu/eli/reg/2015/107/oj (March 03, 2022)

72. https://eur-lex.europa.eu/legal-content/EN/ALL/?uri=CELEX%3A32009R0661 (March 03, 2022)
73. https://eur-lex.europa.eu/legal-content/EN/TXT/?uri=CELEX%3A32012R0347 (March 03, 2022)
74. https://eur-lex.europa.eu/legal-content/EN/TXT/?uri=CELEX%3A32012R0351 (March 03, 2022)
75. https://eur-lex.europa.eu/legal-content/EN/ALL/?uri=CELEX%3A42011X0302%2801%29 (March 03, 2022)
76. https://www.nhtsa.gov/sites/nhtsa.dot.gov/files/documents/13069a-ads2.0_090617_v9a_tag.pdf (March 03, 2022)
77. https://www.eckertseamans.com/app/uploads/PLAC_Eckert_Seamans_AV_Survey-1.pdf (March 20, 2020)
78. https://www.ghsa.org/state-laws/issues/Autonomous%20Vehicles (April 04, 2022)
79. https://www.ghsa.org/taxonomy/term/521?page=0 (March 04, 2022)
80. https://www.ncsl.org/research/transportation/autonomous-vehicles-legislative-database.aspx (March 05, 2022)
81. https://legiscan.com/AL/bill/SB47/2019 (March 05, 2022)
82. https://legiscan.com/AL/bill/SB125/2018 (March 05, 2022)
83. https://legiscan.com/AZ/bill/HB2813/2021 (March 05, 2022)
84. https://legiscan.com/AR/bill/HB1562/2021 (March 05, 2022)
85. https://legiscan.com/AR/bill/HB1561/2019 (March 05, 2022)
86. https://legiscan.com/AR/bill/HB1822/2019 (March 05, 2022)
87. https://legiscan.com/AR/bill/HB1754/2017 (March 05, 2022)
88. https://legiscan.com/CA/bill/SB500/2022 (March 05, 2022)
89. https://legiscan.com/CA/bill/SB570/2022 (March 05, 2022)
90. https://legiscan.com/CA/bill/AB87/2018 (March 05, 2022)
91. https://legiscan.com/CA/bill/SB145/2018 (March 05, 2022)
92. https://legiscan.com/CA/bill/AB669/2018 (March 05, 2022)
93. https://legiscan.com/CO/bill/SB260/2021 (March 05, 2022)
94. https://legiscan.com/CO/bill/SB213/2017 (March 05, 2022)
95. https://legiscan.com/CT/bill/SB924/2019 (March 05, 2022)
96. https://legiscan.com/CT/bill/SB260/2017 (March 05, 2022)
97. https://legiscan.com/FL/bill/HB1289/2021 (March 05, 2022)
98. https://legiscan.com/FL/bill/HB107/2019 (March 05, 2022)
99. https://legiscan.com/FL/bill/HB311/2019 (March 05, 2022)
100. https://legiscan.com/FL/bill/HB725/2019 (March 05, 2022)
101. https://legiscan.com/GA/bill/SB165/2022 (March 05, 2022)
102. https://legiscan.com/GA/bill/SB219/2018 (March 05, 2022)
103. https://legiscan.com/GA/bill/HB472/2018 (March 05, 2022)
104. https://legiscan.com/HI/text/HCR220/2019 (March 06, 2022)
105. https://legiscan.com/IL/bill/HB0791/2017 (March 06, 2022)
106. https://legiscan.com/IA/bill/SB302/2020 (March 06, 2022)
107. https://legiscan.com/KY/bill/SB116/2018 (March 06, 2022)
108. https://legiscan.com/LA/bill/HB455/2019 (March 06, 2022)
109. https://legiscan.com/LA/bill/HB308/2018 (March 06, 2022)
110. https://legiscan.com/ME/bill/LD1724/2017 (March 06, 2022)
111. https://legiscan.com/MD/bill/SB291/2019 (March 06, 2022)
112. https://legiscan.com/MS/bill/HB1343/2018 (March 06, 2022)
113. https://legiscan.com/NE/bill/LB989/2018 (March 06, 2022)
114. https://legiscan.com/NV/bill/AB412/2021 (March 06, 2022)
115. https://legiscan.com/NV/bill/AB69/2017 (March 06, 2022)
116. https://legiscan.com/NH/bill/SB216/2020 (March 06, 2022)
117. https://legiscan.com/NJ/bill/AJR164/2019 (March 06, 2022)
118. https://legiscan.com/NM/bill/HB270/2021 (March 06, 2022)

119. https://legiscan.com/NM/bill/SJM3/2018 (March 06, 2022)
120. https://legiscan.com/NC/bill/HB814/2022 (March 06, 2022)
121. https://legiscan.com/NC/bill/HB469/2018 (March 06, 2022)
122. https://legiscan.com/NC/bill/HB716/2018 (March 06, 2022)
123. https://legiscan.com/NC/bill/HB469/2017 (March 06, 2022)
124. https://legiscan.com/ND/bill/HB1199/2019 (March 06, 2022)
125. https://legiscan.com/ND/bill/HB1418/2019 (March 06, 2022)
126. https://legiscan.com/ND/bill/HB1202/2017 (March 06, 2022)
127. https://legiscan.com/OH/bill/HB26/2017 (March 06, 2022)
128. https://legiscan.com/OR/bill/HB4063/2018 (March 06, 2022)
129. https://legiscan.com/PA/bill/HB1958/2018 (March 06, 2022)
130. https://legiscan.com/SC/bill/HB3289/2018 (March 06, 2022)
131. https://legiscan.com/TN/bill/SB151/2018 (March 06, 2022)
132. https://legiscan.com/TN/bill/SB676/2018 (March 06, 2022)
133. https://legiscan.com/TX/bill/SB1308/2021 (March 06, 2022)
134. https://legiscan.com/TX/bill/HB3026/2021 (March 06, 2022)
135. https://legiscan.com/TX/bill/HB1791/2017 (March 06, 2022)
136. https://legiscan.com/TX/bill/SB2205/2017 (March 06, 2022)
137. https://legiscan.com/UT/bill/HB101/2019 (March 06, 2022)
138. https://legiscan.com/UT/bill/SB56/2018 (March 06, 2022)
139. https://legiscan.com/VT/bill/SB149/2020 (March 06, 2022)
140. https://legiscan.com/WA/bill/SB5460/2022 (March 06, 2022)
141. https://legiscan.com/WA/bill/HB2676/2020 (March 06, 2022)
142. https://legiscan.com/WA/bill/HB2970/2018 (March 06, 2022)
143. https://legiscan.com/WI/bill/SB695/2018 (March 06, 2022)
144. https://www.ncsl.org/research/transportation/autonomous-vehicles-self-driving-vehicles-enacted-legislation.aspx#enacted (July 20, 2020)
145. https://upload.wikimedia.org/wikipedia/commons/a/a4/Map_of_USA_with_state_and_territory_names_2.png (March 05, 2022)
146. https://www.dentons.com/en/insights/guides-reports-and-whitepapers/2021/january/28/global-guide-to-autonomous-vehicles-2021. (March 05, 2022)
147. http://www.conventuslaw.com/report/china-national-administrative-rules-of-road/ (April 20, 2020)
148. http://www.npc.gov.cn/zgrdw/englishnpc/Law/Frameset-index.html (March 08, 2022)
149. Ziyan, C., Shiguo, L., China's self-driving car legislation study, Computer Law & Security Review, Volume 41, 2021, ISSN 0267–3649, https://doi.org/10.1016/j.clsr.2021.105555 (March 08, 2022)
150. https://cms.law/en/int/expert-guides/cms-expert-guide-to-autonomous-vehicles-avs/china (March 08, 2022)
151. https://wiki.unece.org/download/attachments/75530396/VMAD-01-04%20attachment%201.pdf?api=v2 (March 08, 2022)
152. https://www.eurasiagroup.net/files/upload/chinese-auto-vehicle-industry-faces-geopolitical-headwinds.pdf (March 08, 2022)
153. https://baijiahao.baidu.com/s?id=1676534150221070852&wfr=spider&for=pc (March 08, 2022)
154. https://www.roadtraffic-technology.com/news/baidu-robotaxi-china-hebei/ (March 08, 2022)
155. http://en.hubei.gov.cn/business/zones/201305/t20130521_450066.shtml (March 08, 2022)
156. http://en.people.cn/n3/2020/1016/c90000-9770120.html (March 08, 2022)
157. http://www.npc.gov.cn/englishnpc/lawsoftheprc/202001/c983fc8d3782438fa775a9d67d6e82d8.shtml (March 08, 2022)
158. https://www.advant-beiten.com/sites/default/files/downloads/China-Data%20Protection%20and%20Localisation,%20Cyber%20Security%20Law,%20VPN%20and%20Encryption.pdf (March 08, 2022)
159. https://www.ntc.gov.au/transport-reform/automated-vehicle-program (March 14, 2022)

160. https://www.ntc.gov.au/codes-and-guidelines/automated-vehicle-trial-guidelines (March 14, 2022)
161. https://austroads.com.au/publications/connected-and-automated-vehicles/ap-r665-22 (March 14, 2022)
162. https://austroads.com.au/publications/connected-and-automated-vehicles/ap-r664-22 (March 14, 2022)
163. https://austroads.com.au/publications/connected-and-automated-vehicles/ap-r654-21 (March 14, 2022)
164. https://www.gtlaw.com.au/insights/legal-issues-driverless-cars-autonomous-vehicles-australia (March 14, 2022)
165. https://www.legislation.gov.au/Details/C2014C00076 (March 14, 2022)
166. https://www.oaic.gov.au/privacy/australian-privacy-principles-guidelines/chapter-3-app-3-collection-of-solicited-personal-information (March 14, 2022)
167. https://www.oaic.gov.au/privacy/australian-privacy-principles-guidelines/chapter-5-app-5-notification-of-the-collection-of-personal-information (March 14, 2022)
168. https://www.legislation.gov.au/Details/C2021C00183 (March 14, 2022)
169. https://trid.trb.org/view/187219 (March 14, 2022)
170. https://www.mlit.go.jp/road/road_e/r1_standard.html (May 22, 2022)
171. https://www.reuters.com/article/us-honda-autonomous-legend-idUSKBN2AW0TZ (March 14, 2022)
172. https://unece.org/wp29-introduction (March 14, 2022)
173. https://unece.org/DAM/trans/doc/2018/wp29grva/GRVA-01-34.pdf (March 14, 2022)
174. https://www.giroj.or.jp/disclosure/pdf/eibunjibaihou201309.pdf (March 14, 2022)
175. https://www.soumu.go.jp/main_sosiki/joho_tsusin/eng/Resources/laws/pdf/090204_5.pdf (March 14, 2022)
176. https://asia.nikkei.com/Business/Automobiles/Japan-to-create-legal-framework-for-level-4-self-driving-cars (March 14, 2022)
177. https://www.accenture.com/_acnmedia/pdf-60/accenture-insurance-autonomous-vehicles-pov.pdf. (April 04, 2022)
178. https://www.statista.com/statistics/1230664/projected-number-autonomous-cars-worldwide/ (March 15, 2022)
179. https://eur-lex.europa.eu/legal-content/EN/TXT/?uri=celex%3A32009L0103 (March 04, 2022)
180. https://www.autonomousvehicleinternational.com/news/safety/aviva-set-to-trial-autonomous-vehicles-to-help-build-future-insurance-model.html (March 15, 2022)
181. https://ec.europa.eu/docsroom/documents/34802/attachments/1/translations/en/renditions/native (March 04, 2022)
182. https://www.insidetechlaw.com/-/media/files/nrf/inside-tech-law/autonomous-vehicles-volume-3.pdf?revision=4c010c82-05ef-4c76-bfe3-312ebb49b248&revision=4c010c82-05ef-4c76-bfe3-312ebb49b248 (March 03, 2022)

Chapter 8
Social Implication

8.1 Social Impact

The social impact of any new technology must be assessed in terms of the efficiency and benefits it brings to society, as well as the direct impact on the environment in which it evolves. Society needs to prepare for these new technologies, first to accept them and then to become familiar with them. There are people who will be directly affected by these new technologies and for whom society needs to prepare retraining or alternative plans for their current jobs or profession. Once solutions are found to cover people affected by vocational retraining and social security, all steps must be taken to implement these new technologies on a large scale [1, 2].

The global report on road safety published by the World Health Organization (WHO) in 2018 shows a number of 1.35 million deaths (for 2016) as a result of road accidents, respectively 18.2 deaths per 100,000 people. Unfortunately, this ratio has remained constant since the 2000s (Fig. 8.1). Figures indicate that over 90% of road accidents are due to driver errors [3].

Autonomous vehicles will have a massive positive or negative impact on society. The advantages of this technology are obvious, but because there are human subjects directly involved, they must learn to accept that a computer will make decisions for them. Thus, car manufacturers need to present valid and transparent risk scenarios to all those interested in understanding how the relevant decisions are made. The key element in this acceptance process is transparency. Using this approach, the fear of the unknown and the uncertainty of the new will begin to disappear. The millions of test kilometers driven by autonomous vehicles will eventually become millions of kilometers of daily use. Only then will this innovative product be considered accepted by society [1, 2, 4].

The mobility of people and goods is a fundamental issue for our current society. There are many entities interested and working together to make this safer, more efficient, and more environmentally friendly (businesses, scientific organizations, and government institutions). One of the most important stages in the development of this sector is the automation of road transport. The introduction of autonomous

C. Iclodean et al., *Autonomous Vehicles for Public Transportation*, Green Energy and Technology, https://doi.org/10.1007/978-3-031-14678-7_8

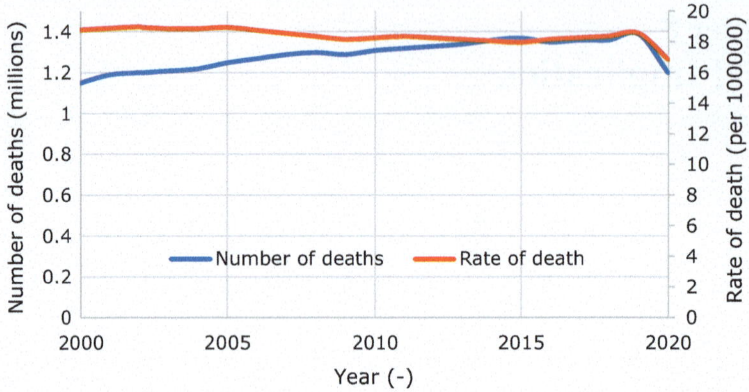

Fig. 8.1 Global report on road safety (2000–2020)

functions in the command and control of vehicles began with the implementation of intelligent systems (Embedded Systems).

According to Directive 2010/40/EU of the European Parliament and of the Council of 7 July 2010 on the framework for the deployment of Intelligent Transport Systems in the field of road transport and for interfaces with other modes of transport [5–7], the following fields are pursued:

(1) Optimal collection of road traffic data.
(2) ITS traffic and freight management services.
(3) ITS applications for road safety and security.
(4) Establishing the link between the vehicles and the infrastructure.

In Villeurbanne, France, on July 9, 2020, the company Navya launched the Autonom® Shuttle Evo with a level 4 autonomy (according to SAE J3016™), which travels in traffic without an on-board safety operator. This model of autonomous bus is in active service at the National Sports Shooting Center (NSSC) in Châteauroux, France and it operates in partnership with the transportation company Keolis [8]. The route implemented within the center has a length of 1.5 km, and the maximum velocity on this route is 18 km/h.

According to Public Relation service of Navya [8], the launch of Autonom® Shuttle Evo is the result of six years of experimentation and is the result of approximately 200 implementations of Navya level 3 autonomous shuttle bus models worldwide (see Chap. 1, Table 1.2). The new version of Autonom® Shuttle Evo includes key enhancements to achieving level 4 autonomy, namely:

(1) Automatic data connection and transfer function via Remote Control to the monitoring center.
(2) An improvement and optimization of the sensor architecture that equips the autonomous shuttle bus.

(3) A revised and improved version that offers an improved package to the user (passenger information screen, optimal access for people with reduced mobility etc.).
(4) Improved connectivity to remotely monitor autonomous shuttle bus status in real time.

A study on the implementation of Navya Autonom® Shuttle Evo worldwide shows that the management plan applied by the integrating companies included the following criteria [9]:

(1) Operational safety of the autonomous shuttle bus in traffic.
(2) Autonomous shuttle bus interaction with all traffic participants.
(3) Risk assessment for all traffic participants.
(4) Assessment of potential system errors.
(5) Ensuring the transition processes from level 3 to level 4 and preparing for the transition to level 5.
(6) Autonomous shuttle bus operation in traffic with/without monitoring.
(7) Interaction of autonomous shuttle bus with the ITS through V2X technology.

ITS are used for the maximum integration of autonomous shuttle buses in urban traffic and their interconnection with all traffic participants who are equipped with intelligent communication systems (Fig. 8.2). The central element of the ITS architecture is the autonomous vehicle which is accompanied by a human operator who can transmit requests to the autonomous driving system (request) and receive information from it (status) using the vehicle as an interface. The autonomous vehicle communicates via V2V technology with other nearby intelligent vehicles.

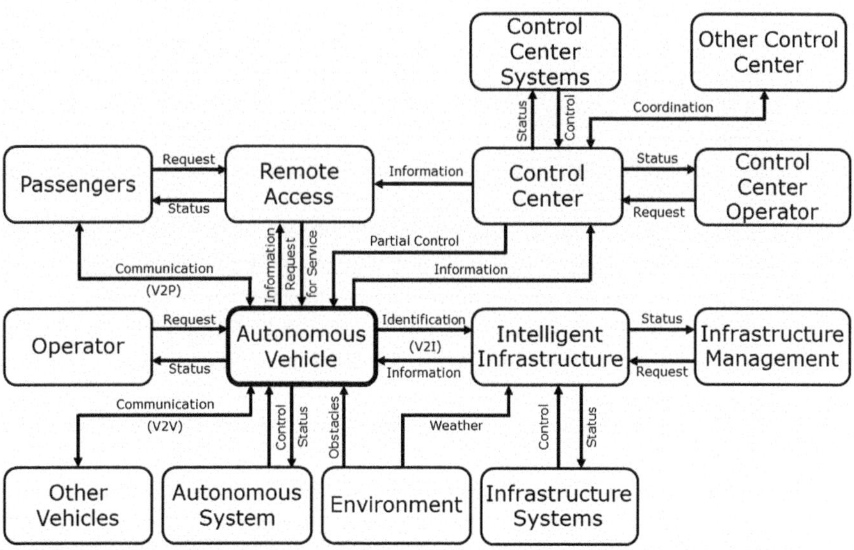

Fig. 8.2 Integration of the autonomous vehicle in an intelligent transport system

The identification of the autonomous vehicle is made through V2I technology with the intelligent infrastructure on which it runs, from which it receives information. Intelligent infrastructure is controlled by infrastructure systems and is operated by infrastructure management. Autonomous vehicle collects information about static and dynamic objects in the environment through sensors, and intelligent infrastructure collects information about the climatic conditions for the entire route contained in the infrastructure.

The autonomous vehicle is connected to the control center which is controlled by the control center system and operated by the control center operator. The control center, which monitors the activity of the autonomous vehicle, permanently receiving information from it, has partial control over the movement, being able to issue alerts to the operator or in extreme cases to act on the braking system to stop the vehicle. The control center can communicate with other control centers to transmit the coordinates of autonomous vehicle's movement.

Passengers, both those on board the autonomous vehicle and those at the embarkation points can request and obtain information via remote access about autonomous vehicle's movement through the intelligent information screens, respectively through the information applications on mobile devices such as smartphones (example: https://shuttlebus.cavcloud.com/shuttle).

8.2 Advantages and Benefits

The advantages of introducing autonomous driving for public transportation, especially for fleets of autonomous shuttle buses at the level of urban traffic are as follows [1, 10]:

(1) Reduction of up to 90% of deaths due to traffic accidents (according to the USDOT report [11] over 90% of traffic accidents are due to human errors in road traffic).
(2) Reduction of up to 60% of polluting emissions from road traffic (according to The Future of Driving report, Ohio University March 02, 2021 [12]).
(3) Elimination of vehicles stopping in traffic (Stop-and-Go) and reduction of traffic jams, especially during peak hours thanks to optimal travel scheduling algorithms consistent with the movement of other traffic participants.
(4) Up to 10% reduction in fuel consumption due to the information mentioned in (3) (according to Autonomous Vehicle Technology Guide [13]).
(5) Increase of the capacity of lanes by up to 100% due to the movement of autonomous vehicles along a predetermined route and the optimal use of the width of the lane.
(6) Reduction of up to 40% of travel time due to the information mentioned in (3) (according to the KPMG Connected and Autonomous Vehicles Economic Opportunity study [14]).

(7) Reduction of up to 15% of the area required for parking spaces due to entry/exit from the parking lot in autonomous driving mode, without passengers on board, which means keeping a shorter distance between parked vehicles (according to study [12]).
(8) Increase of the functional safety in road traffic through perception-prediction-planning intervention made by the autonomous driving system.
(9) Extended operating time especially for urban public transportation after a pre-established schedule (permanent, 24/24 h, 7/7 days).
(10) Unlike a human driver, in the case of a permanent operating regime for the autonomous driving system, fatigue does not occur (driver fatigue).
(11) Unlike a human driver, the autonomous driving system does not make inadvertent mistakes in the driving maneuvers in road traffic.
(12) Unlike a human driver, the autonomous driving system has no emotions that influence driving maneuvers in the road traffic.
(13) Some of the traffic accidents are due to driving under the influence of alcohol or drugs. Autonomous driving system excludes this risk.
(14) Optical sensors that equip autonomous vehicles offer a 360° circular view of the environment both horizontally and vertically, double visibility compared to human visual perception which covers an angle of 180° horizontally [15].
(15) Unlike a human driver, the attention of the autonomous driving system when driving cannot be distracted. The sensors that equip an autonomous vehicle through the functions of perception of the environment complement each other. Even if the functionality of some sensors may be affected by external phenomena, others remain functional.
(16) The operating algorithms of the autonomous driving system are set to recognize all traffic rules and indicators, velocity limits, lane keeping and other elements of preventive driving.
(17) The operating algorithms of the autonomous driving system optimize and reconfigure (as appropriate) the routes for real-time travel, using the information received from the communication systems with intelligent vehicles in traffic (V2V) and intelligent infrastructure (V2I).
(18) Liability insurance costs will be lower for an autonomous vehicle due to a lower likelihood of being involved in a road traffic event [16].
(19) Autonomous vehicles can generate useful periods for drivers (they can rest, relax, or carry out other activities) during the entire journey.
(20) Due to the activation of the autonomous driving system (password, fingerprint, face recognition) and due to the predetermined routes for travel, the possibilities of theft on these vehicles are reduced to a minimum.
(21) Autonomous vehicles provide access for persons with disabilities and persons with reduced mobility as passengers with a human operator who meets the legal requirements for monitoring (generally a driving license with a category similar to the category to which the autonomous vehicle belongs) for level 3 autonomous vehicles, respectively individually for level 4 autonomous vehicles that do not require the presence of a human operator on board the vehicle.

(22) Truck platooning, regulated by the legislation of some US federal states, which consists of grouping vehicles (especially freight trucks) in columns connected by electrical connections, which allow simultaneous braking and acceleration, ensures additional safety in road traffic.

(23) Autonomous vehicles can be used to transport patients to/between hospitals avoiding human interaction, or to transport samples or tests between different investigation centers (during a pandemic).

Studies by Yuen et al. in [7] on the different behavioral theories of a representative sample of urban road traffic participants to identify their intention to use SAV (Shared Autonomous Vehicle) have identified their following perceptions:

(A) The use of shared autonomous vehicles for public passenger transportation, through the superior performance of this technology, ensures the following advantages: increased safety of passengers in road traffic, reduction of the travel time, reduction of the travel costs, reduction of pollutant emissions etc.

(B) Understanding and accommodating passengers with autonomous technology requires minimal effort, the shared autonomous vehicles ensuring an intuitive and logical interaction from users.

(C) Habit is defined as a measure in which travelers tend to automatically exhibit behaviors that are developed by learning, and which is especially common for travelers who commute frequently to the same destination (home-work and return, for example).

(D) The cost of travel is a compromise between the benefits that travelers get from using shared autonomous vehicles and the monetary cost of this technology.

(E) Hedonic motivation is defined by the satisfaction obtained by passengers through the use of shared autonomous vehicles, due to the innovative technology brought by them.

(F) The attitude of the passengers towards shared autonomous vehicles are influenced by relative advantages (the extent to which the new technology is better than the precursor), the degree of complexity (the extent to which the new technology is perceived as difficult to understand, learn and operate) and the degree of compatibility (the extent to which the new technology is perceived to be in line with existing values, past experiences, and current needs).

(G) Subjective rules refer to the extent to which significant reference groups decide to perform or avoid a behavioral model with respect to shared autonomous vehicles.

(H) Behavioral control of passengers regarding the presence or absence of resources, the means, and opportunities necessary to access public transportation with shared autonomous vehicles.

8.3 Disadvantages and Solutions

The disadvantages of introducing autonomous driving for public transportation, especially for fleets of autonomous shuttle buses at the level of urban traffic are as follows [1, 4, 16]:

(1) Rising unemployment among professional drivers (a public transportation fleet in a metropolis has over 1000 drivers). There are possibilities for professional retraining of a part of this category of employees for monitoring, maintenance, upkeep etc.
(2) The high costs to replace the existing transportation fleets (some consisting of clean electric vehicles) with autonomous vehicles and the communication infrastructure (V2X), together with monitoring and operation. The possibilities for the purchase of autonomous vehicles through non-reimbursable funds and/or possibilities for the implementation of autonomous technologies on the existing electric buses.
(3) The high costs for the maintenance and upkeep of the autonomous vehicles. High costs for the spare parts and consumables. The possibility of assistance under procurement contracts from the manufacturer/supplier to reduce these costs.
(4) Cyber security risks, regarding unauthorized access to the computer systems of the autonomous vehicle. It is absolutely necessary that the computer systems and software applications are constantly updated with the latest version available from the manufacturer and that encrypted communication channels, VPN connections, tunnel etc. are used.
(5) The defects in some systems or general defects difficult to fix due to new, innovative technology. It is absolutely necessary that service and support contracts be concluded with the manufacturer's service representatives throughout the warranty and post-warranty period. Through professional retraining some former drivers can be trained/certified for maintenance and upkeep operations.
(6) Extreme weather conditions or unfavorable weather may affect the functionality of the autonomous driving system due to the vulnerability of the sensors equipping the autonomous vehicles to these conditions. Strict compliance with the manufacturer's specifications regarding the environmental conditions regarding the operation of the autonomous vehicles is recommended.
(7) The reluctance and/or restraint of the population regarding the acceptance of the integration of autonomous vehicles as means of public transportation. It is recommended to publicly promote this innovative technology and to allow periods of time for accommodation among the population by conducting free demonstration rides with passengers, highlighting the benefits, visibility etc.
(8) Issues of national legislation, in particular due to the lack of legislation, norms and regulations necessary for the registration and entry into service of autonomous vehicles on public roads. Close cooperation and communication between the manufacturer and the authorities is recommended to supplement and harmonize the legislation necessary for the registration and entry into service

of autonomous vehicles on public roads, and in situations where this requires lengthy procedures, the entry into service of autonomous vehicles on private circuits for testing (airports, campuses, industrial parks etc.).

(9) Moral and ethical issues that in extreme situations may require a momentary decision of the autonomous driving system and moreover an analysis of the consequences of this decision, for example if a last minute maneuver by the autonomous vehicle endangers the safety of passengers in the vehicle. The legislative packages that create the legal framework for the implementation of autonomous vehicles in operation on public roads in many states of the world, expressly require simulations performed by universities/research institutes on virtual models built based on the real models, to analyze all possible driving scenarios on the routes on which the autonomous vehicles are to be implemented.

There is no right or wrong approach. It is not right or wrong. There is the only approach to move forward. Forward may sometimes involve staying in the same place or moving to the right or left of that place; as long as there is movement, technology developers will have/will find the catalyst they need. New technology is not always accepted by society. Sometimes the ancestral human need for security makes individuals reluctant to change, but there have been times when curiosity has pushed people out of their proverbial caves into the unknown. And then humanity evolved.

8.4 Autonomous Vehicles in Use. Case Studies

According to the locations specified in Chap. 1, Table 1.2 Autonomous shuttle buses for public transportation, the following are some case studies on the implementation of autonomous shuttle buses for public transportation in road traffic.

8.4.1 Europe

8.4.1.1 Modellregion Oberfranken (Germany)

Autonomous shuttle buses are implemented in a public transportation project in Germany, Bavaria, in three locations: Hof, Kronach and Rehau. According to the data published on the project's website [17], the transportation service has been suspended due to the pandemic and will be available on the route from Rehau starting with March 2022 (Fig. 8.3). The transportation service is free (in Kronach and Hof the service was available to all citizens, in Rehau it was available to Rehau AG employees), but the operation of autonomous shuttle buses is conditioned by the weather and the availability of seats in the vehicles. In bad weather (rain, snow, ice, frost etc.) autonomous shuttle buses do not operate.

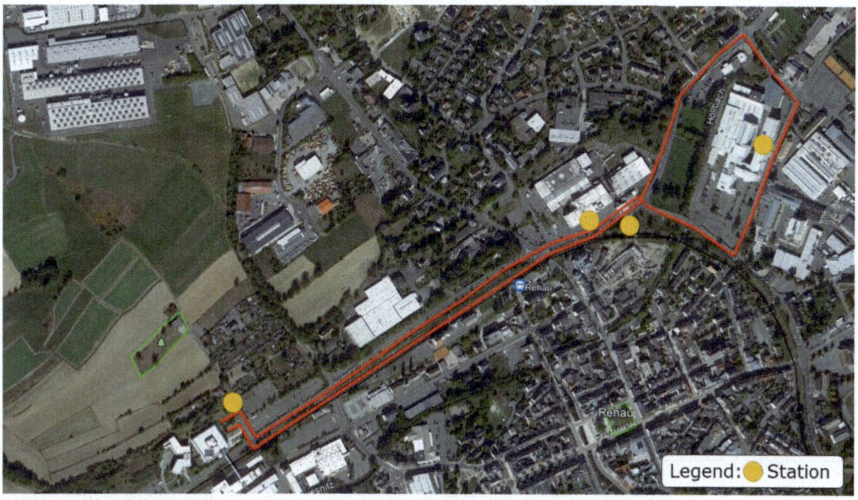

Fig. 8.3 Modellregion Oberfranken (Rehau, Germany)

The SMO (Shuttle Modellregion Oberfranken) research project started in 2021 with six Navya autonomous shuttle buses and was funded by the Bundesministerium für Digitales und Verkehr (BMDV), and the tests will run until June 2022. In the three months of operation the Navya autonomous shuttle buses will have traveled about 4500 km and will have transported over 4000 passengers (about 50 passengers per day).

A survey of the population of the three localities by Coburg University showed that over 80% of the more than 500 respondents supported the project and were interested in the new means of transportation [17].

In addition to public transportation operations (tourist routes in Kronach, the connection between the train station and the city center in Hof) or employee transportation in Rehau, the SMO project also serves scientific research purposes carried out in collaboration with Coburg University and Hof University. The movement of autonomous shuttle buses is monitored by an operator located on board the vehicles, respectively from a control center which is integrated in the public transportation company in the area, and which has the possibility to monitor the operation of level 4 vehicles without an operator on board the vehicles. Control center operators perform the following basic tasks:

(1) Simultaneous monitoring of several autonomous shuttle buses.
(2) Monitoring the functions and functional parameters of the vehicles.
(3) Permanent communication by radio connection with the vehicle.
(4) Contact and communication with passengers in emergency situations.

The mobility sector in Germany, as elsewhere in the world, is facing major challenges due to climate change, social events, and the political context. Combinations of innovative technologies and vehicles equipped with clean powertrains are becoming increasingly important and are increasingly used in metropolitan areas and beyond, especially as part of pilot projects for autonomous driving. New technologies have the potential to strengthen public transportation and make it more efficient, flexible, and optimally tailored to the needs of long-term users [17].

8.4.1.2 Holo (Denmark)

The Holo organization was founded in 2016 by Semler Group, one of the largest automotive companies in Denmark [18], and it aims to develop an efficient and sustainable transportation system based on electric and autonomous vehicles.

Since 2018, Holo has collaborated on the development of projects to test autonomous shuttle buses, using models delivered by the company Navya in the following Nordic countries: Denmark, Norway, Sweden, Finland, and Estonia. Currently the project implemented in Zealand, Denmark is the only one operational.

The first phase of the project started in 2018 with an autonomous shuttle bus Navya level 3, with a maximum capacity of 14 passengers, which was implemented on an operational route with a length of 0.4 km, carried out inside the campus of the University Hospital in Køge Zealand (Sjællands Universitetshospital) (Fig. 8.4). The first phase of the project lasted three months, during which the autonomous shuttle bus covered a distance of 842 km, transporting 5800 passengers (an average of 13 km traveled daily with 75 transported passengers, at an average velocity of 3.6 km/h). The budget allocated for the project was 605,000 Euros [19, 20].

Because it was a route located on a private area, the operation of autonomous shuttle buses was not subject to Danish legislation on autonomous vehicles, the purpose being to test the operation of autonomous vehicles, to perform functional

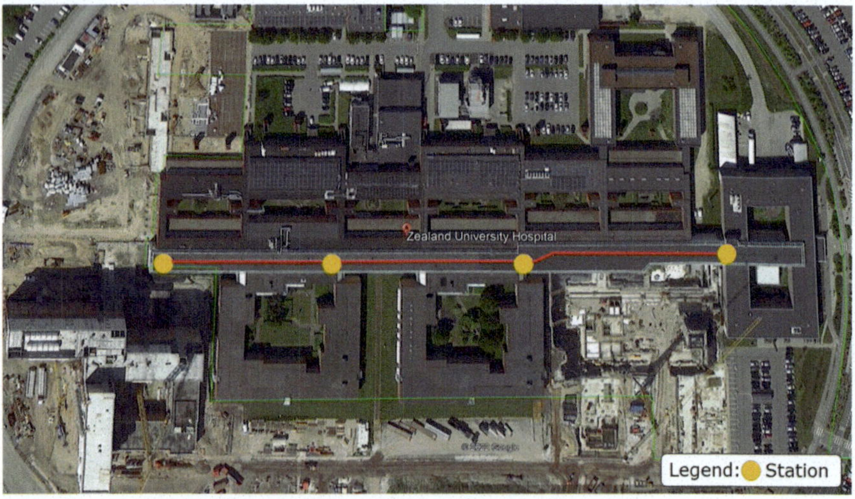

Fig. 8.4 Køge Zealand (Sjællands Universitetshospital)

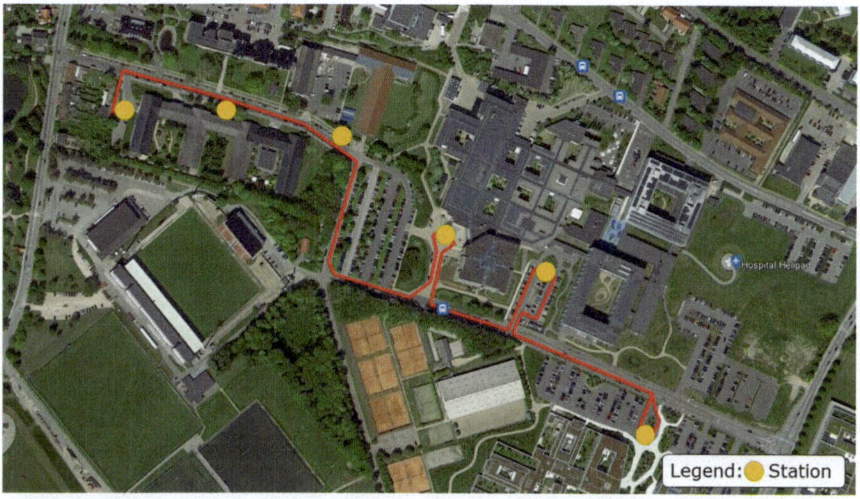

Fig. 8.5 Fælledvej Road (Slagelse Sygehus)

calibrations of the autonomous driving system and to conduct a study regarding the degree of perception and satisfaction of the transported passengers (patients, staff, and visitors).

The two studies on the transported passengers were conducted by the University of Aalborg "Interaction with the build-up area" and the University of Roskilde "Passenger perceptions" [21]. The published results showed a very high degree of passenger satisfaction (98% of users said they were satisfied with the service).

The project implemented on the campus of the university hospital was coordinated by Movia, the Copenhagen public transport authority, and the operation of autonomous shuttle buses was carried out by Holo (Semler Group).

The second phase of the project has as main purpose the implementation of the passenger transportation service on request (continuous movement with stops in the stations arranged on the route only at the request of the passengers) on a road open to public traffic. For the second phase two autonomous shuttle buses were in operation, implemented on a route between two sections of the Slagelse University Hospital in Copenhagen, on a section of the Fælledvej public road (Fig. 8.5). The maximum velocity of the autonomous shuttle buses on public roads is limited to 23 km/h, and the operation of autonomous vehicles will be authorized by the competent authorities on the basis of the Danish legislation in force in relation to the movement of these vehicles on public roads [19, 22].

8.4.1.3 Darwin (Oxfordshire, UK)

Darwin Innovation Group of Oxfordshire, UK is a research-development group in the field of autonomous and connected vehicles. The group has implemented an

Fig. 8.6 Darwin (Oxfordshire, UK) (*Capture* https://shuttlebus.cavcloud.com/shuttle)

autonomous Navya shuttle bus on public roads at the Harwell Science and Innovation Campus [23].

Autonomous shuttle bus travel is available to Harwell Science and Innovation Campus license holders and their partners. The current position of the autonomous shuttle can be tracked publicly in real time (Fig. 8.6) [24].

The tests were launched on November 29, 2021, the autonomous shuttle bus operates daily during the week from morning to evening on two routes, one along Fermi Avenue and one along Eighth Street, routes on which 10 stations are defined for embarkation/disembarkation of passengers. The autonomous shuttle bus is permanently accompanied by an operator who oversees the safe embarkation/disembarkation of passengers and the movement of the vehicle.

The operation of the autonomous shuttle bus in test mode on the routes of Harwell Science and Innovation Campus is carried out according to the specifications of the Code of Practice: Automated Vehicle Trialling (updated January 28, 2022) [25], respectively according to the specifications of PAS 1881:2020 Assuring the Safety of Automated Vehicle Trials and Testing [26]. The civil liability insurance of the autonomous shuttle bus was concluded by the company Aviva, the vehicle being registered with the DVLA (Driver and Vehicle Licensing Agency) for use on public roads and is registered with the number FX21HWO [23].

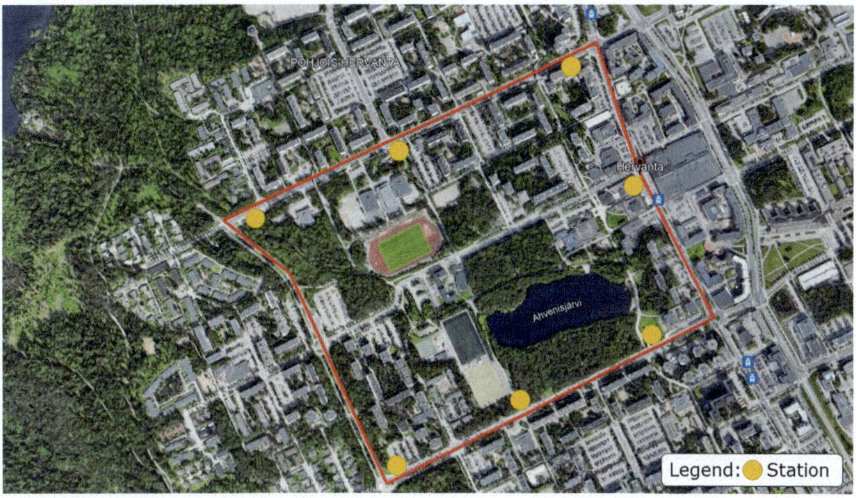

Fig. 8.7 Tampere, Hervanta (Finland)

8.4.1.4 Tampere, Hervanta (Finland)

The Tampere Pilot Project provides public transportation in the Hervantakeskus, Hervanta, Tampere area. The project uses two prototypes of Toyota Proace Verso autonomous vehicles that incorporate Sensible4 autonomous technology and can transport a maximum of four passengers on board on public roads in traffic (Fig. 8.7) [27, 28].

The tests are conducted with passengers, the travel is free, and the vehicles are accompanied by an operator who is prepared to intervene in case of emergencies. The total length of the route is 3.5 km, on the route there are seven stations for embarking/disembarking passengers, and the maximum velocity of the autonomous vehicle is 30 km/h.

8.4.2 United States

8.4.2.1 RoboRide (Peoria, AZ)

For the implementation of the RoboRide autonomous vehicle, City of Peoria, AZ has entered into a partnership with MAG (Maricopa Association of Governments) and Beep Inc., a company that provides autonomous mobility solutions. The implemented autonomous shuttle bus became operational on January 5, 2022 and is part of a six-month pilot program designed to serve local medical facilities in an urban area [29].

Fig. 8.8 RoboRide (Peoria, AZ)

RoboRide is a level 3 electric autonomous shuttle bus built by the company Navya. The shuttle bus travels on a 2.5 mile (4 km) long route with 13 embarkation/disembarkation stations and has top velocity of 15 mph (24 km/h) (Fig. 8.8) [30].

RoboRide trips are free of charge, do not require prior registration and are available to anyone wishing to experience a journey with an autonomous vehicle. The transportation capacity is a maximum of 8 passengers. RoboRide's movements are constantly monitored by an on-board operator and monitored remotely from a control center operated by Beep Inc. A study on the satisfaction of RoboRide public service users has been initiated by Arizona State University and is running for the duration of tests with the passenger. The questionnaire contains the sections [31]:

Section A: Travel Options. This section contains questions about respondents' recent travel options.

Section B: The respondents' perception of autonomous vehicles. This section contains questions about autonomous mobility services and technologies and deals with similar travel experiences, not just with RoboRide.

Section C: General information. This section discusses the needs of the community in relation to the mobility services based on autonomous vehicles.

8.4.2.2 A2GO (Ann Arbor, MI)

A2GO Project (Autonomous to GO) is an experimental project for autonomous driving led by the University of Michigan, May Mobility, and Mcity (research partner in the field of public–private mobility at the University of Michigan).

Fig. 8.9 A2GO (Ann Arbor, MI)

A2GO is a free shuttle service available to the general public that offers an easy way to travel between different locations in Ann Arbor, Michigan starting October 11, 2021. The route travelled by the autonomous vehicles is 2.64-mile (4.25 km) long and has 13 embarkation/disembarkation stations in downtown Ann Arbor (Fig. 8.9). The fleet of autonomous vehicles includes four Lexus RX 450 h hybrid-electric vehicles that can transport three passengers each and a Polaris GEM electric vehicle that can also transport people in wheelchairs [32, 33].

Mcity's primary role is to collect and process data sets from the autonomous vehicles, infrastructure, and other road users for use by researchers at the University of Michigan. This research aims to increase the safety of road users, and/or to improve the decision-making mechanism and perception algorithms for the autonomous vehicles. May Mobility will survey the transported passengers about their experience with the autonomous vehicles from Ann Arbor, and after the end of the testing period will publish the results and analysis of these questionnaires.

The intermediate results of the surveys of passengers transported in the first two months after launch are favorable, indicating a score of 4.81 out of a maximum of 5 possible points [34].

8.4.2.3 AVA (Clearwater Beach, FL)

AVA (Autonomous Vehicle Advantage) Pilot through a partnership with Beep Inc. implemented a Navya autonomous shuttle bus on a route in Tampa Bay, which stops at six embarkation/disembarkation stations, located at the main tourist and commercial attractions in the Clearwater Beach area (Fig. 8.10), in a mixed circulation traffic [35, 36].

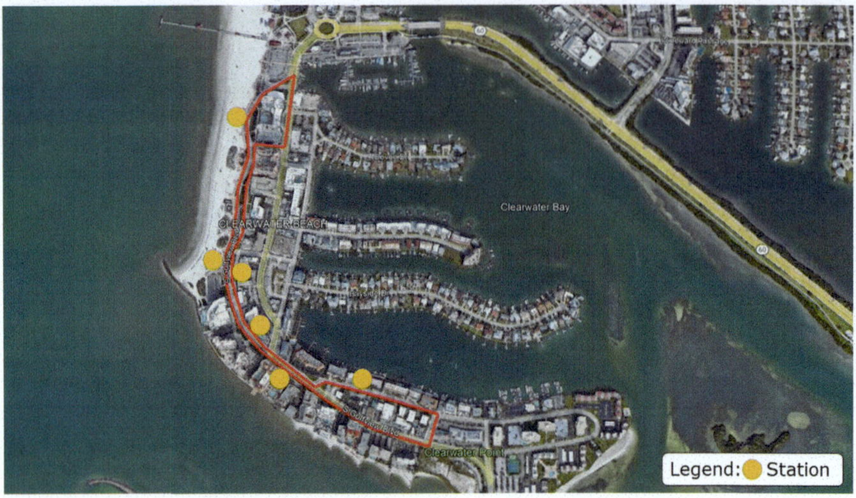

Fig. 8.10 AVA (Clearwater Beach, FL)

The pilot project is managed by PSTA (Pinellas Suncoast Transit Authority), an authority that also manages feedback from transported passengers to understand the users' perception of this project integrated in an intensely lively tourist area throughout the year.

The service is available from Tuesday to Sunday between 10 am and 10 pm for a fee. The travel tickets can be purchased online and cost USD 2.25 per trip.

The Navya autonomous shuttle bus is constantly monitored by an operator on board the vehicle and is monitored remotely from a control center operated by Beep Inc.

Users can access information about the autonomous shuttle bus program and track its movement on a digital map available online [37]. The organizers are taking clear and firm measures regarding passenger transportation both in terms of the sanitary rules imposed by the pandemic and in terms of the number of transported passengers (only passengers seated in the available seats with a fixed seat belt are accepted) [38].

8.4.2.4 RAPID (Arlington, TX)

Arlington, TX is the first city in the United States to implement the operation of autonomous shuttle buses for public transportation, namely the Milo model manufactured by EasyMile, a successful project from August 2017 to August 2018 [39].

In October 2018, the City of Arlington started its second autonomous driving project in partnership with Drive.ai, which ended in May 2019.

In March 2021, the RAPID project (Rideshare, Automation, and Payment Integration Demonstration) was launched, a project that covers downtown Arlington and

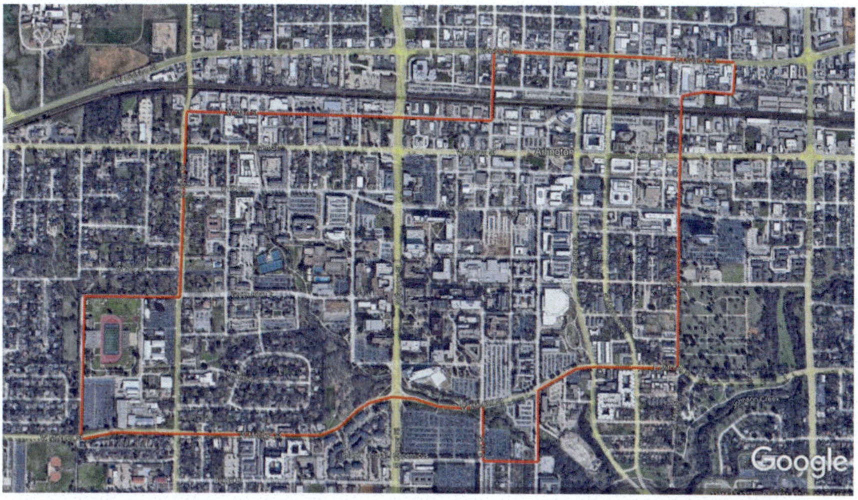

Fig. 8.11 RAPID (Arlington, TX)

the University of Texas at Arlington campus on an 18-mile (29 km) route (Fig. 8.11) and is addressed to the public [40].

In collaboration with the automotive prototype provider—the company May Mobility [41], the management platform application provider—the company VIA [42], and the research-development University of Texas at Arlington [43], the Arlington authorities put four Lexus RX 450 h hybrid-electric vehicles and one Polaris GEM electric vehicle into service.

The RAPID project is part of the Via Rideshare microtransit program and offers a limited number of free rides to University of Texas at Arlington students, and standard rates for the public ranging from USD 3 to USD 5.

After a year of public service, the five autonomous vehicles serving the RAPID project made more than 28,000 journeys and provided a 98% satisfaction rate for the transported passengers (4.9/5 rated based on statistics from the University of Texas at Arlington). According to the monitoring reports of the management platform, the vehicles operated in autonomous driving mode 80% of the total time in use.

8.4.2.5 AVGR (Grand Rapids, MI)

AVGR (Autonomous Vehicle Grand Rapids) is an on-demand autonomous driving service that uses four Lexus RX 450 h hybrid-electric vehicles and covers 4.21-mile (6.77 km) route in Grand Rapids, MI (Fig. 8.12) [44, 45].

The transportation capacity of the autonomous vehicles is a maximum of five passengers or one person in a wheelchair. The movement of autonomous vehicles is supervised by an operator who is permanently on board. The service is free for all passengers depending on availability, and orders for on-demand service are made

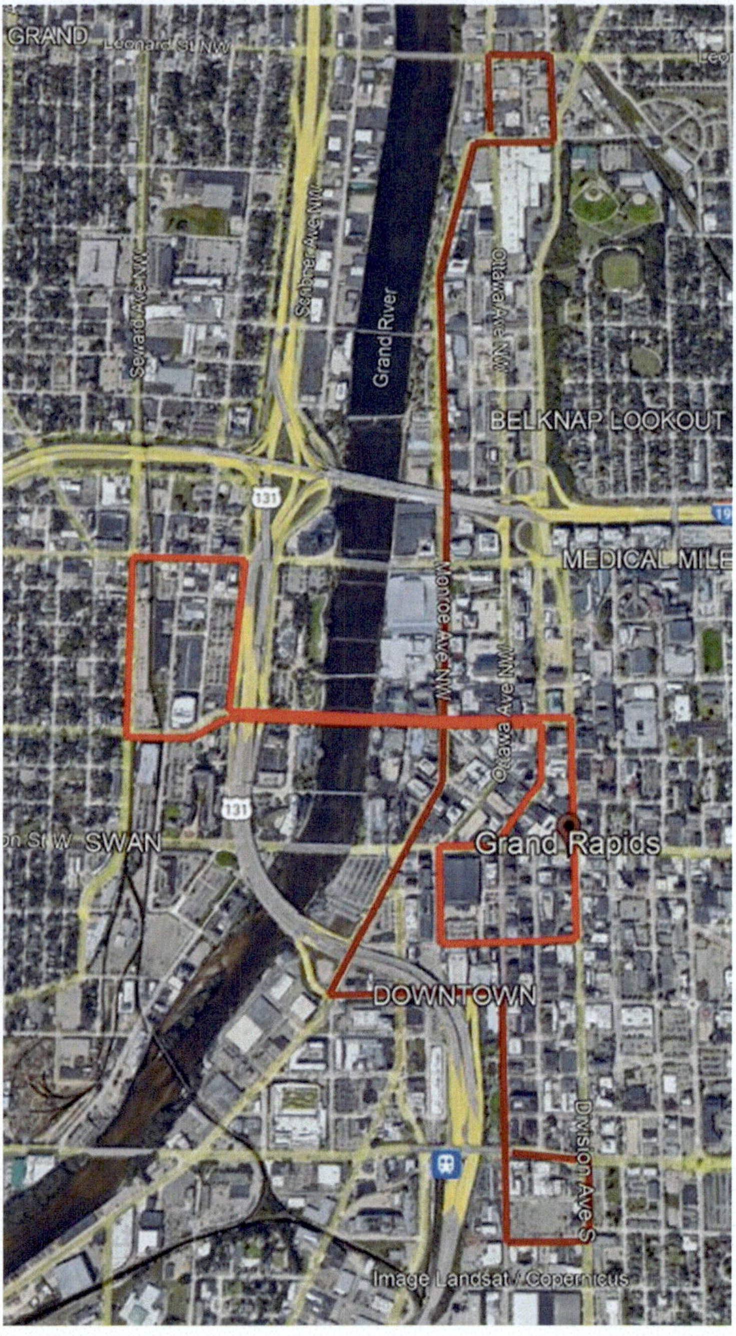

Fig. 8.12 AVGR (Grand Rapids, MI)

from the May Mobility application. The passengers can embark and disembark only in the 20 stations arranged on the route [46].

May Mobility also conducted a survey of the transported passengers who found that 95% of passengers said they felt safe while traveling on autonomous vehicles, although 90% of respondents declared they were on their first trip with an autonomous vehicle [47].

8.4.3 China

8.4.3.1 WeRide (Guangzhou)

In 2019, WeRide launched the first autonomous driving public service Robotaxi in Guangzhou, China (Fig. 8.13), covering an area of more than 144 km^2 with more than 200 embarkation/disembarkation stations, serving a population of over one million. To date, autonomous vehicles have accumulated more than 10 million kilometers on January 17, 2022, of which 2.5 million kilometers in the mode of driving without an operator on board the vehicles. The WeRide fleet has over 300 level 4 autonomous vehicles and consists of various construction models delivered by the Renault-Nissan-Mitsubishi, Yutong and GAC Motor groups [48, 49].

In parallel with the implementation of autonomous vehicles on public roads, WeRide simulated over 100,000 complex driving scenarios using decision-making algorithms implemented in the autonomous driving system. The scenarios were developed based on real-time data generated by vehicles in traffic, on data stored

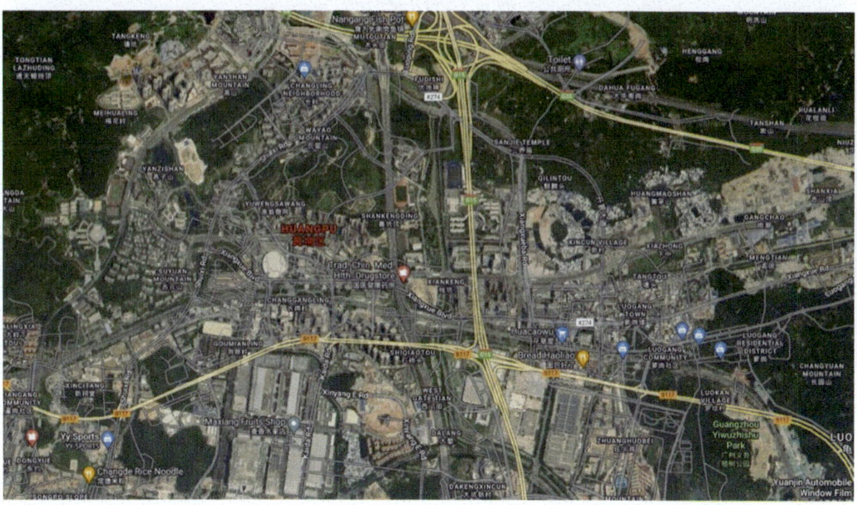

Fig. 8.13 WeRide (Guangzhou)

in a Cloud Computing environment and cover over 8 billion kilometers in simulations, a number equivalent to the mileage of 300 autonomous vehicles that would travel on the public roads for 50 years [50].

The simulation environment used is hosted by the AMAP Alibaba Cloud (AutoNavi) platform, which also contains a Robotaxi on-demand application for the population of the Huangpu District, Guangzhou.

AMAP is a tool for managing urban travel in China, with over 100 million daily active users. The AMAP platform covers information services, navigation of vehicles integrated in the platform, intelligent public transportation etc.

The urban travel management carried out by the AMAP platform is based on a mechanism for generating high-precision real-time travel maps consisting of the following stages [51]:

(1) Collection of data from the sensors of the vehicles in traffic.
(2) Automatic identification of elements of interest on the route:
    (a) Recognition of traffic signs.
    (b) Recognition of POIs (Points Of Interest).
(3) Manual correction of the identified items.
(4) Generation of travel maps.
(5) Route planning and final destination search recommendations.
(6) Virtual positioning of the autonomous vehicle within the map generated based on GPS coordinates.
(7) Virtual navigation of the autonomous vehicle using AR (Augmented Reality) in the generated map (Fig. 8.14).

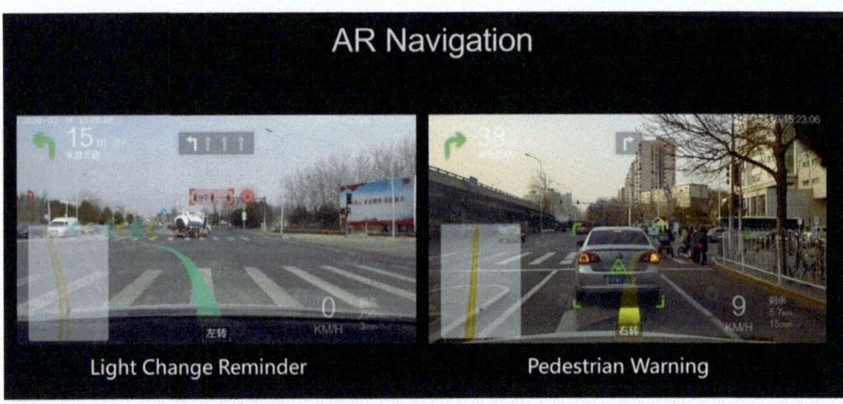

Fig. 8.14 Augmented Reality using AMAP [52]

8.4.3.2 AutoX (Shenzhen)

AutoX operates Robotaxi services in the following cities in China: Shenzhen, Guangzhou, Shanghai, and Beijing. According to official data [53], AutoX owns the largest fleet of Robotaxi in China consisting of over 1000 level 4 autonomous vehicles based on Chrysler Pacifica, Lincoln MKZ, BYD Qin models and the vehicles are authorized to operate on public roads without a driver on board.

The access of the population from the cities served by AutoX for trips with autonomous vehicles can be achieved through the AMAP Alibaba Cloud platform. Compared to other projects, AutoX covers all public roads in its service area, including all smaller and congested streets, main roads, and highways, and it has no predefined route and fixed embarkation/disembarkation stations. The Robotaxi service is available to customers, transporting passengers on request door-to-door [54].

In Shenzhen, Pingshan District, AutoX operates on an area of 168 km^2 offering Robotaxi services without drivers on board [55].

The autonomous vehicles operated by AutoX are equipped with the autonomous driving system AutoX Gen5 which contains 50 sensors, managed by a command and control unit of the vehicle with a computing power of 2200 TOPS (Theoretical Operations Per Second). The vehicle is equipped with 28 video cameras that capture a total of 220 million pixels per second, six high-resolution Lidar sensors that generate 15 million dots per second and eight 4D radar sensors with a resolution of 0.9° and a coverage of 360° around the vehicle. The video camera and Lidar sensors are arranged on the vehicle to cover all blind spots [56].

Permanent monitoring of the operation of the autonomous vehicles is carried out from the operation centers which are connected to each autonomous vehicle by ultra-reliable mobile data connections provided by several internet providers (to ensure connection redundancy) [57].

The virtual models' simulation operation is based on the data collected during the operation in real traffic, and the simulation processes evaluate all possible scenarios for the situations that autonomous vehicles have gone through. Virtual simulations take place on the xSim platform which can recreate, modify, and amplify real-world experiences or generate completely new situations from scratch. Only in situations where a characteristic of the autonomous driving system functionality has been validated by simulation, tested, and certified in a closed and controlled environment, it can be implemented in the operation of the autonomous vehicles driving in real traffic on public roads [58].

8.4.3.3 Alphaba (Shenzhen)

Alphaba Smart Bus is a demonstration project operating on mixed-traffic public roads in Shenzhen and it started in December 2017. This on-demand service uses a fleet of six AnKai, Dongfeng and Scania electric buses with level 3 autonomy and a maximum capacity of 20 passengers (eight seated and 12 standing passengers).

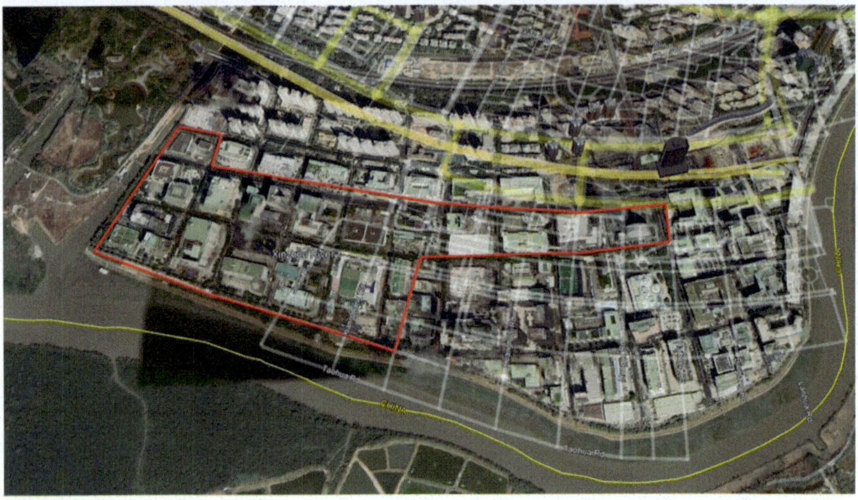

Fig. 8.15 Alphaba (Shenzhen)

The route through the center of Shenzhen (Fig. 8.15) is 3 km long and has four stop stations for embarking/disembarking passengers [59, 60].

The goal of the project is to develop and implement autonomous driving technology in a public transportation service. Thus, from the launch until now, 20,000 passengers have been transported (an average of 55 passengers per day), over a total distance of 120,000 km (an average of 333 km per day) with an average velocity of 15 km/h [59].

The autonomous buses are equipped with V2I communication technology with the intelligent traffic lights in downtown Shenzhen and are connected via 5G internet to the Haylion MaaS management platform [61].

8.4.3.4 Apolong (Beijing)

The Apolong autonomous shuttle bus manufactured by the company Baidu was launched in July 2018, in collaboration with bus manufacturer King Long, and it was implemented on public roads in Beijing, Shenzhen, Xiongan, Wuhan and Pingtan.

Apolong has a level 4 autonomy, it can transport up to 14 passengers (eight seated and four standing passengers) with a velocity between 20 and 40 km/h and is connected for monitoring to the Apollo Fleet Management Platform. During the test period, Apolong traveled more than 10,000 km transporting more than 10,000 passengers [62].

The functionalities of the autonomous driving system are as follows [63]:

(1) Analysis of the ability to drive autonomously in real traffic situations.
(2) Entering and exiting from the parking lots.

Fig. 8.16 RAC Intellibus® (South Perth, Australia)

(3) Autonomous driving mode on the entire established route.
(4) Identification and avoidance of obstacles along the way.
(5) Crossing intersections and giving priority.

8.4.4 Australia

RAC Intellibus® is a pilot project implemented in South Perth, Australia, that uses a Navya autonomous shuttle bus with a level 4 autonomy that travels without a driver on board.

The autonomous shuttle bus travels on a route that is 3.5 km long in a tourist area (between the South Perth Esplanade and the Old Mill) with stops only at both ends (Fig. 8.16).

Since the launch of the service in 2016, 25,084 passengers have been transported over a total distance of 33,338 km. The transported passengers were also surveyed about the degree of satisfaction with the service, resulting in a percentage of 97% of passengers receptive to this mobility service and who consider autonomous vehicles as the means of transport of the future [64, 65].

References

1. Iclodean, C., Cordos, N., Varga, B.O.: Autonomous shuttle bus for public transportation: a review. Energies **13**, 2917 (2020)
2. https://www.sae.org/publications/books/content/pt-158/ (March 16, 2022)

3. https://www.who.int/publications/i/item/9789241565684 (May 22, 2022)
4. Chehria, A., Mouftah, H.T.: Autonomous vehicles in the sustainable cities, the beginning of a green adventure. Sustain. Cities Soc. **51**, 101751 (2019)
5. Pype, P., Daalderop, G., Schulz-Kamm, E., Walters, E., Westermann, S.: Intelligent transport systems: the trials making smart mobility a reality. In: Watzenig, D., Horn, M. (eds.) Automated Driving Safer and More Efficient Future Driving, pp. 595–600. Springer International Publishing, Cham, Switzerland (2017)
6. https://eur-lex.europa.eu/eli/dir/2010/40/oj (March 16, 2022)
7. Yuen, K.F., Huyen, D.T.K., Wang, X., Qi, G.: Factors influencing the adoption of shared autonomous vehicles. Int. J. Environ. Res. Public Health **17**, 4868 (2020)
8. https://www.keolisdowner.com.au/wp-content/uploads/2018/08/KDR-Autonobus-Report.pdf (March 16, 2022)
9. https://navya.tech/en/navya-reaches-a-new-milestone-in-autonomous-mobility-with-the-first-fully-autonomous-level-4-operation-on-a-closed-site/ (March 16, 2022)
10. https://www.thalesgroup.com/en/markets/digital-identity-and-security/iot/magazine/7-benefits-autonomous-cars (March 16, 2022)
11. https://www.nhtsa.gov/technology-innovation/automated-vehicles-safety (March 16, 2022)
12. https://onlinemasters.ohio.edu/blog/the-future-of-driving/ (March 16, 2022)
13. https://www.rand.org/content/dam/rand/pubs/research_reports/RR400/RR443-2/RAND_RR443-2.pdf (March 16, 2022)
14. https://assets.kpmg/content/dam/kpmg/images/2015/05/connected-and-autonomous-vehicles.pdf (March 16, 2022)
15. https://www.bbva.ch/en/news/advantages-and-disadvantages-of-autonomous-vehicles/ (March 17, 2022)
16. https://environmental-conscience.com/self-driving-cars-pros-cons/ (March 16, 2022)
17. https://www.shuttle-modellregion-oberfranken.de/ (March 22, 2022)
18. https://semler.dk/en/ (March 22, 2022)
19. https://space.uitp.org/initiatives/copenhagen-hospital-av-denmark (March 22, 2022)
20. https://space.uitp.org/storage/projects/NkGIQ6fQjB7tBCzUVZyKtR70Q5CWaAfL4cAJPAXy.jpeg (March 22, 2022)
21. https://www.letsholo.com/slagelse (March 22, 2022)
22. https://www.udenhaender.dk/wp-content/uploads/2018/10/slagelse-fase-2.jpg (March 22, 2022)
23. https://darwincav.com/autonomous-shuttle/ (March 22, 2022)
24. https://shuttlebus.cavcloud.com/shuttle (March 22, 2022)
25. https://www.gov.uk/government/publications/trialling-automated-vehicle-technologies-in-public/code-of-practice-automated-vehicle-trialling (March 22, 2022)
26. https://www.bsigroup.com/globalassets/documents/pas/pas1881_final-design-proof.pdf (March 22, 2022)
27. https://sensible4.fi/pilots/ (March 22, 2022)
28. https://sensible4.fi/wp-content/uploads/2022/03/pilot-route-hervanta-1221-1.png (March 22, 2022)
29. https://www.peoriaaz.gov/government/departments/public-works/transportation/autonomous-shuttle (March 23, 2022)
30. https://futuretransport-news.com/wp-content/uploads/sites/3/2022/01/RoboRide-Medical-Shuttle-Route-Map.jpeg (May 22, 2022)
31. https://asu.co1.qualtrics.com/jfe/form/SV_9BPXaSnHQ9syW0K (March 23, 2022)
32. https://record.umich.edu/articles/mcity-supports-new-autonomous-shuttle-service-in-ann-arbor/ (March 23, 2022)
33. https://record.umich.edu/wp-content/uploads/mc-image-cache/2021/10/A2GOServiceAreaMap.jpg (March 23, 2022)
34. https://mcity.umich.edu/a2go-shuttle-off-to-solid-start-draws-hundreds-of-riders-monthly/ (March 24, 2022)
35. https://www.psta.net/programs/ava/ (March 24, 2022)

36. https://www.psta.net/media/5844/final-ava-route.png (March 24, 2022)
37. http://ridepsta.net/bustime/eta/eta.jsp (March 24, 2022)
38. https://www.psta.net/media/5041/ava-rules-of-the-road.pdf (March 24, 2022)
39. https://legistarweb-production.s3.amazonaws.com/uploads/attachment/pdf/264980/Milo_Closeout_Report.pdf (March 24, 2022)
40. https://p1cdn4static.civiclive.com/UserFiles/Servers/Server_14481062/Image/City%20Hall/Departments/Office%20Of%20Strategic%20Initiatives/RAPID/RAPID_Zone_Map.png (March 24, 2022)
41. https://maymobility.com/technology/ (May 22, 2022)
42. https://ridewithvia.com/ (March 24, 2022)
43. https://www.uta.edu/news/news-releases/2022/03/18/rapid-one-year (March 24, 2022)
44. https://www.arlingtontx.gov/city_hall/departments/office_of_strategic_initiatives/rapid (March 24, 2022)
45. https://www.grandrapidsmi.gov/files/assets/public/departments/mobile-gr/images/avgr_service_map.jpg (March 25, 2022)
46. https://www.grandrapidsmi.gov/Government/Departments/Mobile-GR/DASH-the-Downtown-Area-Shuttle (May 31, 2022)
47. https://www.grandrapidsmi.gov/files/assets/public/departments/mobile-gr/files/avgr-survey/avgr-dash-survey-report-022420-w-cover.pdf (March 25, 2022).
48. https://www.weride.ai/en/ (March 26, 2022)
49. https://www.weride.ai/static/64aa5edd77873e4b25bddcaac552cefa/fd753/04.jpg (March 26, 2022)
50. https://www.weride.ai/en/10-mln-kilometers-en/ (March 26, 2022)
51. https://www.alibabacloud.com/blog/connecting-the-real-world-algorithms-and-innovations-behind-amap_596304 (March 26, 2022)
52. https://yqintl.alicdn.com/98e78c1630884a6726e90f5ff6c3047667b688ac.png (March 26, 2022)
53. https://www.autox.ai/en/index.html (March 26, 2022)
54. https://www.autox.ai/blog/20220128.html (March 26, 2022)
55. https://www.autox.ai/download/images/AutoX_65_Driverless_RoboTaxi_Area.jpg (March 26, 2022)
56. https://www.autox.ai/en/mobility.html (March 26, 2022)
57. https://www.autox.ai/blog/20210128.html (March 26, 2022)
58. https://www.autox.ai/blog/20200226.html (March 27, 2022)
59. https://space.uitp.org/initiatives/alphaba-smart-bus-av-pilot-shenzhen-china (March 27, 2022)
60. https://space.uitp.org/storage/projects/VaLp0UjxaGxB2b3HYKhhWo1YtF4Pt0EWXDf9NkxY.png (March 27, 2022)
61. https://www.haylion.cn/alpha_e.jsp (March 27, 2022)
62. https://www.apollo.auto/minibus/ (March 27, 2022)
63. https://iot-automotive.news/baidu-apollo-3-0-x-king-long-l4-minibus-apolong/ (March 27, 2022)
64. https://rac.com.au/about-rac/advocating-change/initiatives/automated-vehicle-program/intellibus (March 27, 2022)
65. https://www-cdn.rac.com.au/-/media/files/rac-website/about-rac/advocacy/intellibus/intellibus-route-south-perth-2021.pdf?modified=20210910022835&_ga=2.151179518.69377094.1647977382-1392404325.1647527504 (March 27, 2022)

Conclusions

Autonomous driving is no longer an exotic idea belonging to the future, but rather belongs to the present for those who dare to connect themselves with the performance of the current technologies. If we started the history of automobiles with a conductor walking in front of the vehicle with a whistle and a flag alarming those present on the road of the presence of a spectacular "machine" limiting the speed to 6 km/h due to possible harmful influence of the speed over the travelers, now after 150 years of engineering evolution we are opening the road to vehicles able to communicate with one another, to vehicles able to make decisions and to vehicles that are aware of the existing pedestrians on the road. At the evolution scale several generations of engineers have been able to prove that the mobility of the people can be not only improved but rather taken to a totally different level. Our estimation is that the autonomous driving is the first stone in the foundation of the most complex transportation system that the humankind will need in the future. We predict a system with no traffic jams with a higher vehicle density and zero traffic collisions, all of the above being possible with the help of autonomous driving.

Autonomous shuttle buses designed for public passenger transportation use a constantly evolving technology, a technology that certainly needs years of testing in order to replace public transportation with conventional buses on a large scale.

The use of autonomous shuttle buses for public passenger transportation, with transportation capacities of up to 15 people, allows the choice and implementation of fixed routes, established according to traffic conditions, route geometry, intersections with other routes, travel schedule etc.

The need for the presence of a human operator who must permanently accompany the autonomous vehicles, the reduced number of passengers transported (maximum 15 people), the reduced travel velocity (maximum 20 km/h), the high price of the autonomous shuttles (approximately 300,000 Euro) etc. are major barriers to the implementation of these vehicles in public passenger transportation. Overcoming these barriers will require years of testing and significant technological advances in autonomous driving.

C. Iclodean et al., *Autonomous Vehicles for Public Transportation*, Green Energy
and Technology, https://doi.org/10.1007/978-3-031-14678-7

On the other hand, niche areas in the field of public passenger transportation can now be identified where autonomous vehicles can be used successfully. For example: routes during the night (when urban traffic is reduced to a minimum), routes through airports, industrial parks, shopping areas, etc.

We can only predict what the future of public passenger transportation will look like in the next 50 years. However, everything related to the public service will certainly be autonomous and independent. Public transportation is a critical form of public service which will face a dramatic change in the years to come. It needs to become improved, modernized, more environmentally friendly and more efficient.

The autonomy of the public transportation fleets is one of the challenges of the years to come. The policies of the public authorities will dictate the quality of life of the residents, giving them the means and reasons to give up private transport. Autonomous public transportation has many advantages in terms of its efficiency (fixed schedule), 24-h operating time with zero downtime, low maintenance costs for vehicle maintenance, maximum energy efficiency for individually optimized routes, greater traffic safety (because independent routes automatically mean fewer accidents) and continuous monitoring (improved personal safety). These are just some of the benefits that autonomous public transportation will provide for the future. Clearly, the current mobility solutions for public transportation will be obsolete in 10 years. In the years to come, a significant proportion of people will work from home, may not need public transportation, but will need public services supported by autonomous transportation (for example, for deliveries of food or medicines, for garbage collection, for delivery of clothes (from shops or cleaners), for street maintenance and cleaning, etc.).

This study provides a comprehensive assessment of the current autonomous driving solutions designed for public transportation. At the same time, it offers the reader, regardless of his/her field of specialization, the means to understand the purpose, technology, legal implications, social implications, usefulness, advantages, and disadvantages of this public transportation solution in direct connection with its primary users: the city residents.

The propulsion solution for the autonomous vehicles is the electric platform that currently equips the present day electric vehicles, with all its advantages and limitations of today. The diversity of solutions is great, and the development of future autonomous platforms is vast and long-term, because only the future can establish appropriate technological directions.

From the perspective of the sensors, we expect the current study to be valid for 3 to 5 years, as technology in this area is so dynamic that future implementation of 5G technology will allow vehicles to outsource their sensors and use some information from an external source. 5G technology will have a major influence in the development of the autonomous driving technology, especially for autonomy Level 5, making autonomous vehicles more connected, smarter, safer, and faster and turning them into real mobile data centers with real-time information, not only for them, but also for the other vehicles they are connected to (V2V).

From an IT perspective, the dynamics of updating will increase exponentially as machine learning and artificial intelligence become a ubiquitous part of the code and of the algorithm that controls the autonomous driving strategies. Currently, this type of code is too expensive for a serial product. However, as this technology advances, such software solutions will become more and more affordable and will guarantee a competitive price for this product.

The legal framework will change along with each active generation of autonomous vehicles, providing a continuous transformation through legal updates and corrections. The social implications will become apparent once the market share of these vehicles reaches 10% of public transportation vehicles.

The current study provides a clear scientific perspective on the current status of autonomous driving in the field of public passenger transportation, as well as a valid insight into the benefits of this technical solution with direct links to the current technology, with a particular focus on the social benefits of this technology.

This scientific study is like a faded light in a dark room: it attracts attention, brings an unbalance in the dark, and gives a direction. The future will for sure provide more robust and more concrete solutions, but for now since everything is at the beginning, we intend to set the direction and put a stone in the bridge between the classical transportation systems and the autonomous driving systems. What we can say for certain is that the direction is set, the tolls are present, and the product exists—the rest is just a question of timing and opportunity.

The future of public transportation will be autonomous, or it won't exist at all.

The manufacturer's authorised representative in the EU is Springer
Nature Customer Service Centre GmbH, Europaplatz 3, 69115 Heidelberg,
Germany. If you have any concerns regarding our products, please
contact ProductSafety@springernature.com

Printed and bound by CPI Group (UK) Ltd, Croydon, CR0 4YY
30/09/2026
02225304-0017